Advances in Intelligent Systems and Computing

Volume 985

The series "Advances in Intelligent Systems and Computing" contains publications on theory, applications, and design methods of Intelligent Systems and Intelligent Computing. Virtually all disciplines such as engineering, natural sciences, computer and information science, ICT, economics, business, e-commerce, environment, healthcare, life science are covered. The list of topics spans all the areas of modern intelligent systems and computing such as: computational intelligence, soft computing including neural networks, fuzzy systems, evolutionary computing and the fusion of these paradigms, social intelligence, ambient intelligence, computational neuroscience, artificial life, virtual worlds and society, cognitive science and systems, Perception and Vision, DNA and immune based systems, self-organizing and adaptive systems, e-Learning and teaching, human-centered and human-centric computing, recommender systems, intelligent control, robotics and mechatronics including human-machine teaming, knowledge-based paradigms, learning paradigms, machine ethics, intelligent data analysis, knowledge management, intelligent agents, intelligent decision making and support, intelligent network security, trust management, interactive entertainment, Web intelligence and multimedia.

The publications within "Advances in Intelligent Systems and Computing" are primarily proceedings of important conferences, symposia and congresses. They cover significant recent developments in the field, both of a foundational and applicable character. An important characteristic feature of the series is the short publication time and world-wide distribution. This permits a rapid and broad dissemination of research results.

** **Indexing: The books of this series are submitted to ISI Proceedings, EI-Compendex, DBLP, SCOPUS, Google Scholar and Springerlink** **

More information about this series at http://www.springer.com/series/11156

Radek Silhavy
Editor

Artificial Intelligence Methods in Intelligent Algorithms

Proceedings of 8th Computer Science
On-line Conference 2019, Vol. 2

 Springer

Editor
Radek Silhavy
Faculty of Applied Informatics
Tomas Bata University in Zlín
Zlín, Czech Republic

ISSN 2194-5357   ISSN 2194-5365   (electronic)
Advances in Intelligent Systems and Computing
ISBN 978-3-030-19809-1   ISBN 978-3-030-19810-7   (eBook)
https://doi.org/10.1007/978-3-030-19810-7

This Springer imprint is published by the registered company Springer Nature Switzerland AG
The registered company address is: Gewerbestrasse 11, 6330 Cham, Switzerland

Preface

Modern trends and approaches of artificial intelligence research and its application to intelligent systems are presented in this book. Paper discuss hybridisation of algorithms, new trends in neural networks, optimisation algorithms and real-life issues related to artificial method application.

This book constitutes the refereed proceedings of the Artificial Intelligence Methods in Intelligent Algorithms section of the 8th Computer Science On-line Conference 2019 (CSOC 2019), held on-line in April 2019.

CSOC 2019 has received (all sections) 198 submissions; 120 of them were accepted for publication. More than 59% of accepted submissions were received from Europe, 34% from Asia, 5% from America and 2% from Africa. Researches from more than 20 countries participated in CSOC 2019 conference.

CSOC 2019 conference intends to provide an international forum for the discussion of the latest high-quality research results in all areas related to computer science. The addressed topics are the theoretical aspects and applications of computer science, artificial intelligence, cybernetics, automation control theory and software engineering.

Computer Science On-line Conference is held on-line, and modern communication technology, which is broadly used, improves the traditional concept of scientific conferences. It brings equal opportunity to all the researchers around the world to participate.

I believe that you will find the following proceedings interesting and useful for your own research work.

March 2019   Radek Silhavy

Organization

Program Committee

Program Committee Chairs

Petr Silhavy	Faculty of Applied Informatics, Tomas Bata University in Zlin
Radek Silhavy	Faculty of Applied Informatics, Tomas Bata University in Zlin
Zdenka Prokopova	Faculty of Applied Informatics, Tomas Bata University in Zlin
Roman Senkerik	Faculty of Applied Informatics, Tomas Bata University in Zlin
Roman Prokop	Faculty of Applied Informatics, Tomas Bata University in Zlin
Viacheslav Zelentsov	Doctor of Engineering Sciences, Chief Researcher of St. Petersburg Institute for Informatics and Automation of Russian Academy of Sciences (SPIIRAS)

Program Committee Members

Boguslaw Cyganek	Department of Computer Science, AGH University of Science and Technology, Krakow, Poland
Krzysztof Okarma	Faculty of Electrical Engineering, West Pomeranian University of Technology, Szczecin, Poland
Monika Bakosova	Institute of Information Engineering, Automation and Mathematics, Slovak University of Technology, Bratislava, Slovak Republic

Pavel Vaclavek	Faculty of Electrical Engineering and Communication, Brno University of Technology, Brno, Czech Republic
Miroslaw Ochodek	Faculty of Computing, Poznan University of Technology, Poznan, Poland
Olga Brovkina	Global Change Research Centre Academy of Science of the Czech Republic, Brno, Czech Republic; Mendel University, Brno, Czech Republic
Elarbi Badidi	College of Information Technology, United Arab Emirates University, Al Ain, United Arab Emirates
Luis Alberto Morales Rosales	Head of the Master Program in Computer Science, Superior Technological Institute of Misantla, Mexico
Mariana Lobato Baes	Superior Technological of Libres, Mexico
Abdessattar Chaâri	Laboratory of Sciences and Techniques of Automatic Control & Computer Engineering, University of Sfax, Tunisian Republic
Gopal Sakarkar	Shri. Ramdeobaba College of Engineering and Management, Republic of India
V. V. Krishna Maddinala	GD Rungta College of Engineering & Technology, Republic of India
Anand N. Khobragade	Maharashtra Remote Sensing Applications Centre, Republic of India
Abdallah Handoura	Computer and Communication Laboratory, Telecom Bretagne, France

Technical Program Committee Members

Ivo Bukovsky	Roman Senkerik
Maciej Majewski	Petr Silhavy
Miroslaw Ochodek	Radek Silhavy
Bronislav Chramcov	Jiri Vojtesek
Eric Afful Dazie	Eva Volna
Michal Bliznak	Janez Brest
Donald Davendra	Ales Zamuda
Radim Farana	Roman Prokop
Martin Kotyrba	Boguslaw Cyganek
Erik Kral	Krzysztof Okarma
David Malanik	Monika Bakosova
Michal Pluhacek	Pavel Vaclavek
Zdenka Prokopova	Olga Brovkina
Martin Sysel	Elarbi Badidi

Organizing Committee Chair

Radek Silhavy   Faculty of Applied Informatics, Tomas Bata
                                       University in Zlin

Conference Organizer (Production)

OpenPublish.eu s.r.o.
Web: http://www.openpublish.eu
Email: csoc@openpublish.eu

Conference Web site, Call for Papers

http://www.openpublish.eu

Contents

The Method of Deductive Inference of Consequences with the Scheme Construction

Anastasia Bardovskaya, Gennadiy Chistyakov$^{(\boxtimes)}$, Maria Dolzhenkova, and Dmitry Strabykin

Department of Computers, Vyatka State University,
Moskovskaya, 36, 610000 Kirov, Russia
gennadiychistyakov@gmail.com

Abstract. The paper describes the method of inference in first-order predicate calculus, which allows, besides obtaining the result, to build a special structure—the scheme of inference. This structure represents a special kind of graph and it can be used to interpret the solution both visually and in the analytical way. Besides, the scheme can be applied to evaluate the development of situation in dynamic systems research. The method is based on the operation of division of disjuncts, characterized by a high level of AND-, OR- DCDP-parallelisms, thereby it can be effectively realized on modern multiprocessor and multicore platforms in software mode.

Keywords: Deductive inference · Disjuncts division operation ·
Conclusion of consequences · Inference scheme

1 Introduction

Logic-based modeling of reasoning is one of the promising areas of research, studying artificial intelligence methods and algorithms. Most often it employs sentential calculus or first-order predicate calculus as a formal system. Propositional logic allows to create simple to realize methods of inference and, consequently, to develop high-performance software and software-hardware systems. But the description of knowledge in the real subject area requires a more expressive formal system. On the one hand, reasoning, presented in the form of formulas of first-order predicate calculus, allows to set different relations and causal connections between objects. On the other, this type of presentation seems quite natural for a human. All this makes predicate calculus a convenient means of formalization of applied tasks. Modeling of reasoning in intellectual systems is performed with the help of the inference apparatus—deductive and inductive methods, method of abductive inference, Case-Based Reasoning (CBR) [1]. This apparatus allows to solve logical problems. Nevertheless quite often in the course of modeling complex, multi-step reasoning one has to define the consequences

© Springer Nature Switzerland AG 2019
R. Silhavy (Ed.): CSOC 2019, AISC 985, pp. 1–10, 2019.
https://doi.org/10.1007/978-3-030-19810-7_1

that can be inferred and the new facts, reflecting the conditions of the changing environment having a set of initial premises [2,3]. This problem can be solved with the help of a special type of deductive inference—parallel logical inference of consequences.

2 Logical Inference of Consequences

The task of the logical inference of consequences can be formulated in the following way. There are consistent premises, presented as a set of disjuncts $M = \{D_1, D_2, ..., D_J\}$. Each disjunct contains a literal without inversion. Set M includes a subset of input data M^F. Besides there is a set of new facts $m^F = \{L_1, L_2, ..., L_p, ..., L_P\}$, with set $M \cup m^F$ being also consistent. The task of logical inference of consequences (literals without inversion) is as follows.

1. Set of consequences M^S and set of sets s^H of consequences $s^H = \{e_0, e_1, ..., e_h, ..., e_H\}$ must be found. Set of consequences e_0 consists of initial premises coinciding with new facts: $e_0 = M^F \cap m^F$. Set e_1 includes consequences, inferred in one step from set of new facts m^F on the basis of the set of disjuncts-premises $M : m^F, M \Rightarrow e_1$, but not inferred from M only. Set of consequences $e_{h+1}(h = 1, ..., H - 1)$ includes consequences, inferred in one step from set of consequences e_h, of new facts m^F and the set of consequences obtained in the previous steps, on the basis of the set of disjuncts-premises $M : e_h, m^F, c_h, M \Rightarrow e_{h+1}; c_h = c_{h-1} \cup e_{h-1}, c_0 = \varnothing$. Set of sets of consequences is $s^0 = \{e_0\}$, and set of sets $s^h(h = 1, ..., H)$ is defined in the following way: $s^h = s^{h-1} \cup \{e_h\}$. We obtain the total set of consequences M^S joining sets of set of sets $s^H : M^S = e_0 \cup e_1 \cup ... \cup e_H$.

2. It is required to give a description O of the scheme of inference as set of sets $O = \{f^1, f^2, ..., f^h, ..., f^{H-1}\} \cup \{s^+\}$, where f^h is a set of literals, obtained while forming a description of the scheme on step h of the inference, s^+ is a set of finite consequences, from which no new consequences can be inferred $(s^+ \subseteq M^S)$.

A special operation of generalized division of clauses is used to realize logical inference together with forming the description of the scheme.

3 Supporting Operations and Procedures

3.1 Generalized Division of a Disjunct by a Literal

The operation of generalized division of disjunct b, containing literal L, by literal L', resulting in quotient b' and remainder d', can be presented as $b[L]\%L' = <b', d'>$.

Quotient b' is a set of literals, obtained after "gluing" literal L of the dividend and literal L'. By "gluing" we mean transformation of a pair of literals of the same name $L[+, k] \in b$ and $L'[j, +]$ (or $L[k, +] \in b$ and $L'[+, j]$), which have become identical as a result of application of unifying substitution λ

and containing an auxiliary variable "+" as a parameter, appearing in various positions, into literal $L[j, \lambda k]$ (or $L[\lambda k, j]$). If "gluing" the literal doesn't require the unifying substitution λ, it may be omitted in the parameter of the resulting literal.

Resulting remainder d' is calculated with the help of the operation of special joining the sets of literals $\lambda \tilde{b} \sqcup b'$, where $\lambda \tilde{b}$ is a set of literals of disjunct b, to which the unifying substitution λ was applied. The peculiarity of the operation consists in absorption of literals $L[+, k] \in \tilde{b}$ and $L[j, +] \in \tilde{b}$ by literal $L[j, k] \in b' : \{L[+, k]\} \sqcup \{L[j, k]\} = \{L[j, +]\} \sqcup \{L[j, k]\} = \{L[j, k]\}$.

Remainder d' is determined according to the following rules:

- if $b' = \varnothing$, then $d' = 1$;
- if $b' \neq \varnothing$ and $(\lambda \tilde{b} \sqcup b') - b' = \varnothing$, then $d' = 0$;
- if $b' \neq \varnothing$ and $(\lambda \tilde{b} \sqcup b') - b' = \tilde{d}, \tilde{d} \neq \varnothing$, then $d' = L_1 \vee L_2 \vee ... \vee L_s \vee ... \vee L_S$, where $L_s \in \tilde{d}'(s = 1, ..., S)$ and $\tilde{d}' = \{L_1, L_2, ..., L_S\}$.

Example 1. Let $b = P(x, y)[+, 1] \vee O(y, x)[1, +]$ and $L' = P(b, a)[2, +]$, then $b[P(x, y)[+, 1]]\%L' = <b', d'>$, where $b' = \{P(b, a)[2, \{b/x, a/y\}1]\}$, $\lambda = \{b/x, a/y\}$, $d' = O(a, b)[1, +]$, as $\lambda \tilde{b} \sqcup b' = \{P(b, a)[+, 1], O(a, b)[1, +]\} \sqcup \{P(b, a)[2, \lambda 1]\} = \{P(b, a)[2, \lambda 1], O(a, b)[1, +]\}, (\lambda \tilde{b} \sqcup b') - b' = \{O(a, b)[1, +]\}$.

3.2   Partial Division of Disjuncts

Partial division of disjuncts is performed with the help of a special procedure of formation of remainders—one of the main procedures used in the method of inference of consequences in predicate calculus. The procedure assumes that the premise and inference should be presented in the form of disjuncts. The initial expressions of premises and inference in calculus are brought to the necessary form with the help of algorithms [4]. For convenience of description of this procedure let us introduce a number of notations. $\omega = <b, d, g, q, n, s, g'>$ is the procedure of partial division, in which: b is the remainder-dividend (disjunct of the premise), used to obtain remainders; d is the remainder-divisor (disjunct of the inference), participating in forming the remainders; g is an intermediate set of literals in the description of the scheme of inference; q is a partial attribute of the continuation of division of disjuncts: "0"—further division is possible; "1"— further division is impossible; $n = \{< b_t, d_t, g_t >, t = 1, ..., T\}$—a set of threes, consisting of the new remainder-dividend b_t and the corresponding remainder-divisor d_t, formed in the result of applying the procedure ω, and also set g_t of the obtained literals of the description of the inference scheme; s is a set of obtained consequences; g' is a set of sets of literals of the description of the inference scheme, with the help of which consequences of set s were obtained.

Let us define "the derivative" $\frac{\partial b[L]}{\partial L'}$ of disjunct $b[L]$, containing literal L, with respect to literal L' as remainder d', obtained while performing the operation of generalized division $b[L]\%L'$. Besides if remainder d' is not equal to zero or one and contains only one literal, the right parameter of which is the symbol of the auxiliary variable "+", this literal is the consequence.

Let us determine the matrix of "the derivatives" of disjunct b with respect to disjunct d in the following way:

$$\mu[b, d] = |\frac{\partial b[L_j]}{\partial L'_k}| = |\Delta_{kj}|,$$

where $j = 1, ..., J$ and $k = 1, ..., K$, besides J is the number of literals in disjunct b, and K is the number of literals in disjunct d.

Before performing the procedure of partial division of disjuncts it is assumed, that $s = \varnothing$ and $g' = \varnothing$.

The following operations are performed in the procedure.

1. The matrix of "derivatives" $\mu[b, d]$ is calculated. The condition of forming the remainders is checked. If all the "derivatives" in matrix $\mu[b, d]$ are equal to one, it is assumed, that $q = 1$, $n = \{<1, 1, 1>\}$ and point 4 is performed, otherwise it is assumed, that $q = 0$ and the next point is performed.

2. The presence of consequences is checked. In case of their absence the next point is passed on to. If there are consequences, for each consequence s_t a set of literals of the description of the inference scheme $g'_t = \{L[\lambda j, \lambda_i k]$, $i = 1, ..., I\}$ is formed, where T is the total number of consequences, λ is some substitution (possibly empty), λ_i is a substitution, obtained in the course of performing operations of generalized division. Set g'_t is formed by means of joining set g with quotient b', formed in the course of performing the operation of generalized division, as a result of which the corresponding consequence $s_t : g'_t = g \cup \{b'\}$ was obtained. Besides, for each consequence s_t the unified substitution $\lambda_t = \cup_{i=1}^{I} \lambda_i$, where λ_i are substitutions from the right parts of literals in set g'_t, is calculated. Set s is formed as $s = \{s_t[\lambda_t j, +], t = 1, ..., T\}$, and set of sets g' is formed as $g' = \{g'_t, t = 1, ..., T\}$. It is assumed that $q = 1$, $n = \{<1, 1, 1>\}$, and point 4 is performed.

3. Set $n = \{<b_t, d_t, g_t>, t = 1, ..., T\}$ of threes, consisting of the new remainders-dividends b_t, remainders-divisors d_t and corresponding sets of literals of the description of the inference scheme g_t is determined. For each of the remainders-dividends b_t the corresponding remainder-divisor is calculated: $d_t = \lambda_t(d \div w)$ with the help of substituting λ_t, used in its formation, where w is an auxiliary disjunct, containing excluded from remainder d literals. For calculating remainder d_t the literals, for which in matrix $\mu[b, d]$ one of the listed below conditions are met, are excluded from remainder d. The conditions are as follows:

   - $\frac{\partial b[L_j]}{\partial L'_h}$ for all $j = 1, ..., J$; i.e. the line of ones in the matrix corresponds to the literal;
   - $\frac{\partial b[L_j]}{\partial L'_h}$ for all j, except $j = u$, such, that $\frac{\partial b[L_u]}{\partial L'_h} = b_t$; i.e. in the line of the matrix, corresponding to the literal, all the "derivatives" are equal to one, except one of them, representing the remainder under consideration—remainder b_t.

The set of literals of the description of the inference scheme g_t is formed by means of joining set g with quotient b', obtained in the course of performing the operation of generalized division, in which the corresponding remainder-dividend $b_t : g_t = g \cup \{b'\}$ was obtained. It is assumed that $q = 0$ and $n = \{<b_t, d_t, g_t>, t = 1, ..., T\}$ and the next point is performed.
4. The results of calculating procedure ω are recorded.

Example 2. Let us consider the example of calculating matrix $\mu[b, d]$, where $b = P(x, y)[+, 1] \vee O(y, x)[1, +]$, and $d = P(b, a)[2, +]$ to illustrate the construction of the matrix of "derivatives" and forming the remainders.

1. (Point 1) Matrix of "derivatives" $\mu[b, d]$ is calculated.

$$\begin{array}{cc} & P(x, y)[+, 1] \quad O(y, x)[1, +] \\ P(b, a)[2, +] \; (& \Delta_{11} \qquad\qquad 1 \qquad) \end{array}$$

In the matrix "derivative" Δ_{11} is determined with the help of unifying substitution $\lambda_{11} = \{b/x, a/y\} : \Delta_{11} = O(a, b)[1, +]$, and "derivative" $\Delta_{12} = 1$. The condition of forming the remainders is checked. As not all the "derivatives" in the matrix are equal to one, it is assumed that $q = 0$ and the next point is performed.

2. (Point 2) The presence of consequences is checked. All the remainders, different from zero and one and containing only one literal, the right parameter of which is the symbol of the auxiliary variable "+", are consequences. In the example under consideration $\Delta_{11} = O(a, b)[1, +]$ is the only consequence, for which the condition of forming the set of literals of the description of the inference scheme $g_1' = \{P(b, a)[2, \lambda_{11} 1]\}$ (assuming $g = \varnothing$), $\lambda_t = \lambda_{11}$ is met. Set of consequences $s = \{O(a, b)[\lambda_{11} 1, +]\}$ and set of sets g' as $g' = \{g_1'\}$ are formed. It is assumed that $q = 1$, $n = \{<1, 1, 1>\}$ and point 4 is performed.

3. (Point 4) The results of calculating procedure ω are fixed. Set of consequences $s = \{O(a, b)[\lambda_{11} 1, +]\}$ and set of sets of literals of the description of the inference scheme $g' = \{g_1'\}$ are obtained, continuation of partial division of disjuncts is not possible ($q = 1$, $n = \{<1, 1, 1>\}$).

3.3 Complete Division of Disjuncts

Complete division of disjuncts aims at obtaining all the consequences from clause d, set of new facts m^F, set of premises M^S and set of previously obtained consequences c on the basis of disjunct-premise D with the help of the procedure considered below.

Let us introduce the following notations: $\Omega = <D, d, m^F, c, Q, S, G>$ is the procedure of forming set S of consequences and set of sets G of sets of literals of the description of the inference scheme by means of dividing the disjunct of premise D by disjunct d taking account of sets of new facts m^F and premises M^F and set of previously obtained consequences c, in which Q is an attribute of solution, having two values: "0"—consequences are found, "1"—disjunct d has no consequences on the basis of disjunct D.

The set of consequences is formed by means of multiple application of ω-procedures in a number of steps. On each step ω-procedures are applied to the present remainders-dividends and remainders-dividers, forming new remainders-dividends and new remainders-dividers, used as input data on the next step. The process ends, when it is detected on the next step, that in all the ω-procedures of this step the attributes, demonstrating impossibility of continuation of division of disjuncts ($q = 1$), are formed.

The parallel execution of ω-procedures is described with the help of a special index function, providing a unique identification of each procedure and its parameters. Let us introduce index function $i(h)$ for index of size h, which we will determine with the help of induction for index variable t in the following way: $i(1) = t, t = 1, ..., T$; $i(2) = i(1).t_{i(1)}, t_{i(1)} = 1, ..., T_{i(1)}$ ($t.t_t = 1.t_1, t_1 = 1, ..., T_1, 2.t_2, t_2 = 1, ..., T_2, ..., T.t_T, t_T = 1, ..., T_T$); $i(3) = t.t_t.t_E$ ($E = t.t_t$), $i(3) = i(2).t_{i(2)}, t_{i(2)} = 1, ..., T_{i(2)}$; and so on. In the general case: $i(h) = i(h - 1).t_{i(h-1)}, t_{i(h-1)} = 1, ..., T_{i(h-1)}$. Let us assume, that $i(0)$ means the absence of index of the indexed variable, e.g., $T_{i(0)} = T$, and also that $i(1) = i(0).t_{i(0)} = t$.

Index function $i(h)$ sets an index, which (when $h>1$) consists of a constant, formed on the basis of the value of function $i(h - 1)$, and corresponding to this constant by variable $t_{i(h-1)}$. The set of indexes, described with the help of the function $i(h)$, is formed on the basis of the values of function $i(h-1)$ and values of variables $t_{i(h-1)} = 1, ..., T_{i(h-1)}$, corresponding to them. E.g., if $T = T_1 = 3$, $T_2 = 2$ we obtain $i(1) = t, t = 1, 2, 3; i(2) = 1.t_1, t_1 = 1, 2, 3, 2.t_2, t_2 = 1, 2$. That is $i(2) = 1.1, 1.2, 1.3, 2.1, 2.2$.

1. Preparatory step. On the preparatory step of the procedure of complete division of disjuncts the procedure of forming remainders $\omega = <D, d, g, q, n, s, g'>$ ($g = \varnothing$) is applied to the disjunct of premise D and disjunct d, and the partial attribute of the continuation of division of disjuncts q is analyzed. If $q = 1$ (continuation of division is not possible), then it is accepted that $S = s, G = g'$, besides, if $s \neq \varnothing$, then $Q = 0$, otherwise $Q = 1$, is determined, and the final step (point 3) is performed. If $q = 0$, then new pairs of remainders, formed in set $n = \{<b_t, d*_t, g_t>, t = 1, ..., T\}$, appear as input data for ω-procedures on the main step. Note also, that during the first implementation of the main step disjuncts d_t are used as remainders-dividers instead of disjuncts $d*_t$. Disjunct d_t is formed by means of complementing disjunct $d*_t$ with literals of the set of initial facts M^F, the set of new facts m^F and the set of previously obtained consequences c. Besides, during the first implementation of the main step it is accepted that $S_0 = s, G_0 = g'$.

Using the index function, set n can be presented as follows: $n_{i(h)} = \{<b_{i(h+1)}, d_{i(h+1)}, g_{i(h+1)}>, i(h+1) = i(h).t_{i(h)}; t_{i(h)} = 1, ..., T_{i(h)}\}$, where $h = 0$.

2. The main step. During k-th implementation of the main step ($k = 1, 2, ..., K$) for each three $<b_{i(h+1)}, d_{i(h+1)}, g_{i(h+1)}>$ of all sets $n_{i(h)}$ of such threes ($h = k - 1$), obtained on the previous step: $n_{i(h)} = \{<b_{i(h+1)}, d_{i(h+1)}, g_{i(h+1)}>, i(h + 1) = i(h).t_{i(h)}; t_i(h) = 1, ..., T_{i(h)}\}$, ω-procedure is performed:

$\omega_{i(h+1)} = <b_{i(h+1)}, d_{i(h+1)}, g_{i(h+1)}, q_{i(h+1)}, n_{i(h+1)}, s_{i(h+1)}, g'_{i(h+1)}>, i(h+1) = i(h).t_{i(h)}; t_{i(h)} = 1, ..., T_{i(h)}$. The set of consequences and set of sets of literals of the description of the inference scheme are complemented: $S_{k+1} = S_k \cup \cup_{v=1}^V S_{i(h).l}, G_{k+1} = G_k \cup \cup_{v=1}^V G_{i(h).l}$, where $v = t_{i(h)}$, $V = T_{i(h)}$.

If $S_{h+1} \neq \varnothing$, then $Q_{h+1} = 0$, otherwise $Q_{h+1} = 1$, is determined. Partial attributes of solution $q_{i(h+1)}$ are analyzed. If there is no $q_{i(h+1)} = 0$, then it is accepted that $Q = Q_{h+1}$, $S = S_{h+1}$, $G = G_{h+1}$ and point 3 is performed. Otherwise new sets of pairs are determined, which were obtained during the current implementation of the step: $n_{i(h+1)} = \{<b_{i(h+2)}, d_{i(h+2)}, g_{i(h+2)}>, i(h+2) = i(h+1).t_{i(h+1)}; t_{i(h+1)} = 1, ..., T_{i(h+1)}\}$, and we pass on to $(k+1)$ implementation of the main step (point 2), for which these sets appear as initial ones.

3. Final step. The results of the calculation of procedure Ω are recorded: the attribute of solution Q, the set of consequences S, set of sets of literals of the description of inference scheme G.

Let us illustrate the complete division of disjuncts.

Example 3. A detailed example of the procedure using is presented in the paper [5].

3.4   The Procedure of Inference of Consequences

The procedure of inference of consequences is analogous to the procedure of division of disjuncts, used in propositional logic [6]. The procedure allows to perform the step of inference by transformation of the inferred disjunct into the new inferred disjunct, necessary for continuation of the inference on the next step. Let us introduce the following notifications.

To infer the consequences on the current step a procedure $\nu = <M, R, m^F, o, p, R_1, e, f>$ is used, in which: M is a set of disjuncts of the input sequents; R is the inferred disjunct, consisting of literals L_k $(k = 1, ..., K)$ of the previously obtained consequences and of the new facts on the first step; m^F is a set of new facts; $o = <c, C>$ is a pair of sets of the current consequences, consisting of sets of consequences, formed before performing (c) and after performing (C) the procedure; p is an attribute of continuation of the inference: "0"—further inference is not possible; "1"—inference is finished; R_1 is a new inferred disjunct; e is a set of consequences for the inferred disjunct; f is a set of literals of the description of the inference scheme, obtained while forming consequences e.

As a sub-procedure the procedure of inference uses the previously considered Ω-procedure of complete division of disjuncts.

The procedure of inference can be applied, if $M \neq \varnothing$ and $K \geq 1$, otherwise attribute $p = 1$, $e = \varnothing$, $f = \varnothing$, $C = c$ is immediately determined and point 5 is passed on to. The following actions are performed in the procedure.

1. Complete division of disjuncts is performed.

1. The disjuncts of initial sequences, which are not facts, are divided by the inferred disjunct $\Omega_i = <D_i, R, m^F, M^F, c, Q_i, S_i, G_i>(i = 1, ..., I)$ with the help of Ω-procedures.

2. Attributes of solutions Q_i, performed Ω-procedures are analyzed. If all the attributes of solutions are equal to one, then it is accepted that $p = 1$, $e = \varnothing$, $f = \varnothing$, $C = c$ and point 5 is passed on to, otherwise the next action is performed.

2. Set of consequences e and set of literals of the description of the inference scheme f for the inferred disjunct are formed. Set e is formed with the help of joining sets of consequences S_i, obtained while performing Ω-procedures. If $e \neq \varnothing$, then the absorbed consequences are excluded from the obtained set of consequences. Literal b is absorbed by literal a then and only then, when there is such a substitution λ, that $b \subseteq \lambda a$. Not only consequences from set e, but also consequences from set c, obtained on the previous step of inference of consequences, are used as absorbing literals. Set f is formed by means of including in it literals from elements of sets G_i, corresponding to the unabsorbed elements of set e. If in the process of absorption of literals from the set all the literals are excluded, then it is accepted that $e = \varnothing$, $f = \varnothing$, $C = c$, $p = 1$ and point 5 is passed on to, otherwise it is determined that $p = 0$, and the next action is performed.

3. A new inferred disjunct is formed. Inferred disjunct R_1 represents the disjunction of literals of set of consequences e.

4. A new set of consequences is formed. In the pair of the current consequences $o = <c, C>$ set C is formed in the result of joining set of consequences c, obtained before performing the procedure, and set of consequences e, obtained after excluding the absorbed literals: $C = c \cup e$.

5. The results of the procedure are recorded. If attribute $p = 1$, then further inference of consequences is not possible, and if $p = 0$, then inference can be continued. If set of consequences e is not empty, a new inferred disjunct R_1, a new set of consequences C and a set of literals of the description of the inference scheme f will be formed in the process of performing the procedure.

It should be mentioned that in procedures of inference division of the disjuncts of the initial data by disjunct R can be performed in parallel with the matrix-like way [7].

4   Method of Inference of Consequences

Method of inference of consequences is based on the procedure of inference of consequences and it consists of a number of steps, on each of which the procedure of inference ν is performed, and the results of the procedure become initial data for the procedure on the next step. The process ends, if further inference of consequences is not possible (the value of attribute $p = 1$ is obtained).

Let us denote the number of the step of inference by h, and the general attribute of the continuation of inference by P ($P = 0$—continuation of inference is possible, $P = 1$—continuation of inference is not possible). Then the description of the method can be presented as follows.

1. Determination of the initial values: $h = 1$, $M \neq \varnothing$, $m^F \neq \varnothing$. Forming inferred disjunct R_1, consisting of literals L_k, forming set m^F. Determination of the set of consequences e_0, coinciding with facts M^F, having in the premise: $e_0 = M^F \cap m^F, s^0 = \{e_0\}, c_1 = e_0$. Determination of the initial value of the general attribute of the continuation of inference: $P_0 = 0$.

  2. Performing h-th procedure of inference

1. On the first step of inference procedure $\nu_h = <M, R_h, \varnothing, o_h, p_h, R_{h+1}, e_h, f_h>$ is performed.
2. Otherwise procedure $\nu_h = <M, R_h, m^F, o_h, p_h, R_{h+1}, e_h, f_h>$ is performed.

  3. Forming the set of consequences, the set of sets of literals of the description of the inference scheme and checking the attributes. The set of sets of consequences $s^h = s^{h-1} \cup \{e_h\}$ and the set of sets of literals of the description of the inference scheme $O = O \cup \{f_h\}$ are formed. The general attribute of the continuation of inference $P_h = P_{h-1} \vee p_h$ is formed. If $P_h = 0$, then inference continues: h is increased by one, it is accepted that $c_{h+1} = C_h$ and point 2 is passed on to, otherwise inference ends ($h = H$). The obtained consequences are contained in sets of set of sets s^H, and the general set of consequences is calculated by means of joining these sets: $M^S = e_0 \cup e_1 \cup e_2 \cup ... \cup e_H$.

  The description of the scheme of inference of consequences represents a set of sets of descriptions of inference $O = \{f^1, ..., f^h, ..., f^H\}$, formed on the final step, complemented by the set of finite consequences $\{s^+\}$. This set of sets consists of sets of literals with parameters. The edge of the scheme is marked by a literal, the first parameter of the literal being the vertex of the scheme, out of which the edge goes out, and the second being the vertex the edge comes into. The scheme is built in accordance with the steps of inference: in the beginning vertices and edges, described in set of literals f_1, are marked, then connections and vertices, described in set of literals f_2, are added and so on. Te final step of building the inference scheme consists in marking the edges, corresponding to the finite consequences and having no terminal vertices. The set of finite consequences is determined as follows: $O' = f_1 \cup f_2 \cup ... \cup f_H, s^+ = (M^S \sqcup O') - O'$, and the peculiarity of the operation of special joining the sets of literals "$\sqcup$" is the absorption of literal $L(j, +) \in M^S$ by literal $L(j, k) \in O'$.

  Let us illustrate the usage of the method of inference of consequences by the example about filiation from the work [8].

Example 4. A detailed example of the method using is presented in the paper [5].

5   Conclusion

Initially the task of deductive inference of consequences with building a scheme arose in the course of development of an intellectual system of logical prognosis of situations [9]. The description of schemes, obtained in the process of inference, allows us to trace the process of solution and can be used to evaluate the development of the situation, assuming that the system under analysis is in the dynamic state, reflected on the scheme. Thus the offered method of inference of consequences with building a scheme widens a scope of tasks, which are reasonable to be solved with the help of intellectual systems [10–12].

An important merit of the offered method consists in parallel performing operations of division of disjuncts in the procedure of inference. Application of the high-performance method of inference of consequences, optimized for modern multiprocessor and multicore computing systems and technologies of parallel programming, will allow to reduce the time for solving the tasks of inference.

References

1. Norvig, P., Russell, S.: Artificial Intelligence: A Modern Approach, Global edn. Pearson Education Limited, Edinburgh (2011)
2. Kakas, A.C., Kowalski, R.A., Toni, F.: Abductive logic programming. J. Log. Comput. **2**(6), 719–770 (1992)
3. Zakrevskij, A.: Integrated model of inductive-deductive inference based on finite predicates and implicative regularities. In: Diagnostic Test Approaches to Machine Learning and Commonsense Reasoning Systems, pp. 1–12 (2013)
4. Vagin, D.: Dostovernyy i pravdopodobnyy vyvod v intellektualnykh sistemakh (Reliable and Plausible Conclusion in Intelligent Systems). FizMatLit, Moscow (2008)
5. Bardovskaya, A., Chistyakov, G., Dolzhenkova, M., Strabykin, D.: Examples of the Method of Deductive Inference of Consequences with the Scheme Construction Implements (2018). https://zenodo.org/record/1482457
6. Strabykin, D.A.: Logical method for predicting situation development based on abductive inference. J. Comput. Syst. Sci. Int. **52**(5), 759–763 (2013)
7. Sato, T., Inoue, K., Sakama, C.: Abducing relations in continuous spaces. In: 27th International Joint Conference on Artificial Intelligence, IJCAI 2018, Stockholm, Sweden, pp. 1956–1962, July 2018
8. Ceri, S., Gottlob, G., Tanca, L.: Logic Programming and Databases. Springer-Verlag, Berlin Heidelberg (1990)
9. Strabykin, D.A.: Logicheskiy vyvod v sistemakh obrabotki znaniy (Inference in knowledge processing systems). St. Petersburg State Electrotechnical University LETI, St. Petersburg (1998)
10. Caferra, R.: Logic for Computer Science and Artificial Intelligence. ISTE, London (2011)
11. Bollacker, K., Evans, C., Paritosh, P., Sturge, T., Taylor, J.: Freebase: a collaboratively created graph database for structuring human knowledge. In: 2008 ACM SIGMOD International Conference on Management of Data, SIGMOD'08, Vancouver, Canada, pp. 1247–1249, June 2008
12. Rahman, S.A., Haron, H., Nordin, S., Bakar, A.A., Rahmad, F., Amin, Z.M., Seman, M.R.: The decision processes of deductive inference. Adv. Sci. Lett. **23**(1), 532–536 (2017)

Novel Optimized Filter Design for Filtered-OFDM to Enhance 5G Communication Spectral Efficiency

K. P. Nagapushpa[1(✉)] and N. Chitra Kiran[2]

[1] Visvesvaraya Technological University, Belagavi, Karnataka, India
pushpakiralu@gmail.com
[2] Department of ECE, Alliance College of Engineering and Design Alliance
University, Bengaluru, Karnataka, India
chitrakiran.n@alliance.edu.in

Abstract. The era of 5G communication has to offer a futuristic way of technological advancement in mobile communication and also it has been said that 5G can able to provide higher transmission rate, user-friendly experience, resource utilization, etc. To achieve this 5G technology needs to have support towards small data packets and narrow bands with low power consumption. However, the existing Orthogonal Frequency Division Multiplexing (OFDM) have issues with high side lobes causing undesired leakage, channel interference and high peak to average ratio (PAPR). This paper deals with contextual applicability of modulation schemes especially OFDM as well as filtered OFDM (F-OFDM) to be synchronized with 5G communication scenario. In the time domain OFDM symbol, to improvise the out of band (OfB) radiation of the sub band signal during the process of management of complex domain orthogonally of symbols, a specially designed filter is proposed. The performance evaluation is done among optimized filtered OFDM (F-OFDM) with Cyclic Prefix OFDM (CP-OFDM). From the outcomes, it is found that the streamlined F-OFDM system has got better PAPR value than conventional OFDM system with least Bit Error Rate (BER) at 20 dB of SNR. Through the optimized F-OFDM system spectrum efficiency is enhanced.

Keywords: Bit Error Rate (BER) · Cyclic prefix · Filtered-OFDM ·
Out of band radiation · 5G communication · PAPR · Signal to noise ratio (SNR)

1 Introduction

The era of 5G communication has been predicted as the futuristic technology for mobile communication by next two years [1]. The studies have pretended that 5G can bring evolution into mobile communication as it can able to offer higher transmission rate, user-friendly experience, resource utilization, etc. [2, 3]. Recent researches were presented four technological scenarios for 5G communication that involves wider coverage, higher hotspot capacity, high connectivity with low power utilization and low delay in service [4]. The higher hotspot connectivity needs user friendliness experience of about 1Gbps rate and to have this a wide range of bandwidth is requires

© Springer Nature Switzerland AG 2019
R. Silhavy (Ed.): CSOC 2019, AISC 985, pp. 11–20, 2019.
https://doi.org/10.1007/978-3-030-19810-7_2

which can support high data rate [5]. The 5G communication exhibits the frequency band of over 6 GHz [6]. But a nonstop spectrum bandwidth can be attained only during low-frequency band. Instead, different unconnected free spectrum fragments exist. Thus, to meet higher data rate, spectrum efficiency in 5G communication, it needs to have significant capabilities supporting small data packets and narrow bands with low power consumption [7]. The existing multicarrier systems uses Orthogonal frequency division multiplexing (OFDM) as it has got good features but exhibits disadvantages of cyclic prefix (CP) leading to band resources consumption, needs strict synchronization and the high side lobs of the carrier spectrum which causes undesirable leakage, a high peak-to-average ratio (PAPR), and even severe adjacent channel interference (ACI) [8]. Thus, 5G communication requires an advanced multi-carrier transmission technology.

This paper deals with contextual applicability of modulation schemes specially OFDM as well as filtered OFDM (F-OFDM) to be synchronized with 5G communication scenario. The performance evaluation is done among filtered OFDM (F-OFDM) with Cyclic Prefix OFDM (CP-OFDM). In the time domain OFDM symbol, to improvise the out of band radiation of the sub band signal during the process of management of complex domain orthogonally of symbols, a specially designed filter is proposed. This paper is categorized with different sections like Related work dealing with 5G communication (Sect. 2), System Model (Sect. 3), algorithm implementation (Sect. 4), results and analysis (Sect. 5) and conclusion (Sect. 6).

2 Related Work

This section discusses some of the serious contributory researches in the domain of 5G communication. The consideration of F-OFDM in 5G communication has gained a lot of interest as it offers a multi-service model and also provides the greater spectrum efficiency. However, there is a lack in addressing the systematic analysis of F-OFDM systems, and this concern is been addressed in Zhang et al. [9]. In this regard, [9] have established a mathematical model that has derived the conditions from achieving zero interference channel equalization. Also, low complex analytical expressions were derived from removing the intersub-band (IsB) interference at low cost. Through performance analysis, it has been found that [9] work can be used as technical guidelines for the 5G communication system design and it can mitigate the IsB interferences to a greater extent with least increment in complexity. In order to avoid Adjacent-channel interference (ACI), the existing OFDM systems consider wide guard band which leads to lowering of spectral efficiency. Hence, the upcoming (5G) mobile communication system needs low out of band technique. In that regard, An et al. [10] have presented a widowing F-OFDM (WF-OFDM) system that enhances the spectral efficiency by adapting windowing and filtering technique in OFDM. The performance analysis of [10] suggests that it has got better spectral efficiency by lowering out of band spectrum characteristics.

The recent year of research in efficient utilization of spectrum and offering flexible waveforms for 5G communication and come up with F-OFDM. The design and implementation perspective of F-OFDM have discussed in Guan et al. [11] and performed a field test on it. The test analysis was suggested that the F-OFDM can reduce the spectral leakage and by which the spectral efficiency was enhanced than conventional OFDM systems. Also, identifying a way for an upcoming wireless network is the biggest concern which needs a new form of OFDM waveform. A comparative analysis towards addressing the out of the band and spectral efficiency is found in Stasio et al. [12] where F-OFDM and WF-OFDM were considered with a cyclic prefix for OFDM signals against traditional OFDM signals.

The discussion on the standardized tries for the OFDM inspired waveform was found in Farhang-Boroujeny and Moradi [13]. In this, a unique and common mechanism was built to derive all the different kinds of waveforms which helps to understand the channel equalization and application of each waveform. Through this work [13] was able to identify the limitations and significances of universal F-OFDM and Filter bank multicarrier (FBMC) based OFDM system. In order to overcome the issues of currently existing 4G communication system, Demmer et al. [14] have presented an interesting modulation scheme called block F-OFDM system which relies on cyclic prefix based OFDM system receiver. The block F-OFDM addresses the various challenges in the OFDM system having a legacy in spectral localization, spectral efficiency, etc. The performance analysis [14] gives that block F-OFDM system was able to achieve better scalability, flexibility and slight increment in system complexity. A unique idea of FBMC mechanism significances over futuristic 5G communication was addressed in Ibrahim and Abdullah [15]. The FBMC has got the immunity against multipath fading and also against intersymbol interference. This review analysis of [15] helps to gain the idea about the needs of futuristic communication via bringing FBMC scheme over the OFDM scheme. A comparative study on machine type of communications for 5G is described in Medjahdi et al. [16]. In this various replacement of OFDM scheme were discussed that addresses the challenges involved in 5G communication. The waveforms design of different modulation was compared which gives a better idea of understanding of suitable waveform for 5G communication. Towards same concern, Hu and Armada [17] have presented signal to noise ratio (SNR) for interference analysis using F-OFDM.

The promising comparison of performance among the various multicarrier waveforms for 5G air interfaces is discussed in Eeckhaute et al. [18]. The emerging waveforms include FBMC, resource block F-OFDM and universal F-OFDM. These waveforms were compared each other by considering spectral efficiency, robustness, and numeral complexity. Similarly, the performance with respect to complexity in 5G is presented in Gerzaguet et al. [19]. Towards technologies of the air interface and management is discussed in Demestichas et al. [20]. The simulation analysis of versatile 5G communication is presented in Pratschner et al. [21]. This offers available simulation tool for 4G, 5G or beyond that mobile communications. The work of Zhao et al. [22] has presented a multi-carrier transmission mechanism with resource aware block OFDM for 5G communication. The outcomes of [22] suggest the improvement in performance with respect to intercarrier interference than existing modulation schemes. The adaptive modulation scheme was having OFDM with index modulation, and dual mode OFDM is presented in Çolak et al. [23] that overcomes decrement in data rate and improvement in spectral efficiency. The recent work of Nagarathna and Chitra [24] discussed various feasibility prospective of OFDM in 5G communication and suggested that the existing waveforms were not fully in support of 5G communication. The paradigm of secure mobile payment systems by Chitra et al. [25–29], which needs to be evolved in synchronous to the 5G communication.

From all the recent researches it is observed that most of the researches were focused on conventional OFDM system, less emphasis is given on improving spectral efficiency, complicated usage of Cyclic Prefix and also provided the partial solution to the mobile communication with 5G technique. Thus, there is a need for an optimized F-OFDM system which can resolve the above-stated problems.

3   System Model

The proposed system model considers optimal filter design along with the mechanism of a transceiver for F-OFDM. In order to generate a predictive sequence of number non-negative integer is used as a seed to the random number generation. The control seed state (δ_s) provides a structure. The initialization of seed factor (k) increments with its default value. The proposed OFDM model namely Optimized F-OFDM model for 5G communication (as Shown in Fig. 1) with essential characteristic includes the optimal value of frequency resolution as FFT point ranges between 1024 to 2024, this shall be tested for lower and higher points. In a 5G communication system the dynamic sharing of the spectrum along with support for high data transfer rate in highly heterogeneous collaborative networks require optimal use of frequency resources while using frequency resources optimally the interference with neighbor node should be avoided to achieve suitable waveforms. The effect of high side lobes in OFDM is compensated with a filtering technique and achieve minimization of out of band emission. The use of resource blocks $(\Re_b)$ based filter is gaining popularity. The number of resource block initialization along with a number of subcarrier/resource block (ψ) with the length of a cyclic prefix in each sample is parameterized. The number of data subcarrier in a particular sub-band (η) with the filter length (f_ℓ), $\Re_b$ and ψ is computed by Eq. 1.

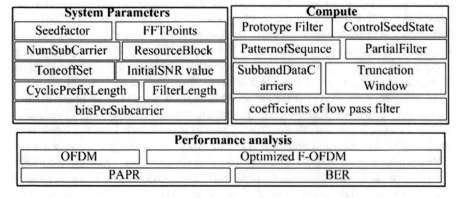

System Parameters		Compute	
Seedfactor	FFTPoints	Prototype Filter	ControlSeedState
NumSubCarrier	ResourceBlock	PatternofSequnce	PartialFilter
ToneoffSet	InitialSNR value	SubbandDataCarriers	Truncation Window
CyclicPrefixLength	FilterLength		
bitsPerSubcarrier		coefficients of low pass filter	

Performance analysis	
OFDM	Optimized F-OFDM
PAPR	BER

Fig. 1. Architectural model of an Optimized F-OFDM model for 5G communication

$$\eta = \Re_b \times \psi \qquad (1)$$

A partial filter (χ) is considered with half of the filter length (f_ℓ) and based on which a pattern of sequence (ρ) is arranged and is represented in Eqs. 2 and 3.

$$\chi \leftarrow \text{floor}\,(f_\ell/2) \qquad (2)$$

$$\rho \leftarrow -\chi, +\chi \qquad (3)$$

Later, the length of cyclic prefix (CP) is provided for each signal samples. In this system, a 64-QAM scheme is utilized which helps in carrying more bits of information over each symbol. The 64-QAM has 6-bits per symbol which can help to enhance the data rate of a link. An initial value of Signal to Noise Ratio (iSNR) and excess bandwidth or tone offset (τ) is adjusted.

The filter design for F-OFDM involves various factors into consideration, i.e., sampling or sinc function for prototype filter (Pf), truncation window (ς), coefficients of low pass filter (Cop), the impulse response of the filter, filtering objects, a matched filter for the receiver, and QAM symbol mapper. The computation of the prototype filter (Pf), truncation window (ς) and Cp is shown in Eqs. 4, 5 and 6 respectively.

$$Pf \leftarrow \text{sinc}((\eta + 2 \times \tau) \times \rho/\varphi) \qquad (4)$$

$$\varsigma = \left(\frac{1}{2} \times \left(1 + cos\left(2\pi \times \frac{\rho}{(f_\ell - 1)} \right) \right) \right)^{\times\, 0.6} \qquad (5)$$

$$Cop = \frac{(Pf \times \zeta)}{sum(Pf \times \zeta)} \qquad (6)$$

The transmit processing of F-OFDM can be initialized with the generation of data symbols by considering η and bits per subcarriers. Further, the data will be packed into OFDM symbols and applied CP. Then, the filter with zero padding will be considered to get transmit signal and computed power spectral density (PSD) and peak to average power ratio (PAPR) of the transmit signal. An Additive White Gaussian Noise (AWGN) is added to this transmit signal and received the matched filter for corresponding f_ℓ, computed filter delay, removed CP, performed FFT and extracted data subcarriers. However, the channel equalization is not necessary as no channel is modeled. Further demapping is applied to compute BER. Finally measured the errors before restoring its control seed state (δ_s). The significance of proposed optimized F-OFDM over existing OFDM is that: in optimized F-OFDM, the sub band cyclic prefix OFDM signal is transmitted through the filter designed above. The filter having passband relevant to the bandwidth of the signal and hence very few subcarriers near to the edges get affected. Also, the filter length (f_ℓ) increases the length of cyclic prefix (Cop) for F-OFDM, and the ISI were are minimized with this widowing (transition) based design of the filter.

4 Algorithm Implementation

The modeling of optimized F-OFDM considers the implementation of the algorithm given below. The modeling begins with initialization of the system parameters like k, φ, $\Re_b$, ψ, f_ℓ, iSNR, τ, iSNR, σ, v with their respective default values (step: 1). To have the control over random number generation, the seed factor (k) is initialized with a default value of 1, and it is then incremented to a standard seeding value of 210 which gives control seed state (δ_s). The number of sub-band data carriers (η) is computed by using Eq. 1. Further, partial filter (χ) is framed as per Eq. 2 and set a pattern sequence range (ρ) by using Eq. 3. The filtering design for F-OFDM has considered flat passband scenario, desired passband attenuation, sinc frequency response, coefficients of low pass filter, etc. This frequency response is computed by adapting Eq. 4. The truncation window can be realized with coefficients of low pass filter truncates responses and eases the smooth transition response and is obtained by using Eq. 6. The digital signal processing filter objects were used for filtering and obtained a matched filter for receiver and QAM mapper is used. The transmission of the signal with F-OFDM is performed with passband having signal bandwidth, least subcarrier at the edge. The used of truncation window reduces the inter-symbol response helps in minimizing the inter-symbol interference (ISI). The data symbols were generated by considering scalar integer of sub-band data carriers (η), bits/Subcarrier (v). Later, packet data will be converted as OFDM symbol by considering FFT points (φ) and η. The cyclic prefix is added into transmitted signal by using inverse FFT and cyclic prefix length (σ). To acquire the transmit signal, the filter with zero padding is considered. At the receiver end, the F-OFDM with no channel is considered but passed a noisy factor into the received signal to acquire respective SNR value. The reverse operation can lead to control seed state. The performance factors like PSNR and BER were computed at respective SNR value.

Algorithm for optimized F-OFDM model

Input: k, φ, $\Re_b$, ψ, f_ℓ, iSNR, τ, iSNR, σ, v
Output: δ_s
Start
Init $\rightarrow$ k, φ, $\Re_b$, ψ, f_ℓ, iSNR, τ, iSNR, σ, v
Increment k $\leftarrow$ k+210, Compute $\delta_s \leftarrow$ rng(k+210)
Compute η $\leftarrow \Re_b \times \psi$, Calculate χ $\leftarrow$ floor $\left(f_\ell / 2 \right)$
Set ρ $\leftarrow -\chi, +\chi$
  Compute $Pf \leftarrow$ sinc$((\eta + 2 \times \tau) \times \rho / \varphi)$
Calculate $\varsigma = \left(\frac{1}{2} \times \left(1 + cos \left(2\pi \times \frac{\rho}{(f_\ell - 1)} \right) \right) \right)^{\times 0.6}$
Matched filter $\leftarrow$ DSP objects (filtering)
Generate data-symbols $\leftarrow$ Scalar (η, v)
Convert packet data (φ, η) $\rightarrow$ OFDM Symbols
Retrieve δ_s , ***End***

---

5 Results and Analysis

The behavioral characteristics of the proposed (optimized) F-OFDM system are being analyzed with the various graphs that includes (1) Magnitude/phase response Vs. Frequency, Power Spectral Density (PSD) Vs. A normalized frequency for (2) OFDM system and (3) optimized an F-OFDM system. For experimentation, system parameters were adjusted for particular values as shown in Table 1. During comparative analysis, the existing OFDM modulation scheme is fully occupied band having similar Cyclic Prefix Length (σ).

Table 1. Values of System parameters/parameter initialization

Parameter	Notation	Value
Seed factor	k	1–211
FFT points	φ	2024
Resource blocks	$\Re_b$	75
Num sub carrier	ψ	24
Filter length	f_ℓ	600
Cyclic prefix length	σ	84
Bits per subcarrier	v	6 (64QAM)
Tone offset	τ	2.0
Initial SNR value	iSNR	20 dB

Figure 2 illustrates the variation of magnitude/phase with varying range of frequency, in which phase response (measured in radians) is the relation among the phase of input and output sinusoidal signal and magnitude response (measured in MHz) is characterized by system magnitude. In Fig. 2, the magnitude is declined towards an increase in Frequency (MHz) and becomes steady at a particular frequency of 7 MHz while the phase of the signal is steady till 7 MHz and starts decreasing gradually with little spikes.

The Fig. 3 represents the change in the power spectral density (PSD) which is a measure of power content against normalized frequency for OFDM system having 1800 subcarriers and are specialized with broadband signals. Through Fig. 3, it is observed that the normalization of frequency, i.e., a division of magnitude at each spectral line remains unchanged between −0.4 Hz to +0.4 Hz of frequency range with PSD of range −30 to −60 dBW/Hz which directly helps to have higher utilization of spectrum allocated and hence spectral efficiency is enhanced than existing OFDM system. The Fig. 4 represents the change in the power spectral density (PSD) against normalized frequency for an optimized F-OFDM system having 24 subcarriers each 75 resource blocks and are specialized with broadband signals. Through Fig. 4, it is observed that the normalization of frequency, i.e., division of magnitude at each spectral line remains unchanged between −0.4 Hz to +0.4 Hz of frequency range with PSD of range −30 to −180 dBW/Hz. From comparing both the Figs. 3 and 4, it is observed that the spectral density of optimized F-OFDM has lower side lobes than OFDM which directly helps to have higher utilization of spectrum allocated and hence spectral efficiency is enhanced than existing OFDM system. Finally, the respective PAPR for both OFDM and optimized F-OFDM is computed as 9.3401 dB and 10.4366 dB respectively at SNR of 20 dB. Similarly, BER rate of F-OFDM is calculated as 0.41546.

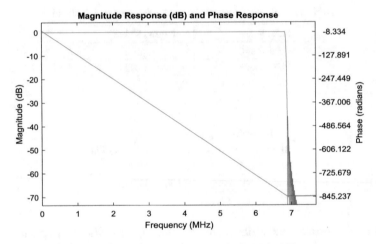

Fig. 2. Magnitude/phase response analysis against Frequency

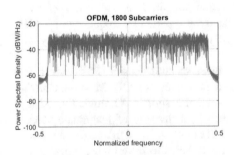

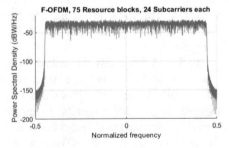

Fig. 3. Power spectral Density analysis against normalized Frequency for OFDM system

Fig. 4. Power spectral Density analysis against normalized Frequency for optimized F-OFDM system

6 Conclusion

This paper deals with the issues of existing researches which are failed to improve spectral efficiency, complicated usage of Cyclic Prefix and partial solution to the mobile communication with 5G technique. The paper presented an optimized filtered OFDM (F-OFDM) to be synchronized with 5G communication scenario. The performance evaluation is done among filtered OFDM (F-OFDM) with Cyclic Prefix OFDM (CP-OFDM). The proposed system model considers optimal filter design along with the mechanism of a transceiver for F-OFDM. The outcomes suggest that the system can offer a higher spectrum for utilization and enhanced spectrum efficiency. From performance analysis, it is found that the proposed F-OFDM is computed as 9.3401 dB and 10.4366 dB respectively at SNR of 20 dB. Similarly, BER rate of F-OFDM is calculated as 0.41546.

References

1. Alliance, N.G.M.N.: 5G white paper. Next generation mobile networks, white paper, pp. 1–125 (2015)
2. Wunder, G., Kasparick, M., Wild, T., Schaich, F., Chen, Y., Dryjanski, M., Buczkowski, M., Pietrzyk, S., Michailow, N., Matthé, M., Gaspar, I., Mendes, L., Festag, A., Fettweis, G., Doré, J.-B., Cassiau, N., Kténas, D., Berg, V., Eged, B., Vago, P.: 5GNOW: intermediate frame structure and transceiver concepts. In: Globecom Workshops (GC Wkshps), Austin, pp. 565–570 (2014)
3. Schaich, F., Wild, T., Chen, Y.: Waveform contenders for 5G - suitability for a short packet and low latency transmissions. In: 2014 IEEE 79th Vehicular Technology Conference (VTC Spring), pp. 1–5 (2014)
4. Lema, M.A., et al.: Flexible dual-connectivity spectrum aggregation for decoupled uplink and downlink access in 5G heterogeneous systems. IEEE J. Sel. Areas Commun. **34**(11), 2851–2865 (2016)
5. Niknam, S., et al.: A multiband OFDMA heterogeneous network for millimeter wave 5G wireless applications. IEEE Access **4**, 5640–5648 (2016)
6. Chen, S., Zhang, P., Tafazolli, R.: Enabling technologies for beyond TD-LTE-Advanced and 5G wireless communications. China Commun. **13**(6), iv–v (2016)
7. Moongilan, D.: 5G wireless communications (60 GHz band) for smart grid-An EMC perspective. In: 2016 IEEE International Symposium on Electromagnetic Compatibility (EMC). IEEE (2016)
8. Wang, X., et al.: Universal filtered multi-carrier with leakage-based filter optimization. In: European Wireless Conference (2014)
9. Zhang, L., et al.: Filtered OFDM systems, algorithms, and performance analysis for 5G and beyond. IEEE Trans. Commun. **66**(3), 1205–1218 (2018)
10. An, C., Kim, B., Ryu, H.-G.: WF-OFDM (windowing and filtering OFDM) system for the 5G new radio waveform. In: 2017 IEEE XXIV International Conference on Electronics, Electrical Engineering and Computing (INTERCON). IEEE (2017)
11. Guan, P., Wu, D., Tian, T., Zhou, J., Zhang, X., Gu, L., Benjebbour, A., Iwabuchi, M., Kishiyama, Y.: 5G field trials: OFDM-based waveforms and mixed numerologies. IEEE J. Sel. Areas Commun. **35**(6), 1234–1243 (2017)

12. Di Stasio, F., Mondin, M., Daneshgaran, F.: Multirate 5G downlink performance comparison for f-OFDM and w-OFDM schemes with different numerologies. In: 2018 International Symposium on Networks, Computers and Communications (ISNCC). IEEE (2018)

13. Farhang-Boroujeny, B., Moradi, H.: OFDM inspired waveforms for 5G. IEEE Commun. Surv. Tutor. **18**(4), 2474–2492 (2016)

14. Demmer, D., et al.: Block-Filtered OFDM: a novel waveform for future wireless technologies. In: 2017 IEEE International Conference on Communications (ICC). IEEE (2017)

15. Ibrahim, A.N., Abdullah, M.F.L.: The potential of FBMC over OFDM for the future 5G mobile communication technology. In: AIP Conference Proceedings, vol. 1883, no. 1. AIP Publishing (2017)

16. Medjahdi, Y., et al.: On the road to 5G: comparative study of physical layer in MTC context. IEEE Access **5**, 26556–26581 (2017)

17. Hu, K.C., Armada, A.G.: SINR analysis of OFDM and f-OFDM for machine type communications. In: 2016 IEEE 27th Annual International Symposium on Personal, Indoor, and Mobile Radio Communications (PIMRC). IEEE (2016)

18. Van Eeckhaute, M., et al.: Performance of emerging multi-carrier waveforms for 5G asynchronous communications. EURASIP J. Wirel. Commun. Netw. **2017**(1), 29 (2017)

19. Gerzaguet, R., et al.: The 5G candidate waveform race: a comparison of complexity and performance. EURASIP J. Wirel. Commun. Netw. **2017**(1), 13 (2017)

20. Demestichas, P., et al.: Emerging air interfaces and management technologies for the 5G era, p. 184 (2017)

21. Pratschner, S., et al.: Versatile mobile communications simulation: the Vienna 5G link level simulator. arXiv preprint arXiv:1806.03929 (2018)

22. Zhao, Y., et al.: Resource block filtered-OFDM as a multi-carrier transmission scheme for 5G. Comput. Electr. Eng. (2017)

23. Çolak, S.A., Acar, Y., Basar, E.: Adaptive dual-mode OFDM with index modulation. Phys. Commun. **30**, 15–25 (2018)

24. Nagapushpa, K.P., Chitra Kiran, N.: Studying applicability feasibility of OFDM in upcoming 5G network. Int. J. Adv. Comput. Sci. Appl. (IJACSA) **8**(1), 216–220 (2017)

25. Chitra Kiran, N., Kumar, G.N.: Modelling efficient process oriented architecture for secure mobile commerce using hybrid routing protocol in mobile adhoc network. Int. J. Comput. Sci. Issues (IJCSI) **9**(1), 311 (2012)

26. Chitra Kiran, N., Kumar, G.N.: A robust client verification in cloud enabled m-commerce using gaining protocol. Int. J. Comput. Sci. Issues 8(6) (2012)

27. Chitra Kiran, N., Kumar, G.N.: Reliable OSPM schema for secure transaction using mobile agent in micropayment system. In: 2013 Fourth International Conference on Computing, Communications and Networking Technologies (ICCCNT). IEEE (2013)

28. Chitra Kiran, N., Kumar, G.N.: Building robust m-commerce payment system on offline wireless network. In: 2011 IEEE 5th International Conference on Advanced Networks and Telecommunication Systems (ANTS). IEEE (2011)

29. Chitra Kiran, N., Kumar, G.N.: Implication of secure micropayment system using process oriented structural design by hash chain in mobile network. IJCSI Int. J. Comput. Sci. Issues **9**(1), 329 (2012)

Multi-agent Modeling of the Socio-Technical System Taking into Account the Risk Assessment

Natalya Bereza[1]([⊠]), Andrey Bereza[2], Maxim Lyashov[2], and Juliia Alekseenko[2]

[1] Alternative Solutions Limited Liability Company, Shakhty, Russia
nvbereza@bk.ru
[2] Institute of Service and Business (Branch), Don State Technical University, Shakhty, Russia
{anbirch,maxl85}@mail.ru, ajulyav@gmail.com

Abstract. The principles of constructing a risk function within the framework of the parametric model of an intelligent agent are considered. The risk function includes the significance of the individual risk, risk assessment (consequences of risks), susceptibility or vulnerability to risk, interaction degree of risks, frequency of risk occurrence, duration of the risk factors over time (it is important in case if the duration of some risks can change and exacerbate the damage from other risks over time). The algorithm to assess the risk is proposed, risks are determined when buying goods online, and the methodology to determine the risk function is developed.

Keywords: Intelligent agent · Risk assessment · Multi-agent system

1 Introduction

The increase in the volume of incoming information and rapidly changes in the environment state become a serious challenge for the capabilities of the human intellect. Thus, it explains the relevance of informative DSSR and recommender systems with elements of artificial intelligence. One of the most promising approaches to increase the efficiency and performance in developing such systems is to use multi-agent technologies. An intelligent agent is able to imitate human behavior, make decisions and choose strategies for behavior in the environment which constantly changes. The inherent element of human activity is a risk associated with the uncertainty and lack of information, the presence of NO-factors of the environment and the counteraction from other active elements of the system, and therefore, in the modeling multi-agent systems, risk assessment is an essential element of the agent behavior.

© Springer Nature Switzerland AG 2019
R. Silhavy (Ed.): CSOC 2019, AISC 985, pp. 21–31, 2019.
https://doi.org/10.1007/978-3-030-19810-7_3

2 Literature Review and Problem Statement

Having analyzed research papers in the area of modeling multi-agent systems, the authors have concluded that no enough attention has been paid to the topic of risk accounting in the process of decision making under uncertainty. The risk accounting problem on the step of multi-agent system design in the electric power industry is considered in [1]. The authors in [2] describe risk accounting while designing a multi-agent navigation system. We can mention the article [3] by Busby, Onggo, Liu about agent-based computational modelling of social risk responses, [4] by Pasman about new improved process, plant risk and resilience analysis tools, [5] by Pham on modelling multi-agent decision support systems (DSS) working in risk environments.

Risks can arise at all levels of agent interaction:

1. Micro (Internal) Environment (agent subsystems);
2. Micro (External) Environment (counterparties which the agent interacts with). The agent interacts with the micro (external) environment and, thus, it can influence this environment to some extent;
3. Macro (External) Environment (factors that do not depend on the agent's actions, the agent cannot influence them, but they to some extent influence the agent).

The parametric description of the agent includes [6]

$$AgS_i = <B_i, G_i, PL_i, Sn_i, Ev_i >; i = 1, n, \qquad (1)$$

where $B_i = \{b_{i1}, b_{i2}, \ldots\}$—knowledge base of i-th agent; b_{ij}—j-th knowledge area;

$G_i = \{g_{i1}, g_{i2}, \ldots\}$—set of goals g_{ik} (k-th goal of the i-th agent);

$PL_i = \{pt_{i11}, pt_{i12}, \ldots\}$—bank of behavior models (plans) pt_{ikl} (l-st plan to achieve the k-th goal of the i-th agent);

$Sn_i = \{p_{i1}, p_{i2}, \ldots\}$—structure of intentions (list of behavior plans p_{ik} chosen by the agent ag_i to achieve the goal $g_{ik} \in G$);

$Ev_i = \{Ev_i(ag_1), Ev_i(ag_2), \ldots, Ev_i(ag_j), \ldots\}$—description of external relations with agents interacting with ag_i.

Obviously, such parameters of the agent's state as the agent's knowledge base, the bank of behavior models, and the description of external relations with agents imply the presence of some indicators related to risk.

Therefore, the extended parametric description of the agent looks as follows:

$$AgS_i = <B_i, G_i, PL_i, E_i, Sn_i, Ev_i, R_i >; i = 1, n, \qquad (2)$$

where $R_i = \{r_{i11}, r_{i12}, \ldots\}$—bank of risk assessment models $r_{ik\,l}$ (for the l-st plan to achieve the k-th goal of the i-th agent).

The risks of each agent are reflected, in general, by the set of objective functions. For the i-th agent, the j-th component of its objective function we will define as $f_{ij}(\cdot), j = \overline{1, m_j}$, where m_j – «dimension of risks»; the set of components of the agent's objective risk function will be as follows $R_i = \{1, 2, \ldots, r_i\}, i \in I$.

Obviously, it is quite difficult to formulate the unified approach to risk assessment in various areas and to select one common set of criteria. In each area there is a specific set of risks, NO-factors, parameters that need to be monitored for risk assessment, and, also, risk assessment criteria.

3   MAC Architecture for Risk Assessment in Online-Trade

Most often, the risk in the scientific literature is defined as the product of some risk damage amount and the probability of its occurrence. This occurrence probability is based on the statistics of observations and is defined as the ratio of unfavorable outcomes (the occurrence of risk situations) to the total sample size. However, there are also approaches based on the features of technical or social systems functioning. Also, one can identify risk assessment (consequences of risks), vulnerability to risk, and the degree of interaction of risks [9, 10] in risk assessment in terms of system interaction and a decision maker (DM) with the environment. Also in several works like [10] the frequency of risk occurrence is considered. As an indicator assessing the consequences of risks both quantitative indicators, for example, the level of demand for enterprise products or goods [7] and the stage of the life cycle of the market [8] for the economic system (enterprise) and qualitative or linguistic indicators, for example, the level of risk may assessed in terms of the danger of consequences and frequency [10].

In terms of the system approach, the following assessment criteria can be distinguished as part of the risk function:

- significance of the individual risk;
- risk assessment (consequences of risks);
- susceptibility or vulnerability to risk (sensitivity to risk, i.e. the extent to which the risk affects the system indicators);
- interaction of risks (presence or absence of relations between risks that can increase/weaken risks);
- frequency of risk occurrence (probability of occurrence);
- duration of the risk factors over time (it is important in case if the duration of some risks can change and exacerbate the damage from other risks over time).

These risk assessment criteria are universal enough, allow the use of various computational methods, they can be applied in such areas as the effectiveness assessment of innovative projects (innovative management), e-commerce, information security, ecology, energetics, logistics, infrastructure management of socio-technical and technical systems.

The determination of the composition of the objective risk function in the development of search agents on the Internet is associated with the features of the search object and the purposes of the subject performing this search. Consider the principles of constructing the risk function for the search agent that performs the selection of products (goods) in an online store. Obviously, the risk function must tend to a minimum. At the same time, the purchase price (that is, the agent goal is to find the product (goods) of the given name at the lowest price) must also tend to a minimum. The most frequent risks when buying goods online include: 1 fraud/deception/loss of goods, 2 poor quality goods/defect, 3 discrepancy with description/size/color, 4 delivery time violation, 5 the impossibility of partial redemption or return of goods in case when a part of goods for some reasons is not suitable (actual when ordering several items), but it is possible to estimate the cost of the return.

This list can be extended depending on the type of goods, for example, for clothing, it is possible that the delivered product is in accordance with the color, size, and description, has no defect, but does not fit.

Modern search engines and filter systems existing at most sites of online shops allow a buyer to choose goods by its name, model, price, color, size, seller or model rating, popularity and some other criteria. However, the real quality of delivered goods and delivery services can only be assessed by customer feedback. Respectively, to collect the necessary information about the risk of online purchases, buyers need to analyze the reviews and perform a risk calculation.

In order to avoid unnecessarily complicating the structure of agents, it is logical to present a multi-agent system in the format of a multi-level distributed structure which helps to speed up the solution search and make it easier for the user to work with the service (Fig. 1):

Level 1 – the search agent performs the function of searching for the necessary product or service on the Internet. This search can be carried out in the form of a test request (product or service type, model name, color, size, material or other characteristics of the product) or by using a photograph of the desired product. Also, the user specifies the price range of the purchase and some priorities (price, delivery time, seller rating, etc. [11]).

Level 2 – based on the search results of the decomposition agent, the task is decomposed in such a way that each assessment agent works on its trading platform, where it performs risk assessment of purchasing products according to the task it received.

Also, the decomposition agent informs the coordinating agent about the number of links found. If the number of links exceeds a certain reasonable limit, it set a limit on the number of resources processed.

Level 3 – the assessment agents evaluate the risk of buying a product on the selected trading platform and select one or more options from the local resource in terms of minimizing the price/risk criterion.

Level 4 – the coordinating agent collects data from the agents and transfers it to the aggregating agent. If the number of links found on a request exceeds the number of processed resources, the coordinating agent directs the free assessment agent to the decomposition agent for further performing this task.

Level 5 – the aggregating agent chooses from the data provided by the coordinating agent one (optimal) and several (up to 5 reserve) options for the buyer (user).

First, the user sends the initial search task to the search agent in the system, and then the search for suitable resources on the Internet (online shops) starts.

In addition to the product name (model), the maximum and minimum purchase prices, some priorities are determined when choosing the product (price, delivery time, product and seller rating, opportunity and ease of return or partial repurchase). Next, the decomposition is performed depending on the results obtained, the necessary number of assessment agents is determined. Each of the found trading platforms is distributed among the assessment agents and calculated in parallel. The assessment agents also receive data on the price range and priorities which the user sets.

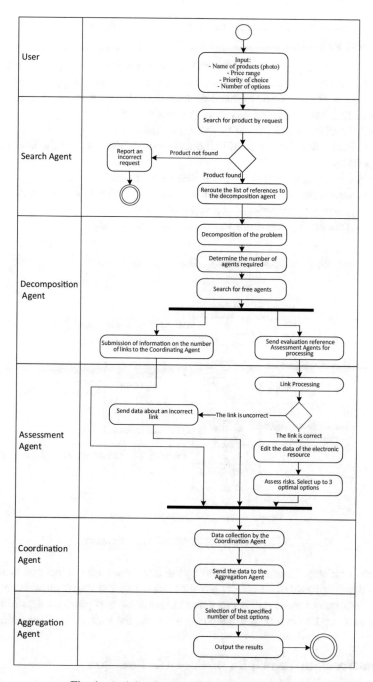

Fig. 1. Activity diagram of multi-agent system.

The algorithm for determining the risk by the assessment agent includes the following steps (Fig. 2)

Step 1. Determining the need for risk assessment. To do this, it is necessary to search and compare the maximum and minimum of possible proposed purchase price options on the selected resource with the value of the price range.
Step 2. Risk ranking according to user's preferences.
Step 3. Determining the value of individual risks.
Step 4. Integrating the values of individual risks and determining the level of risk of the purchase.
Step 5. Choosing the seller by the price/risk criterion.
Step 6. Developing purchase recommendations (the offer of one optimal or several (2–3) options for purchasing the desired product).
Step 7. Transferring the result to the coordinating agent

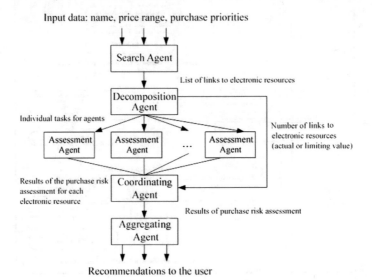

Fig. 2. Architecture of multi-agent system.

At the next step, all results obtained by the assessment agents must be provided to the coordinating agent. The results are then passed to the aggregating agent, which again selects one or more options to make the purchase and provides data to the user. The selection criteria can be the price/risk ratio and the user scale of priorities.

4   Construction Principles of the Risk Function in Online-Trade

Consider the process of risk assessment by the assessment agent in more details. The significance of the individual risk or its priority is determined by the buyer's preferences or purchase characteristics. For example, if the order includes several goods, there is a risk of loss as the result of the impossibility of partial redemption or return. The delivery time violation may have an initial value, if the goods must be purchased by a certain date. The risk significance can be represented in the form of the ordinal scale of risks. When designing a system, it is necessary to provide the ability to rank risks before starting a search.

The risk consequences can be proportional to the order amount (for the first three or fifth risk types) or it can be also associated with the priority of risks (for example, to meet the deadlines by purchasing some goods as a gift). It is set by the user based on personal preferences in the form of linguistic meanings.

The susceptibility or vulnerability to risk (sensitivity to risk, i.e. the influence degree of this risk on the overall value of risk) is due to previous experience with this seller or the product delivery terms. It is the equivalent of an uncertainty measure. In the presence of a large number of reviews, we believe that risk exposure can be considered low; in the absence of feedback there is a high risk possibility.

The interaction of individual risks (the presence or absence of relations between risks) can lead to both increasing and weakening the final risk value. Not all listed risks are interrelated. Risks №1 (fraud/deception/loss of goods) and №5 (the impossibility of partial redemption or return of goods in case when a part of goods for some reasons is not suitable) intensify the negative consequences of risks №2 (poor quality goods/defect) and №3 (discrepancy with description/size/color), while the interdependence of risk №4 (delivery time violation) with the others cannot be called univocal.

The influence of risks on each other can be represented as follows:

Risk №1 increases №2, №3, №4, №5.
Risk №2 increases №3, №5.
Risk №3 increases №2, №5.
Risk №5 increases №1, №2, №3.

The possibility of fitting which some trading platforms provide at their pickup points or when the product is delivered by a courier, as well as partial payment and redemption allow reducing the unpleasant consequences of the risk of receiving a poor-quality or not suitable product. The risk of fraud and deception increases the possibility of obtaining a poor-quality or inappropriate product.

The possibility of delivery time violation can be associated with both the product loss or fraud, and the poor performance of delivery services, which is often impossible to predict. At the same time, the risk of fraud increases the deadline violation possibility.

The frequency of risk occurrence (probability of occurrence) can be estimated by rating the product (the average score or ratio of positive and negative reviews). In each case, the composition of the keywords, by which the ratio of positive and negative reviews will be determined, will depend on the risk type and product characteristics.

The methodology for risk assessment using the risk function is presented in the table below.

Obviously, the most convenient way is using linguistic variables (estimations) in order to assess some criteria. Pre-ranking and prioritization of risks both allow for additive convolution to be used for risk integration; fuzzy inference, neural networks, genetic algorithms, or the hybridization of several methods can also be used [12] (Table 1).

Table 1. Risk assessment methodology of online purchase with risk function

Risk	Assessment criteria as part of the risk function	Means of determining	Markers/indicators/measure
1 fraud/deception/loss of goods	Significance of the individual risk	Risk priority	Ordinal scale, Saati's method (numeric or linguistic value)
	Risk assessment (consequences of risks)	Proportional to the amount of purchase	Seller's rating/work experience on the website
	Susceptibility or vulnerability to risk (sensitivity to risk, i.e. the extent to which the risk affects the system indicators)	Linguistic Estimations	The smaller the number of reviews, the greater the vulnerability to risk Risk increases if the agent's memory lacks the experience of buying goods from this seller
	Interaction of risks (presence or absence of relations between risks that can increase/weaken risks)	Linguistic Estimations	Risks №2, №3, №4, №5 increase
	Frequency of risk occurrence (probability of occurrence)	Linguistic Estimations	Customer Reviews. Search by keywords: did not return money, the goods did not reach
2 poor quality goods/defect	Significance of the individual risk	Risk priority	Numeric or linguistic value
	Risk assessment (consequences of risks)	Proportional to the amount of purchase	Seller's rating/work experience on the website
	Susceptibility or vulnerability to risk (sensitivity to risk, i.e. the extent to which the risk affects the system indicators)	Linguistic Estimations	The smaller the number of reviews, the greater the vulnerability to risk Risk increases if the agent's memory lacks the experience of buying goods from this seller
	Interaction of risks (presence or absence of relations between risks that can increase/weaken risks)	Linguistic Estimations	Risks №3, №5 increase
	Frequency of risk occurrence (probability of occurrence)	Linguistic Estimations	Number of negative/positive reviews. Reviews. Search by words: broken packaging, unpleasant smell, terrible quality, sewn sloppy

(continued)

Table 1. (*continued*)

Risk	Assessment criteria as part of the risk function	Means of determining	Markers/indicators/measure
3 discrepancy with description/size/color	Significance of the individual risk	Risk priority	Ordinal scale, Saati's method (numeric or linguistic value)
	Risk assessment (consequences of risks)	Proportional to the amount of purchase	Cost of purchase
	Susceptibility or vulnerability to risk (sensitivity to risk, i.e. the extent to which the risk affects the system indicators)	Linguistic Estimations	The smaller the number of reviews, the greater the vulnerability to risk Risk increases if the agent's memory lacks the experience of buying goods from this seller
	Interaction of risks (presence or absence of relations between risks that can increase/weaken risks)	Linguistic Estimations	Risks №2, №5 increase
	Frequency of risk occurrence (probability of occurrence)	Linguistic Estimations	Number of negative/positive reviews. Reviews. Search by words: undersized oversized, color/quality does not correspond with the photo, the size does not match
4 delivery time violation	Significance of the individual risk	Risk priority	Ordinal scale, Saati's method (numeric or linguistic value)
	Risk assessment (consequences of risks)	Proportional to risk priority	Duration of delivery/traceability (according to reviews)
	Susceptibility or vulnerability to risk (sensitivity to risk, i.e. the extent to which the risk affects the system indicators)	Linguistic Estimations	The smaller the number of reviews, the greater the vulnerability to risk Risk increases if the agent's memory lacks the experience of buying goods from this seller
	Interaction of risks (presence or absence of relations between risks that can increase/weaken risks)	Linguistic Estimations	Risk №1 increases
	Frequency of risk occurrence (probability of occurrence)	Linguistic Estimations	Number of negative/positive reviews. Reviews. Search by words: delivered on time, delivered quickly, reached for N-days (N shorter than the delivery time)/arrived late, the parcel was delayed, it took a long time

(*continued*)

Table 1. (*continued*)

Risk	Assessment criteria as part of the risk function	Means of determining	Markers/indicators/measure
5 the impossibility of partial redemption or return of goods in case when a part of goods for some reasons is not suitable	Significance of the individual risk	Risk priority	Ordinal scale, Saati's method (numeric or linguistic value)
	Risk assessment (consequences of risks)	Proportional to the amount of purchase	Purchase price/number of items in the order
	Susceptibility or vulnerability to risk (sensitivity to risk, i.e. the extent to which the risk affects the system indicators)	Linguistic Estimations	No prepayment, free return and possibility of inspection/fitting before payment, availability of delivery points
	Interaction of risks (presence or absence of relations between risks that can increase/weaken risks)	Linguistic Estimations	Risks №1, №2, №3 increase
	Frequency of risk occurrence (probability of occurrence)	Linguistic Estimations	Number of negative/positive reviews/number of delivery items.

After the risk assessment of several available options of goods in the online store, the option with the lowest risk value and the lowest price (or 2–3 possible options) is chosen. If all proposed options have a high risk level for the user, the system will give recommendations to refuse purchasing on this resource and search for a suitable product on the site of another online store. Further, the results of the assessment agents are transferred to the coordinating agent for further processing, and the assessment agent can, if necessary, continue the search on another resource.

5   Conclusion

The proposed methodology can be used to develop search and recommender information systems, and Internet services. Risk assessment when shopping in online stores will allow users to make more informed and reasoned decisions, reduce financial losses as the result of inappropriate product purchasing and reduce the time spent on search (reading reviews), and reduce the loading on delivery services.

However, the capabilities of the proposed methodology are not limited to e-commerce. This risk assessment methodology is quite universal, and it can be adapted to solve various applied problems. For example, when adjusting a model in accordance with some features of the particular subject area, it is possible to use the proposed methodology in the design of decision support systems, expert and intelligent recommendation systems based on multi-agent technologies.

The disadvantage of this methodology is that risk assessment is carried out only at the macro (external) environment level, i.e. the agents evaluate the options which were found, choose the most suitable ones and give recommendations to the user. In this case, it does not take into account the interaction at the level of micro (internal) environment (agent subsystems) and at the level of micro (external) environment (interaction with counterparties). Risk assessment of the micro (internal) environment requires evaluation of such factors as the agent type, the internal structure of the agent (the composition of its subsystems is the knowledge base of the agent, the bank of behavior models, the structure of intentions); risk assessment of the micro (external) environment must take into account, besides the factors mentioned above, also the description of external relations with other agents.

The research is supported by Russian Foundation for Basic Research (grants#17-07-01323, 18-07-01054).

References

1. Massel, L.V., Galperov, V.I.: Development of multi-agent systems of the distributed solutions of energy problems using agent scenarios. In: Proceedings of Tomsk Polytechnic University, vol. 326, no. 5, pp. 45–53 (2015)
2. Dmitriev, S.P., Kolesov, N.V., Osipov, A.V.: Safety measures for a ships passing track in the multiagent framework. IFAC Proc. Vol. **33**(21), 373–377 (2000)
3. Busby, J.S., Onggo, B.S.S., Liu, Y.: Agent-based computational modelling of social risk responses. Eur. J. Oper. Res. **251**(3), 1029–1042 (2016)
4. Pasman, H.: New and improved process and plant risk and resilience analysis tools. In: Risk Analysis and Control for Industrial Processes - Gas, Oil and Chemicals, pp. 285–354 (2015)
5. Pham, K.D.: On the determination of cooperative risk-value aware strategies for linear stochastic multi-agent systems. IFAC Proc. Vol. **44**(1), 4198–4205 (2011)
6. Ivashkin, Yu.A., Shcherbakov, A.V.: Multi-agent modeling of poorly formalized conflict. In: Theory of Conflict and its Applications: Proceedings of the International Conference, Voronezh, pp. 7–12 (2006)
7. Arinichev, I.V., Krivko, M.S.: Development of an expert system for quantitative assessment of the risk of bankruptcy of the peasant farming on the basis of a fuzzy-multiple approach. Polytechnical Electronic Scientific Journal of the Kuban State Agrarian University, no. 117, pp. 619–630 (2016)
8. Grushenko, V.I.: Strategy of business management. From Theory to Practical Development and Implementation: Monograph, UNITY-DANA: Law, Moscow, 295 p. (2010)
9. Tukkel, I.L., Surina, A.V., Kultin, N.B. (eds.): Management of Innovative Projects. BHV-Petersburg, St. Petersburg, 416 p. (2011)
10. Chernov, V.G.: Decision support models in investment activity based on fuzzy sets. Goryachaya Liniya -Telecom, Moscow, 312 p. (2007)
11. Bereza, N.V., Beglyarov, V.V., Bereza, A.N., Pavlova, K.A.: Development of principles for constructing a mathematical model of risk assessment for a search intelligent agent in the area of E-commerce. In: Breakthrough Scientific Research: Problems, Patterns, Perspectives: The Collection of Articles of the IX International Scientific and Practical Conference. 2 Parts. Part 1. MTSNS «Science and the Enlightenment», Penza (2017)
12. Piegat, A.: Fuzzy Modeling and Control. Physica-Verlag, New York (2001)

Hybrid Optimization Method Based on the Integration of Evolution Models and Swarm Intelligence in Affine Search Spaces

Boris K. Lebedev, Oleg B. Lebedev$^{(\boxtimes)}$, Elena M. Lebedeva, and Artemy A. Zhiglaty

Southern Federal University, Rostov-on-Don, Russia
lebedev.b.k@gmail.com, lebedev.ob@mail.ru,
clebedeva.el.m@mail.ru, artemiy.zhiglaty@gmail.com

Abstract. A composite architecture of a multi-agent bionic search system based on the integration of swarm intelligence and genetic evolution is proposed. The structure of the affine space of positions is developed, which allows to display and search for interpretations of solutions with integer parameter values. The mechanisms for moving particles in affine space to reduce the weight of affine bonds are considered.

Keywords: Bionic search · Hybridization · Particle swarm ·
Genetic evolution · Affine space · Integer parameters

1 Introduction

Particle Swarm Optimization (PSO) is a stochastic optimization method somewhat similar to evolutionary algorithms. This method models not evolution, but the swarming and gregarious behavior of animals [1]. Unlike population-based methods, PSO works with a single static population, whose members gradually improve with the appearance of information about the search space. This method is a type of **directed mutation**. Solutions in the PSO mutate in the direction of the best solutions found. Particles never die (because there is no selection).

Consider the canonical paradigm of the PSO method, developed by Kennedy and Eberhart [2]. The multidimensional, real, metric search space is populated by a swarm of particles $P = \{p_i | i = 1, 2, \ldots, n\}$ [3]. Each p_i particle is located at position x_i, is connected and can interact with all the particles of the swarm, it is a better solution for the swarm. The process of finding solutions is iterative. At each iteration, the adaptations of the particles are calculated and, if necessary, the information on the best positions found is updated. Each particle moves to a new position. New position is defined as:

$$x_i(t+1) = x_i(t) + v_i(t+1), \tag{1}$$

where $v_i(t+1)$ is the velocity of a particle moving from position $x_i(t)$ to position $x_i(t+1)$. The initial (starting) state is defined as $x_i(0)$, $v_i(0)$. The above formula is

© Springer Nature Switzerland AG 2019
R. Silhavy (Ed.): CSOC 2019, AISC 985, pp. 32–39, 2019.
https://doi.org/10.1007/978-3-030-19810-7_4

presented in vector form. For a separate measurement j of the search space, the formula will take the form:

$$x_{ij}(t+1) = x_{ij}(t) + v_{ij}(t+1), \tag{2}$$

where $x_{ij}(t)$ is the position of the particle p_i in dimension j, $v_{ij}(t+1)$ – is the velocity of the particle p_i dimension j.

We introduce the notation:

- $x_i(t)$ is the current position of the particle, $f_i(t)$ is the value of the objective function of the particle pi at position $x_i(t)$;
- $x_i^*(t)$ is the best position of the particle – x_i^* (t), which she visited from the beginning of the first iteration, and $f_i^*(t)$ is the value of the target function of the particle p_i in this position (the best value since the start);
- $x^*(t)$ is the position of the swarm particle with the best value of the objective function $f^*(t)$ among the swarm particles at the moment of time t.

Then the velocity of the particle p_i in step $(t+1)$ in dimension j is calculated as:

$$v_{ij}(t+1) = w \cdot v_{ij}(t) + k_1 \cdot rnd(0,1) \cdot \left(x_{ij}^*(t) - x_{ij}(t)\right) + k_2 \cdot rnd(0,1) \cdot \left(x_j(t) - x_{ij}(t)\right), \tag{3}$$

where $rnd(0,1)$ is a random number on the interval $(0, 1)$, $(w, k1, k2)$ are some coefficients. The formula for calculating the speed is made up of three components.

By analogy with evolutionary strategies, a swarm of particles can be interpreted as a population, and a particle as an individual (chromosome). This makes it possible to build a hybrid solution search structure based on the integration of genetic search with particle swarm methods [4]. In hybrid algorithms, the advantages of one algorithm can compensate for the shortcomings of another. The link of this approach is the data structure, which describes the solution of the problem in the form of a chromosome [5]. The position of a particle in the search space is equivalent to a genotype in evolutionary algorithms. If a chromosome is used as a particle, then the number of parameters determining the position of the particle in the solution space should be equal to the number of genes in the chromosome. The value of each gene is deposited on the corresponding axis of the solution space. In this case, there are some requirements for the chromosome structure, gene values and search space. A composite architecture of a multi-agent bionic search system based on swarm intelligence and genetic evolution is proposed [6]. The first and simplest approach to hybridization is as follows. From the beginning, the search for a solution is carried out by a genetic algorithm [7]. Then, based on the population obtained at the last iteration of the genetic search, a population is formed for the swarm algorithm. The best, but distinct from each other, chromosomes are included in the formed population. If necessary, the resulting population is equipped with new individuals. After that, a further search for the solution is carried out by a swarm algorithm.

In the second approach, the particle swarm method is used in the process of genetic search and plays a role similar to genetic operators. In this case, at each iteration of the genetic algorithm, the synthesis of new chromosomes on the one hand is carried out with the help of crossing-over and mutation, and on the other hand with the help of operators of the directed mutation of the particle swarm method.

2 Principles of Coding Lists Containing Fixed Amounts of Elements

The canonical paradigm of particle swarm involves the use of real values of parameters in multidimensional, real, metric spaces. However, in most genetic algorithms, genes in chromosomes have integer values. In turn, chromosomes are some interpretations of solutions that transform into solutions by decoding chromosomes.

Analysis of existing methods and algorithms showed that lists of data that are actually interpretations of solutions are most often used as a data structure that carries information about a solution.

This representation is convenient for its use in various metaheuristics (genetic algorithms, ant algorithms), since they work with decision coding sequences. The lists used as interpretations of solutions, depending on the specifics of the problem, are limited.

Consider the principles of coding lists containing fixed amounts of elements. Such lists underlie the interpretation of solutions in coverage problems [8], packaging placements, in distribution problems [9, 10] and in a number of other tasks [11].

The general solution of such problems is represented as a vector $Y = \{y_l | l = 1, 2, 3, \ldots, n\}$ with integer values of the elements that satisfy the constraint.

$$\sum_{l=1}^{n} y_l = b$$

The list corresponds to the chromosome $H = \{h_l | l = 1, 2, 3, \ldots, n\}$.

H is a set of $(n - 1)$ h_l genes whose integer values can vary within the range defined by parameter b:

$$0 \leq h_l \leq b, \; h_l \in H.$$

The values of the genes within H are the reference points inside the segment $[o, b]$ of length b, dividing it into intervals. The length of the interval, defined as $(h_{l+1} - h_l)$, $h_{l+1} \in H$ and $h_l \in H$, is the value of the corresponding element yl of the vector Y (Fig. 1). In Fig. 1, chromosome $H = \; <8, 12, 16, 25>$ for $b = 30$, $n = 5$ corresponds to the list $Y = \; <8, 4, 9, 5>$. The following are analytical expressions for determining the values of y_i on the chromosome H:

$$y_1 = h_1; y_i = h_i - h_{i-1} \text{ for } i = 2, \ldots, n - 1; y_n = b - h_{n-1}. \tag{4}$$

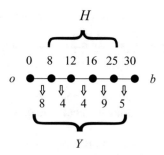

Fig. 1. Chromosome decoding

In the first coding method, first from the H chromosome, they are transferred to the H^* chromosome by ordering the genes by increasing their values in H. Then you can use the analytical expressions (4) applied to H.

In the second coding method, additional constraints are imposed on the gene values, which consist in the fact that within H the genes have values arranged in ascending order, that is, if $h_{i-1} \in H$ and $h_i \in H$, then $h_i \geq h_{i-1}$.

The main genetic operators are crossing over and mutation, using two types of crossing over. The paper proposes an approach to constructing a modified particle swarm paradigm, which makes it possible to simultaneously use chromosomes with integer parameter values in a genetic algorithm and in a particle swarm algorithm.

3 Mechanisms of Moving Particles in Affine Space

The paper proposes an approach to constructing a modified particle swarm paradigm, which makes it possible to simultaneously use chromosomes with integer parameter values in a genetic algorithm and in a particle swarm algorithm.

Let there be a linear vector space (LVS) whose elements are n-dimensional points. To each of any two points p and q of this space, we uniquely associate a unique ordered pair of these points, which we will further call geometric vector (vector). $p, q \in V(p, q)$ is a geometric vector (an ordered pair).

The set of all points of an LVS, supplemented by geometric vectors, is called a point vector or affine space. Affine space is n-dimensional, if the corresponding LVS is also n-dimensional. For any points M, N and P, the equality is feasible:

$$\overline{MN} + \overline{NP} = \overline{MP}$$

Affine-relaxation model (ARM) of a swarm of particles is a graph whose vertices correspond to the positions of a swarm of particles, and the arcs correspond to affine connections between the positions (points) in the affine space. Affinity is a measure of the proximity of two agents (particles). At each iteration, each agent p_i moves in the affine space to a new state (position), at which the weight of the affinity connection between the agent p_i and the base (best) agent p^* decreases. The transfer of the agent p_i from $x_i(t)$ to the new position $x_i(t+1)$ is carried out using a relaxation procedure depending on the type of data structure (chromosomes): vector, matrix, tree and their combination, which is the interpretation of solutions.

The best particles from the point of view of the objective function are declared "center of attraction". The displacement vectors of all particles in affine space rush to these centers.

The transition is possible taking into account the degree of closeness to one basic element or to the group of neighboring elements and taking into account the probability of transition to a new state.

In order to avoid confusion when describing a population (swarm), we will further designate each chromosome describing the I solution of the population as $H_i(t) = \{h_{il} | l = 1, 2, \ldots, n - 1\}$. Moreover, each $H_i(t)$ has the structure described above. In our case, the position $x_i(t)$ corresponds to the solution given by chromosome $H_i(t)$, i.e. $x_i(t) = H_i(t)$. Similarly, $x_i^*(t) = H_i^*(t) = \{h_{il}^* | l = 1, 2, \ldots, n - 1\}$, $x^*(t) = H^*(t) = \{h_l^* | l = 1, 2, \ldots, n - 1\}$. The number of axes in the solution space is equal to the number n of genes in the chromosomes $H_i(t)$, $H_i^*(t)$, $H_i^*(t)$. The starting points on each l axis are the integer values of the genes.

In this paper, the velocity $v_i(t + 1)$ is considered as a means of changing the solution. In contrast to the canonical particle swarm method, in our case the velocity $v_i(t + 1)$ cannot be represented as an analytical expression. The analogue of the $v_i(t + 1)$ rate is the directed mutation operator (DMO), the essence of which is to change the integer values of the genes in the chromosomes $H_i(t)$. Moving the p_i particle to a new position means a transition from the chromosomes $H_i(t)$ to the new one – $H_i(t + 1)$ with the new integer values of the h_{il}, genes obtained after applying the DMO.

As an estimate of the degree of closeness between the two positions $x_i(t)$ and $x_z(t)$, we will use the S_{iz} value of the distance between the chromosomes $H_i(t)$ and $H_z(t)$:

$$S_{iz} = \sum_{l=1}^{n-1} |h_{il} - h_{zl}|. \tag{5}$$

The purpose of moving the chromosome $H_i(t)$ in the direction of the chromosome $H_z(t)$ is to reduce the distance between them.

The essence of the transfer procedure implemented by DMO is to change the difference between the values of each pair of genes (h_{il}, h_{zl}) of the two chromosomes, $l = 1, 2, \ldots, n - 1$.

Moving a particle from the position of $H_i(t)$ to the position of $H_i(t + 1)$ under the influence of attraction to the position of $H_z(t)$ is performed by applying the DMO to $H_i(t)$ as follows. The chromosomes $H_i(t)$ and $H_z(t)$ chromosomes are sequentially viewed (starting from the first), and the corresponding genes are compared. If in the course of sequential viewing of loci in the current locus l with the probability P occurs the event of "mutation", then the $h_{il}(t) \in H_i(t)$ gene mutates.

The probability of mutation P depends on the distance $S_{iz}(t)$ between the positions, and is determined as follows:

$$P = \alpha \cdot S_{iz}(t)/(n-1), \tag{6}$$

where α is the coefficient, $(n-1)$ is the length of the chromosome. Thus, the greater the distance $S_{iz}(t)$ between the positions $H_i(t)$ and $H_z(t)$, the greater the likelihood that the value of $g_{il}(t)$ will be changed.

Gene mutations are performed according to the following rules.

A simple lottery $L(y_1, p, y_2)$ is a probabilistic event that has two possible outcomes y_1 and y_2, the probabilities of which are denoted by p and $(1-p)$, respectively. In other words, with the probability p the lottery is $L(y_1, p, y_2) = y_1$, and with probability $(1-p)$ the lottery is $L(y_1, p, y_2) = y_2$. The expected (or average) price of the lottery is determined by the formula of $py_1 + (1-p)y_2$. As mentioned above, positions are defined by chromosomes. The positions $x_i(t), x^*(t), x_i^*(t) x_i^c(t)$ correspond to the chromosomes $H_i(t) = \{h_{il}(t)|l = 1, 2, \ldots, n-1\}$, $H^*(t) = \{h_l^*(t)|l = 1, 2, \ldots, n-1\}$, $H_i^*(t) = \{h_{il}^*(t)|l = 1, 2, \ldots, n-1\}$, $H_i^c(t) = \{h_{il}^c(t)|l = 1, 2, \ldots, n-1\}$.

The values of the genes of the new position are defined as:

$$h_{il}(t+1) = h_{il}(t) + L(sgn[(h_{zl}(t) - h_{il}(t))], P, 0), \, l = 1, 2, \ldots, n-1. \tag{7}$$

where $sgn[y]$ is the function of the sign of the number:

$$sgn(y) = 1 \text{ if } y > 0; sgn(y) = 0 \text{ if } y = 0; sgn(y) = -1 \text{ if } y < 0.$$

Example. Let $H_z(t) = \,<5, 8, 17, 20>$, $H_i(t) = \,<4, 10, 15, 22>$. In accordance with the expression (7) $S_{iz}(t) = |5-4| + |8-10| + |17-15| + |20-22| = 7$.

Take $P = 1$, then in accordance with the rules of directed mutation:

$$H_i(t+1) = \,<5, 9, 16, 21>\,), \, S_{iz}(t+1) = |5-5| + |8-9| + |17-16| + |20-21|$$
$$= 3.$$

The distance between H_i and H_z positions decreased from 7 to 3.

To account for the simultaneous particle p_i to the $x^*(t)$ and $x_i^*(t)$ positions, a virtual center (position) of attraction $x_i^c(t) = H_i^c(t)$ of the p_i particle is formed. Formation of the virtual position $x_i^c(t)$ is carried out by applying the procedure of virtual movement from the position $x_i^*(t)$ to the virtual position $x_i^c(t)$ towards the position $x^*(t)$.

The values of the genes of the virtual position $x_i^c(t)$ are defined as:

$$h_{il}^c(t) = h_{il}^*(t) + L(sgn[(h_l^*(t) - h_{il}^*(t))], P, 0), l = 1, 2, \ldots, n-1. \tag{8}$$

After determining the center of attraction $x_i^c(t)$, the particle $x_i(t)$ is moved in the direction of the virtual position $x_i^c(t)$ from the position $x_i(t)$ to the position $x_i(t+1)$ using the moving procedure.

$$h_{il}(t+1) = h_{il}(t) + L(sgn[(h_{il}^c(t) - h_{il}(t))], P, 0), l = 1, 2, \ldots, n-1. \qquad (9)$$

After moving the particle p_i to the new position $x_i(t+1)$, the virtual position $x_i^c(t)$ is eliminated.

The local goal of moving the particle p_i is to reach the position with the best value of the objective function. The global goal of a particle swarm is the formation of an optimal solution to the problem.

4 Conclusion

A composite architecture of a multi-agent bionic search system based on the integration of swarm intelligence and genetic evolution is proposed. The link of this approach is a single data structure describing the solution of the problem in the form of a chromosome. The key problem that was solved in this paper is related to the development of the structure of the affine space of positions, which allows displaying and searching for solution interpretations with integer parameter values. In contrast to the canonical particle swarm method, to reduce the weight of affinity bonds, by moving the p_i particle to a new position of the affine solution space, a directed mutation operator has been developed, the essence of which is to change the integer values of genes in the chromosome. New chromosome structures have been developed to represent solutions.

On the basis of the considered paradigm, the program of the planning was developed. For carrying out the experiments of the planning program, the procedure of synthesizing control examples with the known optimum $F_{\text{опт}}$ was used by analogy with the well-known AFEKO method - Floorplanning Examples with Known Optimal area [12]. Quality assessment is the value of $F_{\text{опт}}/F$ – "degree of quality", where F is the evaluation of the solution obtained. Studies have shown that the number of iterations at which the algorithm found the best solution lies within 110–130. The algorithm converges at an average of 125 iterations. As a result of the conducted research, it was established that the quality of the decisions of the hybrid algorithm is 10–15% better than the quality of the solutions of the genetic and swarm algorithms separately. Experiments have shown that an increase in the population of M greater than 100 is impractical, since this does not lead to a noticeable change in quality. The probability of obtaining a global optimum was 0.96. On average, the launch of the program provides a solution that differs from the optimal one by less than 2%. The time complexity of the algorithm with fixed values of M and T lies within $O(n)$. The overall estimate of the time complexity lies within $O(n^2)$–$O(n^3)$.

Acknowledgements. This research is supported by grants of the Russian Foundation for Basic Research of the Russian Federation, the project № 18-07-00737.

References

1. Clerc, M.: Particle Swarm Optimization. ISTE, London (2006)
2. Kennedy, J., Eberhart, R.C.: Particle swarm optimization. In: Proceedings of the IEEE International Conference on Neural Networks, pp. 1942–1948 (1995)

 3. Karpenko, A.P.: Modern search engine optimization algorithms. Algorithms inspired by nature: a tutorial, 448 p. Publishing House MSTU. N.E. Bauman, Moscow (2014)
 4. Wang, X.: Hybrid nature-inspired computational analysis. Doctoral dissertation, Helsinki University of Technology, TKK Dissertations, 161 p. (2009)
 5. Raidl, G.R.: A unified view on hybrid metaheuristics. Lecture Notes in Computer Science, pp. 1–12. Springer (2006)
 6. Blum, C., Roli, A.: Metaheuristics in combinatorial optimization. ACM Comput. Surv. **35**, 268–308 (2003)
 7. Lebedev, B.K., Lebedev, O.B., Lebedev, V.B.: Hybridization of swarm intelligence and genetic evolution on the example of placement. Electron. J. Softw. Prod. Syst. Algorithms **4**, 1–5 (2017)
 8. Lebedev, B.K., Lebedev, V.B.: Coating by the particle swarm method. In: Proceedings of the VIth International Scientific and Practical Conference "Integrated Models and Soft Computing in Artificial Intelligence", pp. 611–619. Fizmatlit, Moscow (2011)
 9. Sha, D.Y., Hsu, C.-Y.: A hybrid particle swarm optimization for job shop scheduling problem. Comput. Ind. Eng. **51**, 791–808 (2006)
10. Lebedev, B.K., Lebedev, O.B., Lebedeva, E.M.: Resource allocation based on hybrid models of swarm intelligence. Sci. Tech. J. Inf. Technol. Mech. Optics **17**(6), 1063–1073 (2017)
11. Lebedev, B.K., Lebedev, V.B.: Tracing based on the particle swarm method. In: Proceedings of the Twelfth National Conference on Artificial Intelligence with International Participation KII-2010, vol. 2, pp. 414–422. Fizmatlit, Moscow (2010)
12. Cong, J., Nataneli, G., Romesis, M., Shinnerl, J.: An area-optimality study of floorplanning In: Proceedings of the International Symposium on Physical Design, Phoenix, AZ, pp. 78–83 (2004)

Applying Context to Handwritten Character Recognition

Richard Fox$^{(\boxtimes)}$ and Steven Brownfield

Northern Kentucky University, Highland Heights, KY 41099, USA
{foxr, brownfiels2}@nku.edu

Abstract. Attempts to automate handwritten character recognition date back to the 1960s, but progress over the past two decades shows extremely accurate recognition of printed characters in English. The most common approaches used today apply a form of machine learning such as support vector machines (SVM) or neural networks. While highly accurate, these forms of machine learning do not attempt to apply higher-level knowledge to improve performance. This paper presents research applying SVM-trained recognizers supplemented with domain knowledge to provide top-down guidance in an attempt to improve recognition accuracy.

Keywords: Handwritten character recognition · Top-down guidance · Support vector machines

1 Introduction

Automated character recognition can be classified into several subtypes such as recognition of machine-produced characters, human-printed characters and human-written (cursive) characters. Character recognition can also be classified by the language being recognized. English is, by far, the most researched language but recent research has explored a wide variety of other languages, including Arabic and Chinese, among others. Attempts to automate handwritten character recognition date back to the 1960s with postal mail sorting [1]. This paper focuses on automated recognition of human-printed English characters.

For printed recognition, handwriting is first digitized, using such steps as noise reduction, sharpening, and thresholding. Segmentation locates distinct characters among the digitized input. Orientation is usually identified to help with segmentation and feature extraction. Feature extraction is applied to supply recognizer(s) with input. Features vary depending on the type of recognizer. Feature extraction techniques range from statistical approaches to mathematical approaches (e.g., topological and geometric analyses) to rule-based approaches to extraction by neural network [2]. Given features, recognition is attempted. Over the years, a wide variety of recognition approaches have been applied to off-line handwritten character recognition. These approaches include the use of rules or some other form of explicitly encoded knowledge base, genetic algorithms, support vector machines, neural networks, and hidden Markov models (HMMs) [3–7]. Each approach may utilize a different set of features (if any).

© Springer Nature Switzerland AG 2019
R. Silhavy (Ed.): CSOC 2019, AISC 985, pp. 40–50, 2019.
https://doi.org/10.1007/978-3-030-19810-7_5

Notably, SVMs, neural networks and statistical forms of processing perform recognition in *isolation*. That is, the recognition task does not utilize other contextual knowledge that might otherwise improve recognition accuracy. With HMMs, grammatical knowledge is available implicitly as transition probabilities [8].

The question examined here is whether additional knowledge can improve accuracy. The most impressive character recognizers available, convoluted or deep neural networks, have recognition accuracy in excess of 99% [9]. With that in mind, why should additional knowledge be utilized? The answer is twofold. First, why not? If such knowledge can be exploited, why shouldn't it? But more significantly, the 99% accuracy occurs with printed characters in English. Recognition accuracy for written (cursive) characters, and recognition of other languages is not as accurate. Although the research described here is for printed character recognition and not written/cursive character recognition, the application of domain knowledge, as demonstrated here, can have a tangible impact on accuracy.

This paper introduces the KASC system. KASC, Knowledge-Augmented SVM Classifier, utilizes trained SVMs to perform an initial recognition and then applies domain-specific knowledge in a top-down way to reexamine characters to improve recognition. With SVMs alone, recognition accuracy for individual characters is in the 60–70% range while accuracy for an entire input is 0%. With top-down guidance, both individual character accuracy and entire input accuracy improves to over 99%. Currently, the only domain implemented is that of recognizing postal addresses (cities, states, zip codes). Work continues to expand KASC to recognize English sentences.

This paper is organized as follows. In Sect. 2, character recognition is explored in more detail, with reference to both past and current approaches. Section 3 introduces the KASC system including training of the SVM classifiers and the implementation of top-down guidance. Section 4 provides experimental results of KASC. Finally, Sect. 5 discusses the enhancements being worked out to move from the postal address domain to the natural language/English sentence domain.

2 Optical Character Recognition Strategies

Automated character recognition can be classified in many ways. First, there is a distinction between real-time recognition and off-line recognition. Second, characters can be machine-produced or human-produced where human-produced characters can be printed or written (cursive). Third, there are many distinctions that arise from differing languages. For instance, recognizing human-printed Arabic has a number of challenges that human-printed English does not [10]. This research concentrates on off-line, human-printed, English characters, so this section concentrates only on that form of recognition.

The first step in off-line character recognition is the digitization of the handwritten message. This step is ignored here as the process involved requires no artificial intelligence. The result of digitization is some binary form of the printed characters, here assumed to be a bitmap. Segmentation is performed to divide the bitmap into distinct characters. With printed characters, this task is simple, handled by looking for sufficient white-space that is assumed to occur between characters. Only with cursive handwriting does this process pose any challenges.

With individual characters located, the next step involves feature extraction of each character. The simplest form of feature extraction is to remove spurious pixels of the bitmap so that the bitmap consists solely of the pen strokes making up the character. Additional processing might try to limit the width of the pen strokes to some uniform size. For neural networks and SVMs, this may be the extent of feature extraction. For other types of recognizers, additional features sought might include aspect ratio (comparison of the width versus height), average distance of pixels from an edge or center of the image, symmetry of the image, and identification of lines or curves of each pen stroke [2]. Another form of feature extraction is to generate a histogram from the bitmap [11]. Histograms identify locations of groups of pixels. Other techniques are contour profiling, Fourier descriptors, curve approximations and gradient features [12]. These are all generated through mathematical analyses of the pixels found in the bitmap.

The identified features are next provided as input to the recognizer(s). Some forms of recognition apply a single recognizer, such as a trained neural network. In such cases, the single recognizer is built, or trained, to generate some form of score for every character that the recognizer was trained to recognize. In a knowledge-based approach, the recognizer may contain a set of descriptors for every character. Based on the features found or not found, the recognizer ranks each character of the language and responds with the most highly matched character, or an ordered list of characters from most to least likely. Aside from neural networks and knowledge-based approaches, HMMs would also fall into this category.

A Support Vector Machine instead is trained to recognize a single class. For character recognition, there must be a trained SVM for each character to be recognized. Given the input, an SVM returns a score of how closely the input matches the expected input for that given character. Thus, there needs to be one SVM for every character to be recognized.

Another approach is to use an *ensemble* of classifiers. In this case, the individual classifiers do not represent different characters but instead they represent different ways of recognizing characters. For instance, an ensemble might contain an HMM and a neural network. Alternatively, different sets of training data may be used to train each classifier. Given output from each classifier in the ensemble, some form of combination function is required to take the results of the classifiers and unite them into a single decision that represents most likely character [13].

Although knowledge-based approaches are seldom used today, KASC builds upon a prior project based on hand-constructed knowledge-based recognizers in the system CHREC [14]. CHREC contained one recognizer per character. Each recognizer sought features of interest within the bitmap. Features were based on horizontal and vertical lines and different shaped curves. Based on the presence or absence of such features and the presence of unexpected features, the recognizer generated a confidence score that the bitmap represented the character sought by that classifier. Highly rated scores formed "islands of certainty" which then permitted domain knowledge to be generated and applied in a top-down way to guide the reexamination of lesser rated characters.

One such domain implemented in CHREC was that of mathematical equations of the form `decimal-value = hexadecimal-value`. CHREC first attempted to recognize as many characters as possible. Given those characters that were considered very likely matches, all possible equations were generated. From those equations, any which held true (that is, the left side equaled the right side) was then used as a "best guess" for CHREC, which used that equation to generate top-down feedback. This feedback was used to guide CHREC which then reexamined mismatched characters, looking for evidence that expected characters (as based on the equation(s) derived) was actually more likely than the originally matched character. This reconsideration caused CHREC to potentially raise scores of some expected characters and lower scores of other unexpected characters.

For instance, if the equation was "285 = 11D" and the highest rated characters recognized were '2', '?', '5', '=', '1', '?', '?', then various equations using these "islands of certainty" were generated, such as "205 = 11B" and "285 = 17D". The first '?' would then be reexamined as a possible '0' or '8'. The resulting reconsideration might cause other characters' scores to be adjusted. CHREC then attempted to perform the recognition character-by-character with the revised scores.

Four different domains were implemented and tested in CHREC: decimal to hexadecimal equations, addresses (city/state/zip codes), bank checks (comparing the numeric dollar amount to the written dollar amount), and short English sentences with a reduced lexicon and simplified grammar. The result of top-down guidance showed a dramatic improvement of recognition from initial character recognition accuracy in the 40–50% range to over 99%.

The research behind CHREC was criticized because the tests were based on data sets generated by one of the researchers rather than by utilizing one of the well-known character banks. In addition, building the individual character recognizers was very time consuming. The next section of this paper describes a different approach whereby the initial recognizers were trained using SVMs and kernel functions.

3   Using Top-Down Guidance to Improve Accuracy

As just noted, most of the time spent constructing CHREC went into hand-constructing classifiers. Features of interest were all generated using algorithms developed from scratch to locate lines and curves. Less time was spent on constructing the top-down processing portion of CHREC. For this reason, the new system, KASC, utilizes trained SVMs to bypass that large time sink. Additionally, the SVM API that was used to train the SVMs includes a handful of feature descriptors. A feature descriptor is an algorithm which takes an image and outputs feature vectors. Feature descriptors encode useful information and act as a numerical "fingerprint" of the given image.

Histogram of Oriented Gradients (HOG) [15] are used as the feature descriptors in KASC. In the HOG feature descriptor, using a sliding window, x and y gradients are recorded and magnitudes and directions (angles) are calculated. Histograms of the magnitudes at orientations ranging from 0 to 180° are created to be used as features. The magnitude of gradients is typically large at the edges and corners of images, and since edges and corners pack a lot of information about handwritten characters, the HOG feature descriptor is an ideal choice for the KASC system. Using this feature, which came with the SVM API, allowed for rapid training of the SVM classifiers, which in turn allowed most of the effort to be spent on the top-down processing portion of the system.

Building KASC has been a two-phased operation. First, SVMs were trained to recognize every character of the domain. There are a total of 62 SVMs, one for each of the 26 upper and lower case letters and 10 digits (a decision was made early on to not include punctuation marks). White space is identified prior to recognition using a simple algorithm looking for sufficient white space between groups of pixels. The second phase, the top-down guidance, differs by domain. The initial domain of implementation, and the one described here, is of addresses (cities, states, zip codes). Top-down guidance is described later in this section.

All characters in the training and testing sets come from the EMNIST database [16]. This database was selected in part because the test data is not of entire handwritten samples but of individual characters. From the EMNIST database, 610 characters were selected to train each SVM. Of those 610 characters, 305 were positive examples and 305 were negative examples. One SVM was trained to recognize each of the 62 characters that KASC would recognize.

At first, SVMs were trained without utilizing a kernel function [17], but the resulting accuracies were very poor. After this failed attempt, numerous SVMs were trained on each character, each using a different kernel function. The SVM with the highest training accuracy for a given character was then selected to be that character's recognizer. Figure 1 illustrates the training process for the character 'a', in which the CH2 kernel function provided the best accuracy. Other kernel functions provided better accuracies for other characters, as shown in Table 1 for 12 of the 62 characters.

In order to test KASC, 1000 addresses were randomly selected from a database of city/state/zip codes. A bitmap of a given address was generated by randomly selecting individual character bitmaps from the ENMIST database. For instance, if the address were "Highland Heights, KY 41076", then the testing data would be generated by randomly selecting an 'H' from the 'H' data set followed by selecting a random 'i' from the 'i' data set followed by a random 'g' from the 'g' data set, and so forth. Section 4 describes three different approaches taken to randomly select characters.

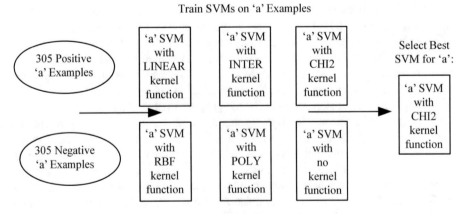

Fig. 1. SVM training process

Table 1. Kernel selected for various characters

Char	Kernel	Accuracy	Char	Kernel	Accuracy
0	CHI2	98.4	A	CHI2	97.2
1	INTER	99.0	B	CHI2	96.9
2	POLY	97.6	C	LINEAR	98.8
3	INTER	98.8	D	RBF	99.5
4	INTER	97.0	E	POLY	97.7
5	CHI2	98.4	F	RBF	97.0

A bitmap was generated from the randomly selected characters to form an address to be used as input for KASC. Note that in generating this address, no punctuation marks were used, the state was a 2-letter abbreviation and the zip code was a 5-digit number. The same database of addresses used to randomly select an address to generate is also used by KASC as the knowledge source for the top-down guidance.

KASC operates as follows. Given the input bitmap (as generated from the description above), each character is isolated. All 62 SVMs analyze each character of the input. Each SVM generates a score predicting that the given character from the input is the one the SVM is trained to recognize. The result for each character of the input is a ranking of the 62 characters that KASC can recognize. The ranked list orders the characters from highest to lowest recognition score as provided by the 62 SVMs. The highest ranked character is selected to form an address, which is KASC's initial guess. For instance, an input address may result in a "recognized" address of `olwrKlwKC m1 4q234` (see run 2 in Sect. 4).

Next, an approximate string matching algorithm [18] is used to compare each address in the addresses database to the address formed out of the top-ranked characters. The string matcher generates a low score for a closely matching string and a high score for a poorly matching string. The 20 lowest scoring addresses (that is, those that most closely matched the initially recognized address) are retained. Each of these 20 addresses is compared to the generated address character-by-character. For any mismatch, the original SVM results are reconsidered. Is the expected character one of the top characters proposed? If so, its score is altered.

Consider the address Crozier VA 23039, which had a score of 4.0 from the original address of `orozICr vA 23039`. By focusing on Crozier VA 23039, the first character is expected to be a 'C'. What score did the 'C' SVM provide for the first character of the input? Was it reasonably high? Alternatively, what score did the 'c' SVM provide? Is there enough evidence in reconsidering 'C' (or 'c') that the score could be improved? Is there any relationship between the selected character, 'o' and the sought character, 'C'? This might be the case if, for instance, the 'o' SVM often mistakes a 'C' for an 'o'. These pieces of information might provide KASC with additional evidence to reconsider 'C' as the appropriate character for the first character in the address. Applying such logic allows KASC to alter the individual character scores as originally generated by the SVMs.

Adjustments to scores take place for each of the top 20 matching addresses as selected by the approximate string matching algorithm. In this way, the sought after 'C' (or possibly 'c') with an adjusted score may cause the score of the address Crozier VA 23039 to be improved (lowered). Another highly matching address might have its score improved because it seeks a 'G' and the 'G' SVM scored reasonably high, or because 'G' is commonly mistaken for 'o' or 'O' by the SVMs. The steps taken during the recognition process are shown in Fig. 2.

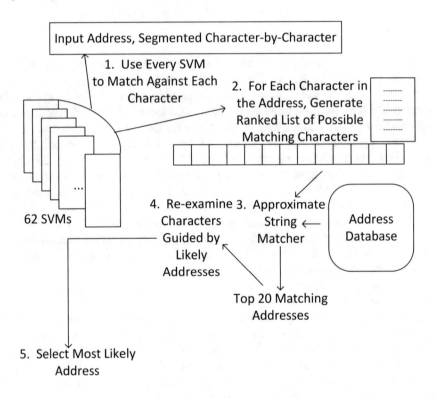

Fig. 2. KASC system architecture

The use of top-down knowledge, in the forms of city/state/zip codes and similarly-shaped characters, dramatically improved recognition accuracy over the SVM accuracies alone. However, a problem still arises for a city with similar zip codes. For instance, Cincinnati, OH, has zip codes of 45222 and 45223. From the approximate string matching algorithm, if both Cincinnati OH 45222 and Cincinnati OH 45223 were equally likely, then the scores for the '2' and the '3' for the final character would both be increased by potentially the same amount and therefore neither address's score would be improved with respect to the other address. The solution here is to select whichever SVM produced a higher score, the '2' SVM or the '3' SVM, for that final character in the input. If both the '2' SVM and the '3' SVM produced equal scores, there is nothing that KASC can do to resolve this problem and both addresses would be considered equally likely.

4 Experimental Results

Before looking at the results, it should be noted that SVMs were selected as classifiers instead of neural networks because it was hoped that the SVMs would not perform so well as to make top-down guidance useless. The idea here is that top-down guidance should be utilized because it *will* improve accuracy. With this in mind, to ensure lower (poorer) accuracy, SVM training used only a limited number of sample characters and features. Had this research wanted the recognizer to achieve high accuracy without top-down guidance, neural networks would probably have been chosen.

Three experiments were run on the system, each using 1000 randomly generated addresses from the address database. Each selected address was pieced together with randomly selected characters from the ENMIST database. In the first run, an index, i, from 1 to 1000 was randomly selected. This index, i, was then used to pull all characters out of the database to form the current address. For instance, if the address were Highland Heights KY 41076, and the random index was 501, then the 501st 'H' was selected, as was the 501st 'i', the 501st 'g', etc. Given the duplication of letters in the address, the second 'H' would be the same as the first 'H', the second 'i', would be the same as the first 'i', the second 'g' would be the same as the first 'g', etc. The second run generated a random index for each character so that, while there was a possibility of selecting the same 'H', it was unlikely. For instance, the 501st 'H' may have been selected for the 'H' in Highland while the 296th 'H' may have been selected for the 'H' in Heights. The third run ensured that there would be no duplicates among the randomly selected characters, so that for instance there would be two distinct 'H' characters selected to form the input address of "Highland Heights".

Table 2 provides the results from running KASC using each of these approaches to randomly generating addresses. The three runs' results are shown in columns 2, 3 and 4. The first row of results shows how many addresses were correctly recognized in their entirety based solely on selecting the most likely characters from the SVMs (that is, without any top-down guidance). Notice that no entire address was accurately recognized. The second row provides the accuracy of each recognized character of the addresses using SVMs alone. SVM accuracies during training were nearly perfect but when applied to the testing set, the accuracies degraded substantially (to 60–69%). Rows 3 and 4 show the improvement after top-down guidance had been applied. Row 3 contains the accuracies of recognizing entire addresses while row 4 contains the accuracies for each individual character.

Table 2. Experimental results

	Run 1	Run 2	Run 3
% correct address by SVM alone	0%	0%	0%
% correct character by SVM alone	60.8%	68.2%	69.0%
% correct address with top-down guidance	99.9%	99.4%	99.3%
% correct character with top-down guidance	99.9%	99.4%	99.3%
Incorrect characters (with the correct character ranked in the top 5)	7 (out of 8)	4 (out of 6)	6 (out of 7)

Notice for all three experiments, the results are the same between accurately recognized addresses and for accurately recognized characters. This is a little misleading in that there were more incorrect characters than incorrect addresses. This is not shown here because only one decimal point is used in the percentages of the table. In run 1, there was only one incorrect address but that one address had eight incorrect characters. In this case, nearly all of the error was caused by recognizing the wrong city while the state and most of the zip code were correct recognized.

Row 5 of the table lists for the incorrectly recognized characters (prior to top-down guidance), how many of the correct characters were ranked within the top 5 according to the initial SVM scorings. This indicates that the SVM rankings were often close enough to permit the incorrectly guessed character to be replaced by the correct character after top-down processing. In conclusion, the SVMs performed poorly in isolation yet correct characters were often scored highly enough that the use of domain knowledge and top-down guidance was able to improve recognition accuracy dramatically.

As noted at the end of Sect. 3, one significant challenge arises when the city has multiple zip codes. One example of this was with the address Houston TX 77080. KASC interpreted the address as Houston TX 77010. The '8' was among the top five characters as recognized by the SVMs alone, but top-down guidance suggested that the incorrect character (which was initially recognized as a 'g') should be corrected to '1' instead of '8'. This was one of only six incorrect cases in run 2.

Below are three example results, one for each of the three experiments. In each case, the actual address is presented followed by the address proposed after SVM recognition but prior to top-down guidance. Also listed is the character recognition accuracy. Next, the address selected by KASC after top-down guidance was applied is shown along with its approximate string matching cost when compared to the SVM-suggested address. Finally, the percentage of correct characters that received a top-5 scoring from their SVM is shown. KASC was able to correctly recognize all three of the addresses after top-down guidance was applied.

> Experiment 1 sample run:
>   Address: Mcdonald NC 28340
>   SVM suggestion: McdOnald NC 2p340
>   Percent character accuracy: 80%
>   Top-down correction: Mcdonald NC 28340
>   Approximate string matching cost: 2.8
>   Percent correct characters in top 5: 100%
> Experiment 2 sample run:
>   Address: Clarklake MI 49234
>   SVM suggestion: olwrKlwKC m1 4q234
>   Percent character accuracy: 43.75%
>   Top-down correction: Clarklake MI 49234
>   Approximate string matching cost: 8.6
>   Percent correct characters in top 5: 87.5%

Experiment 3 sample run:
    Address: `Manassas Park VA 22111`
    SVM suggestion: `mQnQssQs PQrk VA 22111`
    Percent character accuracy: 60%
    Top-down correction: `Manassas Park VA 22111`
    Approximate string matching cost: 7.2
    Percent correct characters in top 5: 100%

The goal of these experiments has been to show that domain-specific knowledge (such as that city names start with capital letters, zip codes are five digits, etc.) can be applied to improve character recognition accuracy. Although the address domain is a restricted one (the lexicon consists only of city names, state 2-letter abbreviations, and 5-digit numbers), the results clearly show a dramatic improvement with top-down guidance.

5 Conclusions

The KASC system, described in Sects. 3 and 4, combine trained SVMs and domain-specific top-down guidance to improve recognition accuracy. As shown in Sect. 4, SVM recognition alone only provided 60–69% accuracy. With top-down guidance however, accuracy improved to 99.3–99.9% accuracy.

As the address domain is too limited to be of practical use, the system is being modified for a second domain, that of natural language English sentences. The top-down knowledge will come in the form of N-grams. This version of the system is only now being created. It is expected that similar results will arise.

References

1. Herbert, H.: The history of OCR, optical character recognition. Recognition Technologies Users Association, Manchester Center, VT (1982)
2. Kumar, G., Bhatia, P.K.: A detailed review of feature extraction in image processing systems. In: IEEE 2014 Fourth International Conference on Advanced Computing & Communication Technologies, India, pp. 5–12. IEEE (2014)
3. Saba, T., Almazyad, A., Rehman, A.: Language independent rule based classification of printed & handwritten text. In: 2015 IEEE International Conference on Evolving and Adaptive Intelligent Systems (EAIS), pp. 1–4. IEEE (2015)
4. De Stefano, C., Fontanella, F., Marrocco, C., Di Freca, A.S.: A GA-based feature selection approach with an application to handwritten character recognition. Pattern Recogn. Lett. **35**, 130–141 (2014)
5. Sabeenian, R., Paramasivam, M., Dinesh, P., Adarsh, R., Kumar, G.: Classification of handwritten Tamil characters in palm leaf manuscripts using SVM based smart zoning strategies. In: Proceedings of the 2nd International Conference on Biomedical Signal and Image Processing, pp. 18–21. ACM (2017)
6. Chen, L., Wang, S., Fan, W., Sun, J., Naoi, S.: Beyond human recognition: a CNN-based framework for handwritten character recognition. In: 3rd IAPR Asian Conference on Pattern Recognition (ACPR), pp. 695–699. IEEE (2015)

7. Ghods, V., Sohrabi, M.: Online Farsi handwritten character recognition using hidden Markov model. J. Comput. **11**(2), 169–175 (2016)
8. Jurafsky, D., Wooters, C., Segal, J., Stolcke, A., Fosler, E., Tajchaman, G., Morgan, N.: Using a stochastic context-free grammar as a language model for speech recognition. In: International Conference on Acoustics, Speech, and Signal Processing, vol. 1, pp. 189–192. IEEE (1995)
9. Lai, S., Jin, L., Yang, W.: Toward high-performance online HCCR: a CNN approach with DropDistortion, path signature and spatial stochastic max-pooling. Pattern Recogn. Lett. **89**, 60–66 (2017)
10. Lawgali, A.: A survey on Arabic character recognition. Int. J. Sig. Process. Image Process. Pattern Recogn. **8**(2), 401–426 (2015)
11. Tian, S., Bhattacharya, U., Lu, S., Su, B., Wang, Q., Wei, X., Lu, Y., Tan, C.: Multilingual scene character recognition with co-occurrence of histogram of oriented gradients. Pattern Recogn. **51**, 125–134 (2015)
12. Santosh, K., Wendling, L.: Character recognition based on non-linear multi-projection profiles measure. Front. Comput. Sci. **9**(5), 678–690 (2015)
13. Wang, S., Mathew, A., Chen, Y., Xi, L., Ma, L., Lee, J.: Empirical analysis of support vector machine ensemble classifiers. Expert Syst. Appl. **36**(3), 6466–6476 (2009)
14. Fox, R., Hartmann, W.: Hand-written character recognition using layered abduction. In: Sobh, T., Elleithy, K. (eds.) Advances in Systems, Computing Sciences and Software Engineering, pp. 141–147. Springer, Dordrecht (2016)
15. Dalal, N., Triggs, B.: Histograms of oriented gradients for human detection. In: Computer Vision and Pattern Recognition, vol. 1, pp. 886–893. IEEE (2005)
16. https://www.nist.gov/itl/iad/image-group/emnist-dataset
17. Amari, S., Wu, S.: Improving support vector machine classifiers by modifying kernel functions. Neural Netw. **12**(6), 783–789 (1999)
18. Ukkonen, E.: Algorithms for approximate string matching. Inf. Control **64**(1–3), 100–118 (1985)

A Cognitive Assistant Functional Model and Architecture for the Social Media Victim Behavior Prevention

Eduard Melnik[1], Iakov Korovin[2], and Anna Klimenko[3(✉)]

[1] Southern Scientific Center of the Russian Academy of Science,
41 Chekhov Street, Rostov-on-Don 344006, Russia
[2] Southern Federal University, Rosdtov-on-Don, Russia
[3] Scientific Research Institute of Multiprocessor Computer Systems of Southern
Federal University, 2 Chekhov Street, GSP-284, Taganrog 347928, Russia
Anna_klimenko@mail.ru

Abstract. Nowadays the facilities of a parental control are not sufficient due to the numerous mobile devices and Internet access points. Another approach to control the children's behavior in the Social Media is to augment the human capabilities by the usage of a cognitive assistant. Cognitive assistants usually provide such features as self-learning, adaptive behavior, big data processing, etc. This paper focuses on the design of the cognitive assistant based on the authorship attribution methods. The major functions of the cognitive assistant as an online service are highlighted, the authorship attribution methods are analyzed and the base model and architecture of the cognitive assistant service are presented. The cognitive map usage as a data analysis intelligent technique is proposed. Also some issues of the service (i.e. scalability, dependability, integration) are considered and discussed.

Keywords: Cognitive assistant · Authorship attribution ·
Intelligent data processing · Intelligent parental control

1 Introduction

Nowadays the Internet and social networking has become an overall communication environment for everybody. In the social context this environment possesses the following important features – anonymity, which allows a person to be whoever he wants, and the absence of physical representation, which lets everyone feel free to be aggressive in relation to anybody without any responsibility. These peculiarities of the socializing in the Internet produce the promising field for a different "victim-predator" interaction, which, firstly, affects the real life of the individuals and their self-estimation, and secondly, can lead to real crimes, from fraud and identity theft, to suicide.

In the context of social networking the question of how to prevent the individual's victim behavior is a topical one. The term "victim behavior" is considered as a kind of behavior which leads to crime. According to the [1], the victim behavior varies from passive to active. The latter presupposes that the victim provokes the perpetrator by

R. Silhavy (Ed.): CSOC 2019, AISC 985, pp. 51–61, 2019.
https://doi.org/10.1007/978-3-030-19810-7_6

carelessness, sharing the individual information, chatting with unknown persons, or even by abusing the potential perpetrator (with aggressive reaction).

Considering the children safety in the Internet and Social media, the parental control must be mentioned like a means to prevent such victim behavior. The examples of the parental control facilities are: Circle with Disney, Router Limits Mini, Linksys AC 1750, and many others. Yet the wide range of personalized devices (smartphones, tablets, etc.) and the Internet access points allows to create multiple fake accounts which are not under parental control, but, at the same time, can be involved in cyberbullying, extremist and suicidal organizations. So, the up-to-date problem of parental control facilities is to identify fake accounts on the social media platforms and other Internet resources [2–5].

The new type of facilities, which help the individual to reach the goals is the cognitive assistant.

The main cognitive assistant difference from the well-known automation is that cognitive assistant enhances the human's possibilities, not substituting them. Within this paper an approach to cognitive assistant for the Social media victim behavior prevention is proposed. It includes the functional requirements to the system, model and the architecture of a cognitive assistant, based on a new cross-cutting technologies (e.g. fog computing concept) and the intelligent data analysis techniques.

The following sections of this paper contain:

- the brief overview of the children victim behavior in the Internet;
- the review of the cognitive assistant field;
- the analysis of the authorship attribution approaches in terms of an online cognitive assistant service;
- cognitive assistant functional requirements, architecture and issues of implementation;
- conclusion.

2   Children Safety in the Social Media

A problem of children safety in the multiple social media is topical nowadays. There are plenty of examples, how the social networking affects the individual's reality, quality of life, relationships, and mental health [6–8].

It must be mentioned, that in case of widespread cyberbullying, bullies are in danger as well as their victims. One study has shown that children who act as cyberbullies are also at increased risk for suicide, although they score lower on measures of suicidal ideation than their victims [9]. Children who are both a cyberbully and a cyberbully victim are at risk for the emotional difficulties associated with being a victim, as well as the behavioural difficulties associated with being a bully [10].

Yet cyberbullying is only one of multiple threats in the Internet and the Social media. A child can communicate with individuals through fake accounts, giving them some private information. He/she may have a date with them offline with absolutely unpredictable consequences, or may become a victim of a fraud, or be involved in a religious or extremist organizations.

Some social media have intelligent facilities to prevent harassment [11], or to remove fake accounts [12, 13]. For example, Twitter and Facebook use intelligent mechanisms based on machine learning to prevent any kind of cyberbullying and to find suicidal posts [14]. Also social media have facilities of reporting to site administration, blocking undesirable accounts and hiding the insulting messages. But, as it was proved in [15], the very little percentage of children do it to defend themselves from perpetrators. Moreover, only few children talk to their parents about such incidents, preferring to create new fake accounts. At the same time, the perpetrator creates new fake accounts too, which make any kind of harassment easy and responsibility-free.

Another type of children safety provisioning in the Internet is the parental control. Nowadays there are lots of applications of parental control, such as: Circle with Disney, Router Limits Mini, Linksys AC 1750, Kaspersky Safe Kids, Quostodio parental control, Symantec Norton family Premier and others. But those facilities are not efficient in case when the child can use multiple network access points, e.g. the smartphone of a friend, etc. So, the major method of control is an instant face-to-face talks with children about the relations and threats in the social media.

In this respect it is important to consider the danger of fake account creating: if the child creates fake accounts, it means that he (she) wants to hide something from the parental control. There is a plenty of reasons to use fake accounts, e.g. to be a cyberbully, or to hide from cyberbullying, or to participate in some religious, or extremist, or suicidal communities. Besides, if the individual uses fake account, it can communicate with other fake accounts, which were created with some uncertain purposes (as is shown in the Fig. 1).

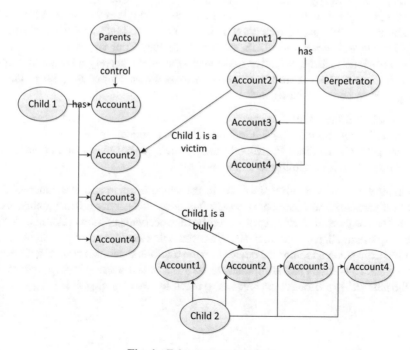

Fig. 1. Fake account interaction

So, the multiple fake accounts creating is the reason to pay attention to the individual in the Social Media. It leads to the problem - how the parental control facilities can find the child's fake accounts in a dynamically changing environment? The promising way to do it is proposed by authorship attribution facilities.

3 Intelligent Cognitive Assistant as a Service

«Cognitive assistant» is a topical term. There are some definitions of a "cognitive assistant" in contemporary science, which were formulated decades ago, for example [16]:

- a software agent "augments human intelligence";
- performs tasks and offer services (assists human in decision making and taking actions);
- complements human by offering capabilities that is beyond the ordinary power and reach of human (intelligence amplification);
- cognitive Assistant offers computational capabilities typically based on Natural Language Processing (NLP), Machine Learning (ML), and reasoning chains, on large amount of data, which provides cognition powers that augment and scale human intelligence.

Nowadays a wide range of cognitive assistants was developed: Apple's Knowledge Navigator System Expert Systems, Virtual Telephone Assistant Portico, Wildfire, Webley; Speech Recognition Voice Controlled, CALO IRIS.

Also Personal Assistants were developed – Siri, Google Now, Microsoft Cortana, Amazon Echo, – Braina, Samsung's S Voice, LG's Voice Mate, DeepDive, OpenCog, YodaQA, OpenSherlock, OpenIRIS, iCub EU projects, Cougaar, Inquire.

It is obvious that the cognitive assistants' area is extensively growing and up-to-date. So, cognitive-assistant-based approach to parental control is quite promising, giving such important features as:

- adaptation to the environment;
- enhancing and augmenting the individual's possibilities;
- more powerful facilities to maintain the parental control than just automated ones because of their possibility to adapt and learn.

Under the term "cognitive assistant" in the scope of this paper we understand the entity (in particular, software entity which functions as a service), which enhances the possibilities of parental control to discover the fake accounts of their children by adding adaptivity and intelligent data analysis. We assume that the fake accounts can be created and used from some uncontrolled devices, so the traditional parental control facilities are not efficient. Hence the general approach to fake accounts discovery is proposed: we will use the authorship attribution techniques to reach the main goal [17, 18].

From this functionality description the most important functional features of the cognitive assistant can be outlined:

- as fake accounts are discovered through the stylometry techniques, there must be possibility to apply these techniques via real-time online service;
- the parents must have the possibility to determine the range of internet resources to begin the fake accounts identification;
- the cognitive assistant must be adaptive and self-teaching to enhance the fake accounts identification process to other internet resources based on its history intelligent analysis.

As the cognitive assistant is considered to be an online service, it must be applicable to the real time data processing. The problem of authorship attribution can be quite time consuming, so at this stage of our research we make a statement, that we use the most simple and fast authorship attribution methods. Yet the problem of service load is a topical one: with a multiple users the learning stage of the authorship attribution can be rather exhaustive for the service (at least 4500 words from each user must be processed and transformed to the text features vector).

4   Analysis and Selection of an Authorship Attribution Approach in Terms of an Online Cognitive Assistant Service

In general, the authorship attribution problem has the following formal description: for document D given and a set of multiple authors A = {A1, A2, …, An} and their written documents, it is needed to determine which one from the set A is the author of D [18]. The most simple scheme of the authorship attribution implementation is as shown in Fig. 2. And named the profile-based approach [20–22].

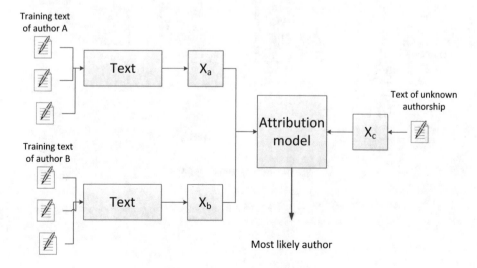

Fig. 2. Profile-based approach

Within this approach the major procedure of authorship attribution can be described as follows:

- training set gathering (the latest research show that in common cases not less than 4500 words per one author are needed);
- feature extraction (the number and semantics of features can vary);
- text representation features vector forming (X_b and X_a in the Fig. 2);
- text representation features vector forming for unknown author text;
- the comparison of the known and unknown text features vectors. Attribution model (Fig. 2) is usually based on a distance function that determines the differences between, e.g. X_a and X_c.

In other words, if *PV(x)* is the feature vector of text of unknown author *x*, and *PV(y)* is the one of author *y* from the author set *A*, the most likely author *x* of a text given is determined as follows: *Author (x) = arg min d(PV(x), PV(y)), y ∈ A*.

Another general approach to the authorship attribution is the instance-based approach (Fig. 3). Within this approach every text of an author must be contributed to the system separately, and each text sample of a known authorship is an instance of a problem. Each text of a training corpus is represented by a vector of attributes (x). Yet it must be mentioned that such classification algorithms need the multiple training instances per class to reach a reliable classification (Their sizes vary from 200 to 1000 words).

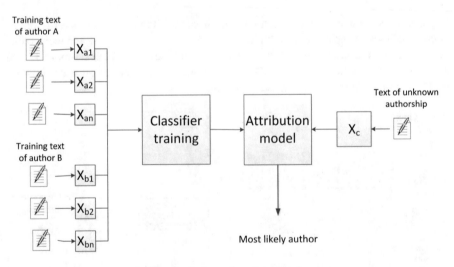

Fig. 3. Instance-based approach

Comparing these two approaches to the authorship attribution problem, the main features of each method are presented in the Table 1:

Table 1. The features of the authorship attribution methods

	Profile-based approach	Instance-based approach
Text representation	One cumulative representation for all training texts per author	Each training text is represented individually
Stylometric features	Difficult to combine different features. Some text-level features are not suitable	Different features can be combined easily
Training time cost	Low	Relatively high
Running time cost	Low	Low

It must be mentioned that within the authorship attribution problem the question of fake author identities discovery is also up-to-date. There are some works which are devoted to this problem, for example the one called Whiteprints method [22]. "Learning by similarity" method [23] performs in the similarity space by creating of similar and dissimilar document sets and comparing the distance between them.

Yet the authors of the work [18] state that the traditional probability-based techniques of authorship attribution does not work properly, when there is a need to group author's fake accounts. The proposed "Doppelganger finder" method presupposes the comparison of all pairs of authors to detect more suspicious ones in terms of fake author identity. It is obvious, that the computational complexity of the full comparison procedure is rather high.

Therefore, considering the general outlined functionality of the cognitive assistant service and the peculiarities of the authorship attribution techniques, the first preliminary conclusion must be made: in conditions of big data processing it is expedient to use the authorship attribution techniques of a lowest computational complexity of the learning and operational stages. Assuming this, it is promising to choose the profile-based approach due to its relative simplicity and low computational costs.

Being based on the authorship attribution approach chosen, the basic model requirements and architecture of a cognitive assistant can be sketched and presented.

5 Cognitive Assistant Architecture and Its Main Issues

Consider the structure and architecture of a cognitive assistant to be designed from the parental control point of view. The basic requirement to the cognitive assistant functionality is producing by the parent's desirable capabilities.

Parent performs as follows:

- connects to the cognitive agent service;
- makes the cognitive agent service to form the text feature vector;

- outlines the initial search space, giving the resource range to identify the children's fake accounts;
- gets the search result.

The generic service's actions are as follows:

- to form the text feature vector and to save it;
- to perform the search procedure in the search space given;
- to propose to widen the search space according to the kernel knowledge base information;
- if there are fake accounts, to report the information found and to adapt the kernel knowledge base.

The kernel knowledge base of the cognitive assistant is the main adaptive mechanism of the system. It is implemented via the fuzzy cognitive map [24] of resources explored. The concepts of the cognitive map are the names of the Internet resources. The links between them are weighed so as if there is a fake account on the resource A, and fake account on the resource B, the weight of the link grows (as is shown in the Fig. 4). So, within the next search, if an individual wants to make a search on the resource A, the resource B will be proposed to expand the search as the probability of the fake account existence on the resource B grows, based on the basis of historical data.

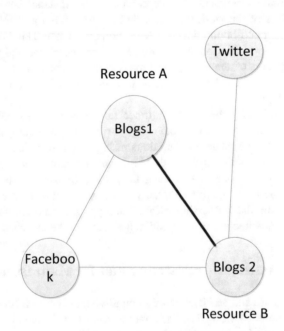

Fig. 4. The example of a kernel knowledge base structure

This is the main feature of the cognitive agent which implements the self-learning of the system and its intelligence.

The basic architecture of a cognitive assistant can be presented as follows (Fig. 5).

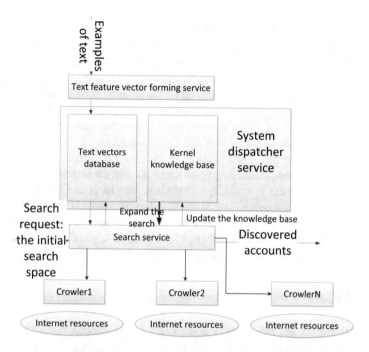

Fig. 5. The basic cognitive agent architecture

According to the structure presented, the services of the cognitive assistant are responsible for the following functions:

- text feature vector forming service receives the text samples and forms the text feature vector. When the vector is formed, it is saved in the vectors database;
- system dispatcher service is responsible for the data- and knowledge bases operability and functionality and for the databases integration in case of distributed cognitive instances operating. Such dispatching can involve the distributed ledger technologies and is out of the scope of this paper.
- the search service performs the search, new text feature vectors forming, based on the information extracted, and the comparison with the sample text feature vector. Search service interacts with multiple crawlers, which crawls the resources given. The text feature vectors, which are formed of crawled information are saved in the database. The results of the comparison are returned to the users (parents).

It is obvious, that the system considered has lots of issues, which are related to the performance, availability of the service, scalability and data integrity.

From the authors' point of view, the major issues are:

- scalability;
- possibility of the real-time operating.

The scalability issue is important in case of centralized assistant implementation, where all data are stored in one data/knowledge base. Besides, with the growth of user's number, the scalability issue transforms to the latency one, e.g. text feature vector forming can become the bottleneck of the system. The scalability problem is solved via decentralization of the assistant, but decentralization itself leads to the new issue, i.e. data integrity one. This problem is related to the distributed ledger field and can be solved by a wide range of data consensus methods.

Another issue is the time-consumption by the major assistant operation, including the relatively high computational complexity of an atomic text features vector comparison operations and the availability of crawlers. The latter is a separate question to consider, because of its being related to software dependability problem. Yet the question of the crawlers availability and the velocity of the crawling process is quite solvable by methods of system reliability provisioning and, possibly, by some heuristic techniques to reduce the time of text feature vector forming.

6 Conclusion

The parental control in social media is a topical question nowadays. Though lots of parental control automated facilities are developed, the problem of a fake account discovery has not been solved so far, while the children's fake accounts relate to victim behavior frequently.

The authorship attribution methods are used widely and give quite accurate results, so this is the way to discover the fake identities through the Internet.

In this paper we propose a basic model and architecture of a cognitive assistant, which augments and expands the parental control in the area of cyberthreats to children and young adults. The cognitive assistant approach is more effective than the simple automation of the process, because of its ability to learn and adapt to the environmental changes.

The architecture of a cognitive assistant is proposed. The authorship attribution approach is chosen from the computational complexity point of view.

Considering the model designed, the major issues of the cognitive assistant implementation are outlined, and the approaches to solve them are presented.

Hence, the future research area is highlighted: the first direction is to minimize the response time of the system by affecting the crawlers dependability, and the second one is to develop robust and dependable techniques to provide the decentralized system functioning and data consensus reaching.

Acknowledgement. The paper has been prepared within the RFBR project 18-29-22093.

References

1. Ilyina, L.V.: Ugolovno-pravovoye znacheniye viktimnosty. Pravovedeniye, № 3, p. 119 (1975)
2. https://www.cybersmile.org/advice-help/category/is-my-child-a-cyberbully
3. https://www.stopbullying.gov/cyberbullying/cyberbullying-tactics/index.html
4. https://www.komando.com/tips/411428/does-your-teen-have-a-fake-account-online
5. https://www.telesign.com/fake-user-impact-report/
6. Cheng, J.: Ars Technica Not an obscene racist after all: 4 flagged for Facebook fake. Ars Technica, 28 September 2009. http://arstechnica.com/tech-policy/2009/09/that-obscene-racist-may-be-fake-4-sued-for-profile-prank/. Accessed 24 Mar 2014
7. Girl, 13, commits suicide after being cyber-bullied by neighbour posing as teenage boy (n.d.). http://www.dailymail.co.uk/news/article-494809/Girl-13-commits-suicide-cyber-bullied-neighbour-posing-teenage-boy.html. Accessed 24 Mar 2014
8. Henry, L.: Man faces cyber-bullying charge in ex-girlfriend's fake adult-date profile, 7 November 2013. 5NEWSOnlinecom. http://5newsonline.com/2013/11/07/man-faces-cyber-bullying-charge-in-ex-girlfriends-fake-adult-date-profile/. Accessed 24 Mar 2014
9. Hinduja, S., Patchin, J.W.: Bullying, cyberbullying, and suicide. Arch. Suicide Res. **14**, 206–221 (2010)
10. Kowalski, R.M., Limber, S.P.: Psychological, physical, and academic correlates of cyberbullying and traditional bullying. J. Adolesc. Health **53**(1 Suppl), S13–S20 (2013)
11. Davis, A.: New tools to prevent harassment, 19 December 2017. https://newsroom.fb.com/news/2017/12/new-tools-to-prevent-harassment/
12. Garreffa, A.: Facebook has deleted 1.3 billion fake accounts. Social Networking News, 10 September 2018. https://www.tweaktown.com/news/63113/facebook-deleted-1-3-billion-fake-accounts/index.html
13. https://blog.hubspot.com/marketing/twitter-harassment-cyberbullying
14. https://www.washingtonpost.com/news/the-switch/wp/2018/04/11/ai-will-solve-facebooks-most-vexing-problems-mark-zuckerberg-says-just-dont-ask-when-or-how/?noredirect=on
15. SafetyNet: Cyberbullying's impact on young people's mental health. Inquiry report. https://www.childrenssociety.org.uk/what-we-do/resources-and-publications/safety-net-the-impact-of-cyberbullying-on-children-and-young
16. https://www.slideshare.net/hrmn/cognitive-assistants-opportunities-and-challenges-slides
17. Koppel, M., Schler, J., Argamon, S.: Authorship attribution in the wild. Lang. Resour. Eval. **45**, 83–94 (2011). https://doi.org/10.1007/s10579-009-9111-2
18. Afroz, S.: Deception in authorship attribution. Ph.D. thesis (2013). https://www1.icsi.berkeley.edu/~sadia/thesis.pdf
19. Coyotl-Morales, R.M., Villaseñor-Pineda, L., Montes-y-Gómez, M., Rosso, P.: Authorship attribution using word sequences. In: Proceedings of the 11th Iberoamerican Congress on Pattern Recognition, pp. 844–853. Springer (2006)
20. Diederich, J., Kindermann, J., Leopold, E., Paass, G.: Authorship attribution with support vector machines. Appl. Intell. **19**(1/2), 109–123 (2003)
21. Holmes, D.I.: Authorship attribution. Comput. Humanit. **28**, 87–106 (1994)
22. Abbasi, A., Chen, H.: Writeprints: a stylometric approach to identity-level identification and similarity detection in cyberspace. ACM Trans. Inf. Syst. **26**(2), 29 (2008). https://doi.org/10.1145/1344411.1344413. http://doi.acm.org/10.1145/1344411.1344413
23. Qian, T.Y., Liu, B., Li, Q., et al.: J. Comput. Sci. Technol. **30**, 200 (2015). https://doi.org/10.1007/s11390-015-1513-6
24. Groumpos, P.P.: Fuzzy cognitive maps: basic theories and their application to complex systems. In: Glykas, M. (ed.) Fuzzy Cognitive Maps. Studies in Fuzziness and Soft Computing, vol. 247. Springer, Heidelberg (2010)

An Ontology-Based Approach to the Workload Distribution Problem Solving in Fog-Computing Environment

Anna Klimenko[1($\boxtimes$)] and Irina Safronenkova[2]

[1] Scientific Research Institute of Multiprocessor Computer Systems of Southern Federal University, Taganrog, Russia
anna_klimenko@mail.ru
[2] Southern Scientific Center of the Russian Academy of Science, Rostov-on-Don, Russia
safronenkova050788@yandex.ru

Abstract. A problem of workload distribution within the heterogeneous computing environment is not new and has been solved frequently and successfully. However, optimization problem models neglect such aspects of fog computing concept as:

- the computation nodes are unequal in terms of desirable workload distribution;
- the cloud layer is the mandatory participant of the computing process.

The current paper focuses on the formalizing of workload distribution optimization problem in terms of fog computing special aspects for device "offload" strategy. The task subgraph relocation to computing devices of fog layer takes place. This is a multicriteria NP-hard optimization problem with multiple constraints, which are determined by the fog-computing peculiarities. So there is a problem to get solutions of an acceptable quality under the restricted time conditions.

In this paper an approach based on the optimization problem search space reduction is proposed. An ontological approach is used for this purpose. It allows to reduce the optimization problem search space and the problem solving time thereby.

Keywords: Workload distribution · Fog computing · Distributed computing · Internet of Things · Ontology · Optimization problem · Search space reduction

1 Introduction

Nowadays fog computing is applied ubiquitously. The general reason for this is the need of big data processing, which are generated daily by numerous end-point devices, sensors, smartphones, etc., integrated into the Internet of Things [1, 2].

© Springer Nature Switzerland AG 2019
R. Silhavy (Ed.): CSOC 2019, AISC 985, pp. 62–72, 2019.
https://doi.org/10.1007/978-3-030-19810-7_7

Device "offload" strategy is one of the well-known models applied in fog environment [3]. It presupposes the partially workload relocation to the other computing resources.

In this case the problem of the workload distribution through some computing fog-layer devices occurs: a subgraph of tasks, connected by the information exchanges, must be mapped onto some new locations. This problem is not new and related closely to the scheduling problem [4, 5] and scheduling for parallel distributed computing systems (for instance, the time-cost option) [6–9].

However, the model of heterogeneous computing system is not compliant with peculiarities of fog computing environment:

- computational nodes are not equal in fog computing environment, so some of them could be more preferred in workload distribution tasks [10, 11];
- fog computing can't exist without the cloud, so a part of the computing process must be performed in the cloud layer.

In the current paper a modified model of workload distribution problem is proposed. The model takes into account the peculiarities of the device "offload" procedure and is formalized as an optimization problem.

Such problem should be solved in restricted time conditions, so there is a need to optimize the solution search process.

The approach to the search process optimization bases on the search space reduction. The key idea of the approach proposed is to reduce the candidate computational units number by the ontological approach application [12, 13]. For this goal a domain ontology which specifies the task subgraph is developed. The proposed ontology forms a set of preliminary requirements for computing nodes, which allow to reduce the search space of the optimization problem considered.

2 The Problem Formalization

A device (end or cloud) "offload" problem is considered as follows: a set of tasks is performed by the set of computational devices, but in a time t_0 a subset of tasks, described by the subgraph, must be relocated to the other computating devices in the fog-layer. The objective functions can be multiple, e.g., the workload dispersion minimization, the communication overheads minimizing, etc.

Assume the objective functions mentioned to be an example of the ones for the problem model instance. These cost function are reasonable for the following: first, device load balancing provides system reliability growth [14–16], second, minimization of communication network workload corresponds to the fog commuting conceptual issues [17].

Given task graph G is a directed and acyclic. Each node of the graph G is weighted by computational complexity x_i, and graph edges are weighted by the amount of data w_{kl} which are transmitted between tasks. The graph G includes a subgraph G' which must be relocated, here $G' \subset G$.

A set of computing devices (CD) is given either, CDs are connected by the communication network, and form a graph P.

Considering the structure of the fog-computing, P partitions into three non-overlapping subgraphs: P_{edge} – subgraph of end devices, P_{fog} – subgraph of fog layer devices, P_{cloud} – subgraph of cloud layer devices. $P_{edge} \cup P_{fog} \cup P_{cloud} = P$; $P_{edge} \cap P_{fog} \cap P_{cloud} = \varnothing$. Each graph node is characterized by performance, graph edges are weighted by network infrastructure capacity.

Assume that an initial distribution of tasks A is given and described via matrix as follows:

$$
A = \begin{vmatrix} <u_{11}, t_{11}> & <u_{xy}, t_{xy}> & <u_{1M}, t_{1M}> \\ <u_{21}, t_{21}> & & \\ <u_{N1}, t_{N1}> & & <u_{NM}, t_{NM}> \end{vmatrix} \tag{1}
$$

where

u_{NM} - computing resource which is occupied by task N from CD M;
t_{NM} - the moment of beginning of task N on DC M.

A new task distribution is determined by A' with the constraint:

$$
B = \begin{vmatrix} b_{11} & b_{12} & b_{1M} \\ b_{21} & & \\ b_{N1} & & b_{NM} \end{vmatrix} \tag{2}
$$

where $b_{ij} = \begin{cases} 1, & \text{if the task can be distributed to the node,} \\ 0, & \text{in other cases.} \end{cases}$

Let's impose a restriction that is specified by the peculiarities of fog computing environment.

$$
\forall j, p_j \in P_{cloud}, \sum_{j=1}^{|P_{cloud}|} u_{ij} > 0 \tag{3}
$$

It means the mandatory participation of cloud layer in the computing process.

Let's also impose an execution time restriction:

$$
\forall i, j \ t_{ij} + \frac{x_i}{u_{ij} p_j} \leq T \tag{4}
$$

Total computational device workload is also an important constraint:

$$
\forall i, j : \sum_{i=1}^{N} u_{ij} \leq 1 \tag{5}
$$

As well, the following constraint must also be provided: executing tasks can't be relocated.

Let us introduce an element comparison operator of two matrixes A and A':

$$comp\left(a_{ij}, a'_{ij}\right) = \begin{cases} 0, & g_j \equiv g'_j \\ 1, & \text{otherwise} \end{cases} \tag{6}$$

The function comp $\left(a_{ij}, a'_{ij}\right) = 0$ only if the task g_i, which is located on CD j, has the same location after workload redistribution procedure.

Let us present it as a constraint for task model:

$$\forall g_i \notin G' : \forall i, j \; comp(a_{ij}, a'_{ij}) = 0 \tag{7}$$

The load balancing cost function appears as follows:

$$\forall j, k : F_1 = \sum_{i=1}^{N} u_{ij} - \sum_{i=1}^{N} u_{ik} \rightarrow \min \tag{8}$$

which is equivalent of minimal CD workload dispersion.

Let us form a communication workload cost function. Consider the following simple example (see Fig. 1).

Fig. 1. A communication workload on the CD1–CD4 route

Assume tasks g_1 and g_2 are connected by data transmission from g_1 to g_2. In the case when g_1 is on the CD1, as is shown in Fig. 1, the data is transmitted through CD2 and CD3, so it goes the whole way up to g_2. Communication workload should be described as follows:

$$L_0 = L_1 + L_2 + L_3 \tag{9}$$

where

L_0 - total communication workload of CD route, in terms of transmit data volume;
L_1 - transmit data volume from CD1 to CD2;
L_2 - transmit data volume from CD2 to CD3;
L_3 - transmit data volume from CD3 to CD4.

In the case when tasks g_1 and g_2 are located in another way we will get the following (see Fig. 2).

Fig. 2. A communication workload on the CD3–CD4 route

Then $L_0 = L_3$, what is less than in the first case.

Consequently, the communication workload must be minimized: for each task where a total flow of input data is more than a total flow of output data, the task should be located in a way to minimize the distance to maximum data flow source.

To formalize this requirement the following functions must be introduced: $j = get\_CU(i)$, that returns a number of CD j, where the task i is allocated; $D_{fh} = Dest\_CU(f, h)$, that returns a distance between CD f and h, [hops].

Then a closeness criterion appears as follows:

$$Task\_dest(g_i, g_l) = Dest\_CU(get\_CU(i), get\_CU(l)) \tag{10}$$

and a cost function (8), taking into account, has the following form:

$$\forall i, g_i \in G, w_{ki} = \max(W_i) : F_2 = Dest\_CU(get\_CU(k), get\_CU(i)) \rightarrow \min \tag{11}$$

where

W_i – set of edge which belongs to a i;
w_{ki} – maximum intensity data flow.

A workload distribution problem model using end device "offload" is shown below. Given G, G', P, A, B it is necessary to find a new distribution A' such as:

$$\forall j, k : F_1 = \sum_{i=1}^{N} u_{ij} - \sum_{i=1}^{N} u_{ik} \rightarrow \min \tag{12}$$

$$\forall i, g_i \in G, w_{ki} = \max(W_i) : F_2 = Dest\_CU(get\_CU(k), get\_CU(i)) \rightarrow \min \tag{13}$$

With the following constraints:

$$\forall j, p_j \in P_{cloud}, \sum_{j=1}^{|P_{cloud}|} u_{ij} > 0 \tag{14}$$

$$\forall i, j \ t_{ij} + \frac{x_i}{u_{ij}p_j} \leq T \tag{15}$$

$$\forall i, j : \sum_{i=1}^{N} u_{ij} \leq 1 \tag{16}$$

$$\forall g_i \notin G' : \forall i,j \; comp(a_{ij}, a'_{ij}) = 0 \tag{17}$$

and the following border conditions:

$$i > 0; j > 0; t_{ij} > 0; \; 0 < u_{ij} < 1 \tag{18}$$

The problem formulated is a multicriteria one and belongs to NP-hard problem class, so there are no efficient algorithms for them, which allow to find an optimal solutions in the restricted time conditions. In the scope of device "offload" the time is rather critical. So, the question of problem solving time reduction is quite relevant.

Nowadays metaheuristics are efficiently applied in different domains [18–21].

Yet if a search space is of a high dimensionality, the good solutions can be reached by the weighty number of iterations, or by the parallel search optimization algorithms usage.

In the current research we propose to use ontological approach to reduce a search space of the optimization problem.

3   Ontology-Based Workload Distribution Problem Solving in Fog Computing Environment

Related works [22–25] analysis has shown that the heuristics application reduces the time of problem solving. The examples of such heuristics are: strongly connected tasks should be placed on the single CD, tasks of a high computational complexity should be located on the CD of a higher performance, and so on. In other words, task allocation should be performed considering task structure and computational complexity.

We propose the ontology usage for candidate node set selection considering the graph structure and allocated subgraph G' particularities.

For this purpose a base ontology should be developed. According to the proposed ontology a decision of some CD inclusion in search space should be made.

First, the task subgraph to be distributed through the system, should be classified by tasks interconnection strength attribute. In case if G' is strongly interconnected by the information exchange, devices that are close or a single node should be examined. In case if G' is weakly interconnected, subtasks can be allocated arbitrarily (considering the time constraint for the set of the tasks). Moreover, communication channel capacity and the current workload of channels and nodes must be considered.

Second, task subgraph G' is proposed to be classified by the amount of data transmissions. If ones are large, subtasks should be located on the nodes, which are close to one another, or on the nodes that are connected by channels with high capacity.

Third, intensity of input and output flow from G' is also must be considered. In case of intensive input data flow, there is no reason to locate G' remotely from input data. Similar reasoning should be applied for output data flow from G'.

Finally, an initial location of G matters: if an initial graph performs in cloud layer there is no reason to shift the workload to the fog layer devices, which are close to edge of the network, and vice versa, if user device "offload" takes place, there is no reason to shift the workload of G' to the cloud.

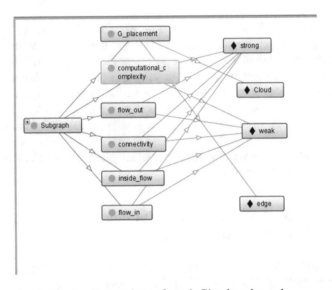

Fig. 3. The ontology of graph G' to be relocated

The ontology based on given classification attributes is developed (see Fig. 3). Some rules, based on the developed ontology, are formulated and given below.

1. If G_placement=cloud, then CD that are closer to cloud layer must be included in search space.
2. If G_placement=edge, then CD that are closer to the edge of network must be included in search space.
3. If Computational_complexity=high, then the nodes with high computational performance must be included in search space.
4. If Computational_complexity=low, then nodes of unspecified complexity must be included in search space.
5. If flow_out=high, then nodes which are close to the data receiver must be included in search space.
6. If flow_out=low, then unspecified nodes can be included in search space.
7. If flow_in=high, then nodes that are close to the data sources, must be included in search space.
8. If flow_out=low, then unspecified nodes can be included in search space.

9. If connectivity=high, then nodes that are near to each other must be included in search space.
10. If connectivity=low, then unspecified nodes can be included in search space.
11. If inside_flow=high, then nodes that are near to each other and connected by the channels of a high capacity must be included in search space.
12. If inside_flow=low, then requirement # 11 is omitted.

By choosing the specific candidate nodes for workload distribution problem a search space is reduced.

4 Ontological Approach Application for the Workload Distribution Problem in Fog Computing Environment

Consider an example of ontological approach application.

For example, assume that a task described by graph G was performing in cloud layer (see Fig. 4). In order to perform the cloud layer "offload" a part of workload should be relocated to the fog layer (a subgraph is highlighted).

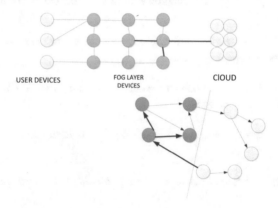

Fig. 4. G' relocation from a cloud to a fog layer

Given: flow_in=high, inside_flow=high, computational_complexity=high, flow_out=low.

If all available nodes are considered as a search space, then solutions for all fog layer nodes must be estimated.

If ontological approach and preliminary analysis of available resources take place, an initial set of candidate nodes is reduced (see Fig. 5).

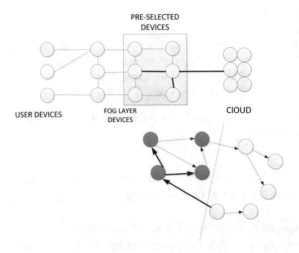

Fig. 5. Subtask relocation to the reduced node set

A complex requirement for candidate nodes is formed according to the rules mentioned above:

flow_in=high&   inside_flow=high&   computational_complexity=high&
flow_out=low ->
{
nodes that are close to the nodes where data source information tasks are located, must be included in search space.
&
nodes that are closely spaced and connected by the high speed channels must be included in search space.
&
high computational complexity nodes must be included in search space.
&
unspecified nodes must be included in search space.
}

So, a CD set for subgraph relocation is reduced by the additional constrains application, as a consequence, the number of possible decisions is also reduced.

It is quite obvious that a proposed technique of candidate node choosing for G′ relocation could be applied as a technique of the initial task location and the initial point for metaheuristic search optimization.

5 Conclusion

The current work deals with the problem of the workload distribution in the fog computing environment. The workload distribution problem in terms of fog environment peculiarities for device "offload" strategy was formalized. The fundamental difference between the proposed model and the classical model of the workload distribution in the heterogeneous computational system is determined by the particular constraints. The workload distribution problem belongs to NP-hard ones. As a result, the search time to get the solution of a satisfactory quality can take an unacceptable period of time, while the time is always limited.

We propose to reduce the search time via the search space reduction. This can be done by the usage of an ontological approach proposed.

The novelty of the current research is the usage of the ontological description of the subgraph to be relocated.

The search space reduction takes place, and so the search time reduces too.

Preselected nodes could be used for initial task location in terms of available computing resources for workload distribution problem solving by metaheuristic search optimization methods.

Further works are supposed to perform experimental studies of the approach proposed.

Acknowledgement. The paper has been prepared within the RFBR project 17-08-01605 and the GZ SSC RAS N GR of project AAAA-A19-119011190173-6.

References

1. Chiang, M., Zhang, T.: Fog and IoT: an overview of research opportunities. IEEE Internet Things J. **3**, 854–864 (2016)
2. Bonomi, F., Milito, R., Zhu, J., Addepalli, S.: Fog computing and its role in the internet of things. In: Proceedings of the 1st Edition of the MCC Workshop on Mobile Cloud Computing (2012)
3. Moysiadis, V., Sarigiannidis, P., Moscholios, I.: Towards distributed data management in fog computing. Wirel. Commun. Mob. Comput. **1**, 1–14 (2018)
4. Pinedo, M.L.: Scheduling: Theory, Algorithms, and Systems, 5th edn. Springer, New York (2016)
5. Konvej, R.V., Maksvell, V.L., Miller, L.V.: Scheduling. Science, Moscow (1975)
6. Barskij, A.B.: Parallel Processes in Computation System. Radio i svjaz', Moscow (1990)
7. Horoshevskij, V.G.: Computer architecture. MGTU imeni N.Je. Baumana, Moscow (2008)
8. Gonchar, D.R., Furugjan, M.G.: Efficient scheduling algorithm on real time multiprocessors. https://cyberleninka.ru/article/n/effektivnye-algoritmy-planirovaniya-vychisleniy-v-mnogoprotsessornyh-sistemah-realnogo-vremeni. Accessed 02 Dec 2018
9. Kostenko, V.A.: Architecture synthesis problems: formalization, particularities and opportunities of different solution methods. In: Software and Tools. Subject Collection, pp. 31–41. MAKS Press, Moscow (2000)

10. Fog computing and the Internet of Things: extend the cloud to where the things are. https://www.cisco.com/c/dam/en_us/solutions/trends/iot/docs/computing-overview.pdf. Accessed 20 Nov 2018

11. Wang, Y., Uehara, T., Sasaki, R.: Fog computing: issues and challenges in security and forensics. In: Proceedings International Computer Software and Applications Conference, pp. 53–59 (2015)

12. Noy, N., McGuinness, D.: Ontology development 101: a guide to creating your first ontology. Stanford knowledge systems laboratory Technical report KSL-01–05 and Stanford Medical Informatics Technical report SMI-2001-0880 (2001)

13. Gavrilova, T.A.: Knowledge engineering. models and methods. https://e.lanbook.com/book/81565. Accessed 22 Nov 2018

14. Melnik, E.V., Klimenko, A.B., Ivanov, D.Y.: Distributed information and control system reliability enhancement by fog-computing concept application. In: IOP Conference Series: Materials Science and Engineering, vol. 327 (2018)

15. Melnik, E.V., Klimenko, A.B.: Informational and control system configuration generation problem with load-balancing optimization. In: Proceedings of the 10th International Conference on Application of Information and Communication Technologies, pp. 492–496 (2017)

16. Klimenko, A.B., Ivanov, D., Melnik, E.V.: The configuration generation problem for the informational and control systems with the performance redundancy. In: 2nd International Conference on Industrial Engineering, Applications and Manufacturing (2016)

17. Linthicum, D.: Edge computing vs. fog computing: definitions and enterprise uses, https://www.cisco.com/c/en/us/solutions/enterprise-networks/edge-computing.html. Accessed 03 Dec 2018

18. Ingber, L.: Very fast simulated re-annealing. Math. Comput. Model. **12**(8), 967–973 (1989)

19. Dorigo, M., Maniezzo, V., Colorni, A.: Ant system: Optimization by a colony of cooperating agents. IEEE Trans. Syst. Man Cybern. Part B Cybern. **26**, 29–41 (1996)

20. Iba, H., Aranha, C.C.: Introduction to genetic algorithms. In: Practical Applications of Evolutionary Computation to Financial Engineering. Adaptation, Learning, and Optimization, vol 11. Springer, Heidelberg (2012)

21. Goldberg, D.E. Holland, J.H.: Machine Learning (1988)

22. Kholod, I.I.: A method for determining the capabilities of parallel execution of data mining algorithm functions. In: Software and Systems, no. 2, pp. 268–274 (2018)

23. Kupriyanov, M.S., Kholod, I.I. Karshiyev, Z.A.: Method of creating parallel algorithms of intellectual analysis of data using flow-independent functional blocks. IZVESTIJa SPbGJeTU «LJeTI», vol. 8, pp. 19–23 (2013)

24. Kholod, I.I.: Models and methods of creating parallel algorithms of analysis of dispersed data. https://etu.ru/assets/files/nauka/dissertacii/2018/holod/avtoreferat_holod-ii.pdf. Accessed 03 Dec 2018

25. Kaliaev, I.A., Melnik, E.V.: Metod multiagentnogo raspredeleniya resursov v intellectualnih mnogoprocessornih vychislitelnih sistemah. In: Kalyaev. Vestnik UNTS RAN, № 12, pp. 40–50 (2007)

Decoupling Channel Contention and Data Transmission in Dense Wireless Infrastructure Network

Jianjun Lei$^{(\boxtimes)}$ ⓘ and Hong Yun ⓘ

School of Computer Science and Technology,
Chongqing University of Posts and Telecommunications,
Chongqing 400065, China
`leijj@cqupt.edu.cn`, `377912964@qq.com`

Abstract. In an infrastructure wireless network, with the increase of the number of wireless nodes, the channel collisions become more intense and the network performance declines sharply. In this paper, we propose a novel distributed Media Access Control (MAC) protocol and attempt to decouple the channel contention and data transmission process. Wherein, the channel utilization time is split into channel contention period and data transmission period. During the contention period, the nodes attempt to transmit a short control frame to compete the channel and they will be piped into the potential queue if they succeed. In the following data transmission period, according to the order of the transmission sequence, the node can be scheduled to transmit successively the data packet, and every node will be noticed by the Acknowledgement (ACK) piggybacking. Furthermore, we design an adaptive scheme to adjust the length of the contention period to improve the channel utilization. The simulation results show that our proposed algorithm can improve significantly the performance in terms of system throughput, transmission delay and channel utilization compared to the IO-MAC algorithm and legacy IEEE 802.11 DCF mechanism.

Keywords: Infrastructure wireless network · DCF · Decoupled MAC · ACK piggybacking

1 Introduction

In recent years, the Wireless Local Area Networks (WLANs) based on the IEEE 802.11 [1] are widely deployed in the home, office and public spaces etc. They mainly adopt Distributed Coordination Function (DCF) with Binary Exponential Backoff (BEB) as the basic media access mechanism to resolve channel collisions. However, the traditional BEB algorithm has some critical problems. One is the big Contention Window (CW) will result in a large amount of idle time being wasted when the network load is light. Inversely, under heavy load scenario, the algorithm resetting the CW to a minimum value after successful transmission will lead to lots of data collisions. Undoubtedly, the DCF mechanism is not a plastic protocol for dynamic of network load, and hence often leads to the extreme deterioration of network performance. To avoid collisions, a centralized channel control algorithm such as the IEEE 802.11 Point

© Springer Nature Switzerland AG 2019
R. Silhavy (Ed.): CSOC 2019, AISC 985, pp. 73–84, 2019.
https://doi.org/10.1007/978-3-030-19810-7_8

Coordination Function (PCF) [1] polls the wireless nodes according its polling list. PCF divides the slot into a contention-free period (CFP) and a contention period (CP), only the polled node is permitted to transmit in CFP period. However, some null-pollings occur when the polled node has no data to transmit, hence waste the channel resource.

Aiming to the potential collision problem in a dense infrastructure network, we propose a novel Distributed Decoupled MAC protocol, referred to as DMAC algorithm in this paper. Different from some previous algorithms that the nodes transmit data immediately after competing the channel successfully, the DMAC divides the time slot into contention period and data transmission period by a decoupling model. Then in the contention period, nodes send DTR (Data Transmission Request) to compete the channel and no response is required from the centralized node AP (Access Point). When a node successfully transmits the DTR, it will be inserted into the sequence queue by the AP. During the data transmission period, the nodes in the sequence queue will be scheduled to transmit the data consecutively and efficiently on a collision-free channel by the ACK piggybacking from the AP. Meanwhile, to improve the channel utilization, the DMAC also adjusts adaptively the length of the contention period.

The rest of this paper is organized as follows. We first summarize the related work in Sect. 2. Then, we describe the DMAC in Sect. 3. In Sect. 4, we validate the DMAC protocol through simulation. Finally, we make a conclusion in Sect. 5.

2 Related Work

Recent literatures develop some MAC protocols and analytical model to attempt to solve the collision problem. In [2], Bianchi analyzes the performance of DCF via a two-dimensional Markov chain model under the assumption of saturated network condition, and derive an optimal CW value which can maximize the network throughput and is related to the number of active nodes in the network. To adjust the CW more precisely, in [3], the authors estimate the number of active nodes by a collision probability algorithm. In [4], the CW is adaptively adjusted to achieve the maximum system throughput according to the network condition, whether saturated or unsaturated. In [5], according to the mathematical model, the relationship between the transmission time of any two nodes and its corresponding CW value is obtained, and under the limitation of time fairness, the CW value that maximizes the system throughput is calculated by an iterative algorithm. In [6], considering the unsaturation of the network, the CW is dynamically adjusted by evaluating the number of active nodes in the network. In [7] and [8], the BEB is modified to improve the performance of the network. In [7], nodes reduce the backoff value by a probability, which is calculated from the available system parameters in the network while the channel is idle. In [8], the CRB algorithm is proposed, the node is assigned a backoff value and a backoff stage by a VBA algorithm running on the AP when the node transmits successfully. Although, the above algorithm can alleviate the collision to some extent, the collision between data packets cannot be eliminated, especial in some dense scenarios.

Some algorithms based on contention/reservation are also proposed to degrade the data transmission collision. In [9], the authors propose a token passing MAC protocol, referred to as Token-DCF. Each node acts as a scheduler with a certain probability to schedule its neighbor that has the longest queue by its MAC header. The Token-DCF can cause the starvation of other's node when a node always has the longest queue length. An IO-MAC (Implicit Ordering) algorithm was proposed in [10] and [11]. In [10], each node regards its previous successful transmission node as a predecessor. When a node's predecessor has no more data to transmit or transmission fails, it will take part in contention with a backoff timer from CWmin to 2CWmin-1. However, the IO-rule can be invalid because that the subsequent nodes may choose a less backoff timer than its predecessor when more than three nodes conform to the IO-rule. Hence, in [11], the authors divide the time slot into the variable contention period and the reservation period. The nodes that successfully enter into the reservation period select a backoff counter value that is equal to its successful transmission order in current cycle. However, a lot of contention time slot can be idle and wasted when all the nodes are in contention-free period. In [12], the authors propose a CSMA/CQ algorithm that uses the SDN (Software Defined Network) to divide the channel into contention channel and transmission channel, which can solve the collision problem but is not suitable to the single channel network. In this paper, we design a novel MAC protocol that is based on the contention/reservation mechanism and is easy to implement in the single channel scenario.

3   Protocol Description

In this section, we describe the DMAC, which is designed for an infrastructure-based WLAN. To better demonstrate the DMAC algorithm, we first present a system model, and then detail the implementation, finally optimize the CW value by an analytical model.

3.1   System Model and Frames Structure

We consider a typical infrastructure network consisted of one AP and several competing nodes. Each node is associated with the AP and shares the same channel with the other nodes. We assume that frame error rate (FER) can be neglected and transmission failure is only caused by the collision.

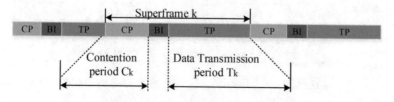

Fig. 1. The superframe structure of DMAC

In DMAC, the time slot consists of a series of superframes, which includes one contention period (CP), one beacon and one data transmission period (TP). As shown in Fig. 1, the C_k and T_k represent the contention period and data transmission period in superframe k. During the CP period, all nodes can transmit a DTR (Data transmission Requirement) frame to the central node (Access Point, AP) and no response is required from AP. The format of DTR is shown in Fig. 2. After the end of the C_k period, the AP broadcasts a beacon to all nodes, which will piggyback the length of the next contention period. During the data transmission period, the nodes are activated consecutively to transmit the data by an acknowledgment (ACK) frame piggybacked an ID of the next transmission node, and its format is shown in Fig. 3.

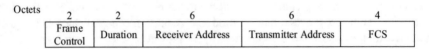

Fig. 2. The format of DTR in DMAC

Fig. 3. The format of ACK in DMAC

As the central node of this infrastructure network, the AP will transmit periodically beacon frames including some control information. The beacon frame activates the data transmission period and provides the sequence information for all potential transmission nodes.

3.2   Protocol Details

In this section, we detail the DMAC, which includes three procedures: contention process, queuing process and data transmission process. To better demonstrate the DMAC, we first present the Fig. 4 and explain the operation details in the following section.

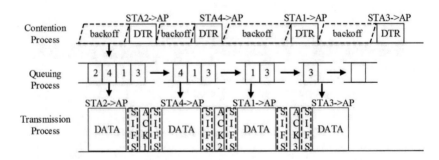

Fig. 4. The procedure of Channel contention and data transmission

3.2.1 Contention Process

In contention process, the node uses CSMA/CA to contend the channel, and it will transmit a DTR frame to the AP once the channel is idle. To achieve the fairness of channel utilization, one node only is permitted to transmit a DTR during a contention period. The CW of nodes is determined by the beacon frame from the AP, which is relevant to the number of nodes in the network. If the collision occurs when the CW is decreased to zero, the node will not attempt to compete the channel again. When the backoff value of the AP reaches zero, The AP indicates the end of the contention period by a beacon when its backoff counter expires. For data transmission period, all frames will be transmitted consecutively and the interval of frame is only a SIFS (Short Inter-frame Space), therefore, all nodes will be able to trigger a new round channel contention process when the channel is idle for a DIFS (Distributed Inter-frame Space) time. The pseudo code of contention process is given as follows:

```
Algorithm 1. Contention Process
 1: for each node do
 2:     if receives Beacon and CQ buffer is empty
 3:        Initialize CW values
 4:     end if
 5:     if idle DIFS
 6:         if backoff value is not zero
 7:            Perform Backoff
 8:         else
 9:            Send a DTR
10:         end if
11:     end if
12: end for
```

3.2.2 Queuing Process

In this process, the AP will perform the queuing and dequeuing operation for of the successful contention nodes. When the AP receives successfully DTR from a node during contention period, it stores the ID of this node into a CQ (contention queuing) buffer. The ID is an association number from the association response management frame and is assigned by the AP during the association procedure.

However, the dequeuing process is operated in the data transmission period. As long as the AP successfully receives the data from a node whose ID is same to the value of the first element in CQ buffer, it will delete the ID from the CQ buffer. As illustrated in Fig. 4, the AP stores sequentially the ID of nodes 2, 4, 1, 3 in contention period.

In data transmission period, AP removes these IDs from its CQ buffer one by one when these nodes have successfully completed the transmission. The pseudo code of queuing process is given as follows:

Algorithm 2. Queuing Process

```
1: if DTR successfully transmits
2:     CQ_Enqueue(DTR.SrcID)
3: end if
4: if Data successfully transmits
5:     CQ_Dequeue(Data.SrcID)
6: end if
```

3.2.3 Data Transmission Process

The data transmission process is initiated by the beacon frame broadcasted by the AP, which will carry on the ID of the first node in the CQ queue. The matching node will transmit a frame after receiving the beacon, and then the AP piggybacks the ID of the next potential transmission node into this ACK frame and broadcasts it. Therefore, the node will transmit a frame after a SIFS time if it overhears this ACK frame and has the same ID piggybacked in it. The data transmission process will keep on until the CQ queue is empty. The pseudo code of data transmission process is given as follows:

Algorithm 3. Data Transmission Process

```
1: for each node do
2:     if myID==ACK_ID
3:         send a data frame
4:     end if
5: end for
```

3.3 Optimal Contention Window

To improve the channel utilization, we design an adaptive scheme to adjust the length of the contention period. Our analytical model borrows a mathematical model studied in [2]. Let τ denote the stationary probability that a node transmits in a randomly selected discrete time slot.

$$\tau = \frac{2}{1+W} \tag{1}$$

Let P_{tr} represent the probability that there is at least one transmission in a time slot. When the n nodes attempt to compete the channel, the P_{tr} is expressed as follows:

$$p_{tr} = 1 - (1 - \tau)^n \tag{2}$$

Let P_s represent the probability that a transmission occurring on the channel is successful, therefore we have

$$P_s = \frac{n\tau(1-\tau)^{n-1}}{P_{tr}} = \frac{n\tau(1-\tau)^{n-1}}{1-(1-\tau)^n} \tag{3}$$

In the contention period, the throughput can be calculated as follows:

$$S = \frac{P_s P_{tr} E[P]}{(1-P_{tr})\delta + P_s P_{tr} T_{succ} + P_{tr}(1-P_s)T_{fail}} \tag{4}$$

Where $E[P]$ is the average packet payload, we use the length of DTR to replace $E[P]$ due to nodes only send DTR in contention period. And δ is the duration of an empty slot time, T_{succ} denotes the average time of successful transmission, and T_{fail} is collision time. The values of T_{succ} and T_{fail} are given by

$$\begin{cases} T_{succ}=T_{DIFS}+T_{DTR} \\ T_{fail}=T_{DIFS}+T_{DTR} \end{cases} \tag{5}$$

Where the T_{DIFS} denotes the time of DIFS, and the T_{DTR} denotes the time required to transmit successfully a DTR frame. We rewrite (4) as

$$S = \frac{E[P]}{T_{succ}-T_{fail}+\frac{(1-P_{tr})\delta/P_{tr}+T_{fail}}{P_s}} \tag{6}$$

Since $E[P]$, δ, T_{succ}, and T_{fail} are constant for all nodes, the throughput S is maximized when the following quantity is maximized:

$$\frac{P_s}{(1-P_{tr})\delta/P_{tr}+T_{fail}} = \frac{n\tau(1-\tau)^{n-1}}{T_{fail}^*-(1-\tau)^n\left(T_{fail}^*-1\right)} \tag{7}$$

Where $T_{fail}^* = T_{fail}/\delta$ is the duration of a collision measured in slot time units δ. After some simplifications for Eq. (7), we have

$$(1-\tau)^n-T_{fail}^*\{n\tau-[1-(1-\tau)^n]\}=0 \tag{8}$$

Under the condition $\tau \ll 1$, the following equation holds.

$$(1-\tau)^n \approx 1-n\tau+\frac{n(n-1)}{2}\tau^2 \tag{9}$$

It also yields the following approximate solution:

$$\tau = \frac{\sqrt{\left[n + 2(n-1)\left(T^*_{fail} - 1\right)\right]/n - 1}}{(n-1)\left(T^*_{fail} - 1\right)} \approx \frac{1}{n\sqrt{T^*_{fail}/2}} \tag{10}$$

Thus, we can get the corresponding optimal contention window size by

$$W \approx n\sqrt{2T^*_{fail}} \tag{11}$$

4 Performance Evaluation

In this section, we evaluate the DMAC via the Matlab simulator, and compare it with to the traditional DCF and IO-MAC [11] in terms of system throughput, delay and channel utilization. We assume that each node in the single hop network communicates directly with the AP and all nodes are within the carrier sense range without the hidden terminal problem. To evaluate the effect of transmission packet size, we choose some different packet size that is 200, 500, 1000 bytes respectively. In addition, we also consider both saturated and unsaturated scenarios. The parameters for MAC and PHY layers are configured as shown in Table 1.

Table 1. Simulation parameters

Parameter	Parameter value	Parameter	Parameter value
Packet payload	200,500,100 bytes	Slot time	20 us
MAC header	224 bits	DIFS	50 us
PHY header	192 bits	SIFS	10 us
Data frame rate	2 Mbps	ACK	128 bits
Simulation time	1800 s	DTR	160 bits

4.1 Saturated Scenario

In saturated scenario, each node has a non-empty packet send queue and the packet interval time is 1 ms. The Figs. 5 and 6 depict the system throughput and channel utilization under various network densities. The DMAC can improve the performance compared to the DCF and IO-MAC regardless of the size of the packet. The reason is that it reduces the idle time and collisions. Although the IO-MAC also can reduce the idle time and collisions, but it will result in much idle when all nodes are in reservation period. Remarkably, when the size of the packet is 1000 bytes and the number of nodes is 100, the system throughput of DMAC is 89.5% and 6.8% higher than that of the DCF and IO-MAC algorithms.

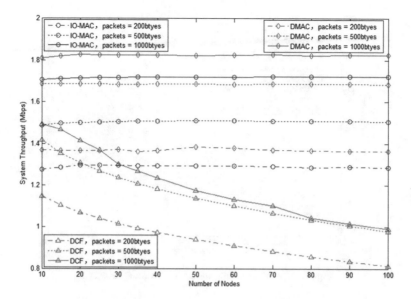

Fig. 5. System throughput under the saturated scenario

The Fig. 7 shows the maximum completion time, minimum completion time and average completion time when all nodes transmit one Mega size file to the AP. It can be noticed that the completion time of three algorithms always increase with the network density increasing. But the DMAC has the lower average completion time than DCF and IO-MAC algorithms. Moreover, the maximum and minimum completion time of the DMAC and IO-MAC are more stable than the DCF, which is due to that their reservation scheme is effective.

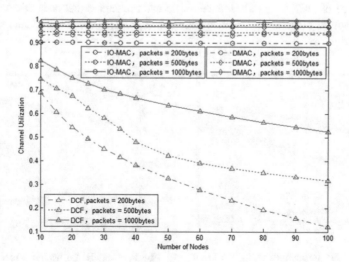

Fig. 6. Channel utilization under the saturated scenario

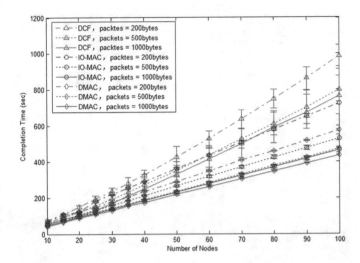

Fig. 7. Completion time under the saturated scenario

4.2   Unsaturated Scenario

To simulate the unsaturated network scenario, we define two types of nodes, the saturation node A has a packet interval time of 10 ms (milliseconds) and the unsaturation type node B has a packet interval time of 100 ms. The number of node A, B and total nodes is represented as N_A, N_B and N, respectively. Here, we choose N as 50 in our simulations. In addition, we only use a packet size of 1000 bytes in the unsaturated environment. The Figs. 8 and 9 show the system throughput and channel utilization under the different network saturated conditions indicated by a variable $\theta = N_A/N$. As the θ increases, the system throughput and channel utilization of DMAC and IO-MAC also increase and the performance of the DMAC is better than IO-MAC. However, the performance of DCF is slightly reduced. The main reason is that both DMAC and IO-MAC adopt a reservation mechanism, but the DCF adopts a contention-based mechanism.

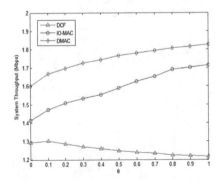

Fig. 8. System throughput under the unsaturated scenario

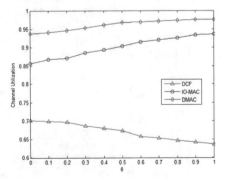

Fig. 9. Channel utilization under the unsaturated scenario

Similar to the saturated scenario, the Fig. 10 also presents the maximum completion time, minimum completion time and average completion time under the different saturated condition when all nodes transmit one 1 Mega file to the AP. With the increasing of θ, the average completion time of DMAC and IO-MAC is gradually decreasing, but the average completion time of DCF are changed slightly.

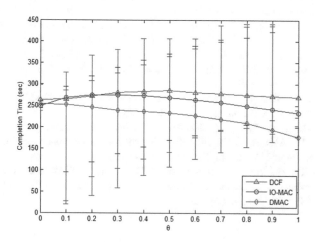

Fig. 10. Completion time under the unsaturated scenario

5 Conclusion

This paper presents a DMAC, which is a decoupled distributed MAC protocol for the single hop infrastructure network. We give the implementation details of the DMAC and propose a derivation model to calculate an optimal length for the contention period. Finally, we verify the effectiveness of our algorithm through intensive simulations. The results show that our algorithm significantly improves the system performance in terms of system throughput, channel utilization and delay under the different network scenarios.

References

1. IEEE Standards Association: IEEE std, 2012, 802: Part 11: Wireless LAN medium access control (MAC) and physical layer (PHY) specifications (2012)
2. Bianchi, G.: Performance analysis of the IEEE 802.11 distributed coordination function. IEEE J. Sel. Areas Commun. **18**(3), 535–547 (2000)
3. Morino, Y., Hiraguri, T., Yoshino, H., et al.: A novel contention window control scheme based on a Markov chain model in dense WLAN environment. In: IEEE International Conference on Artificial Intelligence, pp. 417–421 (2016)
4. Hong, K., Lee, S.K., Kim, K., et al.: Channel condition based contention window adaptation in IEEE 802.11 WLANs. IEEE Trans. Commun. **60**(2), 469–478 (2012)

5. Le, Y., Ma, L., Cheng, W., Cheng, X., Chen, B.: Maximizing throughput when achieving time fairness in multi-rate wireless LANs. In: 2012 Proceedings of IEEE INFOCOM, pp. 2911–2915 (2012)
6. Morino, Y., Hiraguri, T., Yoshino, H., et al.: A novel contention window control scheme based on a Markov chain model in dense WLAN environment. In: 2015 3rd International Conference of IEEE on Artificial Intelligence, Modelling and Simulation (AIMS), pp. 417–421 (2015)
7. Karaca, M., Bastani, S., Landfeldt, B.: Modifying backoff freezing mechanism to optimize dense IEEE 802.11 networks. IEEE Trans. Veh. Technol. **66**(10), 9470–9482 (2017)
8. Kim, J.D., Laurenson, D.I., Thompson, J.S.: Centralized random backoff for collision resolution in Wi-Fi networks. IEEE Trans. Wirel. Commun. **16**(9), 5838–5852 (2017)
9. Hosseinabadi, G., Vaidya, N.: Token-DCF: an opportunistic MAC protocol for wireless networks. In: IEEE 2013 Fifth International Conference on Communication Systems & Networks, pp. 1–9 (2013)
10. Ko, H., Lee, G., Kim, C.: IO–MAC: an enhancement of IEEE 802.11 DCF using implicit ordering. Wirel. Pers. Commun. **72**(2), 1467–1473 (2013)
11. Feng, B., Hong, S., Wang, Z.: IO-MAC: a novel hybrid MAC protocol with implicit ordering in WLANs. In: IEEE 2014 Sixth International Conference on Wireless Communications and Signal Processing (WCSP), pp. 1–5 (2014)
12. Zhao, Q., Xu, F., Yang, J., et al.: CSMA/CQ: a novel SDN-based design to enable concurrent execution of channel contention and data transmission in IEEE 802.11 networks. IEEE Access. **5**, 2534–2549 (2017)

Principal Component Analysis and ReliefF Cascaded with Decision Tree for Credit Scoring

Thitimanan Damrongsakmethee[(✉)] and Victor-Emil Neagoe

Department of Applied Electronics and Information Engineering,
Faculty of Electronics, Telecommunications and Information Technology,
Polytechnic University of Bucharest, Bucharest, Romania
manancc@gmail.com, victoremil@gmail.com

Abstract. The objective of this paper is to propose a credit scoring approval model using a feature selection technique performed by Principal Component Analysis (PCA) and ReliefF algorithm followed by a decision tree classifier. As a reference classifier, we have chosen Support Vector Machine (SVM). The performance of our proposed model has been tested using the German credit dataset. The experimental results of the proposed signal processing cascade for the credit scoring lead to the best accuracy of 91.67%, while classifiers without feature selection show the best accuracy of only 75.35%. On the other side, using the same combination of feature selection (PCA and ReliefF) but cascaded with SVM classifier, one has obtained an accuracy of only 85.15%. The experimental results confirm the accuracy of the proposed model, and at the same time they show the importance of feature selection and its optimization for credit scoring decision systems.

Keywords: Feature selection · Credit scoring · PCA · ReliefF · Decision tree · SVM

1 Introduction

Credit scoring is a bank tool that has been developed for credit approval applications with low credit limits, such as credit cards and cash [4, 13]. Financial institutions are using credit scoring as a tool to measure the willingness of official creditors to repay their debts [3, 7, 11]. Nowadays, credit scoring and credit rating of the consumer are important factors that the bank used for the ability to perform a customer financial in banks [12, 13, 15]. Furthermore, credit scoring is a method using statistical analysis models based on the information about the borrower, allowing one to identify between "good" and "bad" loans and giving an estimate of the probability of default [1, 14]. The priority of credit scoring is performed by the bank based on a judgmental view of credit experts, credit groups or credit bureaus [16]. In addition, credit scoring is a tool that reduces the risk of credit management for financial institutions as well [11]. Moreover, the credit scoring is an essential part of the consumer lending process with the banks [7]. The goal of using credit scoring model is to find a classifier that separates the good credit samples from the bad credit samples [1]. This endeavor is considered as one of the best

© Springer Nature Switzerland AG 2019
R. Silhavy (Ed.): CSOC 2019, AISC 985, pp. 85–95, 2019.
https://doi.org/10.1007/978-3-030-19810-7_9

application fields for both machine learning and operational research techniques [15]. Many models of machine learning classification have been examined to pursue credit scoring accuracy. Classification models for the credit scoring include K-Nearest Neighbors (K-NN) [1], Support Vector Machine (SVM) [20], Multilayer Perceptron (MLP) [14, 17], Naive Bayes, Random Forests, Decision Trees and Logistic [2, 10]. In this study, we have proposed machine learning with feature selection techniques to improve the accuracy in the credit scoring dataset. The main objectives of this paper is to compare the accuracy results of classification techniques including decision trees and SVM (with several kernel functions) with and without feature selection. We used the Principal Component Analysis (PCA) and ReliefF algorithm, because these are popular techniques that researchers expanded and applied for many models and algorithms to analyse the credit scoring risks in banks [6]. According to [9], the feature selection is one of the ways to reduce the dimensionality of the problem and runtime. In evaluating the experimental proposed model, we have used the German credit dataset in a public benchmark from the UCI machine learning repository [21].

The paper is structured as follows: Sect. 2 defines the details of our proposed methods, Sect. 3 presents the experimental results pointing out accuracy improvements of the proposed models, and the conclusions are presented in Sect. 4.

2   The Proposed Model

2.1   Feature Selection

Feature selection is an important processing stage of pattern recognition [16]. The process of feature selection selects the subset of the most significant characteristics from a given set of features and it can simplify the model by reducing the number of parameters [3, 6]. The feature selection method can create an accurate predictive model by choosing significant features that will give a good or better accuracy by comparison to the case of retaining all the features. Moreover, when we have a small number of features, the model becomes more explainable. In this paper, we used two types of feature selection methods: Principal Component Analysis (PCA) and ReliefF algorithm. Each of them will be detailed as follows.

Principal Component Analysis (PCA). PCA is a feature selection method of reducing the data dimensionality by preserving most of the significant information [6, 20]. It is also considered as a factor analysis technique since the components are real factors derived from the correlation matrix [9]. From research literature, it is accepted that PCA is one of the best technique for dimensionality reduction to decrease the number of attributes in the dataset [3]. PCA has as aim to create new components by linear combination of the initial variables in order to describe the information as better as possible [11]. Let X be a random n-dimensional vector. PCA is based on the idea to use an orthogonal transformation to allow the optimal representation of the vector X in relation to the minimal square root error criterion. For a random vector X of N scalar observations, the standard deviation of data set is defined as

$$S = \sqrt{\frac{1}{N-1}\sum_{i=1}^{N}|X_i - \mu|^2}, \tag{1}$$

where μ is the mean of X defined below

$$\mu = \frac{1}{N}\sum_{i=1}^{N} X_i. \tag{2}$$

On German data set, we have calculated the covariance matrix Σ to find the relation between two classes (good credit and bad credit), in which zero value indicates that there is no relations between two dimensions. The covariance matrix [11] is computed using the equation

$$\Sigma = \frac{1}{N}\sum_{i=1}^{N}(x_i - \mu_x)(x_i - \mu_x)^T. \tag{3}$$

Then, we have deduced the eigenvalues and the eigenvectors of the covariance matrix. The eigenvalues are arranged in the descending order, since the largest value of an eigenvalue shows the relative importance of the corresponding principal component. Similarly, the eigenvectors in the matrix Σ must be arranged according to their respective eigenvalues in the diagonal matrix. The eigenvalues correspond to the variances of components in the transformed space and the eigenvectors build the Karhunen-Loeve (PCA) transformation matrix. We have computed the eigenvalues and eigenvectors of the covariance matrix according to relations given below.

$$Eigenvalues: |\Sigma - \lambda I_n| = 0, \lambda_1 \geq \lambda_2 \geq \ldots \geq \lambda_n \tag{4}$$

$$Eigenvectors: \Sigma\,\phi_j = \lambda_j\phi_j, j = 1\ldots n \tag{5}$$

Finally, we have built the K matrix of its eigenvectors as $K = [\phi_1, \phi_2, \ldots, \phi_n]^T$, then matrix K must be truncated as $K' = [\phi_1, \phi_2, \ldots, \phi_m]^T$. Each input vector X from the n-dimensional space is transformed into a m-dimensional space as

$$Z = K'X. \tag{6}$$

The PCA technique generates the set of new variables called *principal components*; they are statistically independent. The first principal component has the highest variance, and next components correspond to decreasing variances according to their order. For example, in our experiments we have selected a number m = 20 principal components, in order to preserve 95% of the total variance in the PCA space.

PCA Algorithm corresponds to the following steps as:

1. Standardize the data to calculate mean (μ) and standard deviation.
2. Calculate the covariance matrix of the standardized data. The covariance matrix contains all necessary information to change the coordinate system.

3. Use the resulted covariance matrix to deduce the eigenvalues, eigenvectors and of the transformation matrix.
4. Sort the eigenvalues in descending order.
5. Choose m components which retain the most amount of variance (information) within the new components (the larger eigenvalue means the feature contains more variance).
6. Compute the transformed data matrix using m components.

Relief Algorithm. Relief algorithm belongs to the class of filtering methods; it is designed to estimate the important of attributes in order to discard noisy features [2]. Relief algorithm estimates attributes to evaluate how well they distinguish between samples which are close to each other in the dataset. It can find two nearest neighbors and rebounds a weight for each attribute [5]. We have used Relief algorithm called *ReliefF*, which improves the original Relief algorithm with a more reliable probability estimation. The main concept of ReliefF algorithm is based on the collection of k nearest hits/misses for each training sample instead of the single hit/miss used for estimation in the original Relief algorithm. ReliefF returns the ranks and weights of predictors for the input data matrix X and response vector Y and predictor ranks and weights usually depend on k. We have explored different settings of k in ReliefF algorithm to know how well the number of nearest neighbors impacts performance, as well as whether setting of k based on a percentage of instances offers a potential alternative to threshold-based neighbor selection [19]. In this study, we have started with $k = 10$ and have investigated the stability and reliability of ReliefF ranks and weights for various values of k. ReliefF starts to set the weights of various values of k to zero ($w_j = 0$). Then, the algorithm iteratively selects a random observation x_r, it finds the k-nearest observations to x_r for each class, and then it updates, for each nearest neighbor x_q, all the weights for the predictors F_j according to next equations.

Case 1: If x_r and x_q are in the same class, then

$$w_{ji} = w_{ji-1} - \frac{\Delta j(x_r, x_q)}{m} * d_{rq} \tag{7}$$

Case 2: If x_r and x_q are in the different classes, then

$$w_{ji} = w_{ji-1} + \frac{p_{y_q}}{1 - p_{y_r}} * \frac{\Delta j(x_r, x_q)}{m} * d_{rq} \tag{8}$$

For the Eqs. (7) and (8) given above, w_{ji} is the weight of the predictor F_j at the *i*th iteration step, p_{y_r} is the a priori probability of the class to which x_r belongs, and p_{y_q} is the a priori probability of the class to which x_q belongs. The parameter m is the number of iterations. Here $\Delta j(x_r, x_q)$ is the difference of the values of the predictor of feature F_j between observations x_r and x_q, the parameter x_{rj} denotes the value of the

jth predictor for observation x_r and the parameter x_{qj} denotes the value of the jth predictor for observation x_q. Then, we have computed the feature difference of the two different samples, according to the equation given below.

For discrete features:

$$\Delta j(x_r, x_q) = \begin{cases} 0, \text{ when } x_{rj} = x_{qj} \\ 1, \text{ when } x_{rj} \neq x_{qj} \end{cases} \tag{9}$$

For continuous features:

$$\Delta j(x_r, x_q) = \frac{|x_{rj} - x_{qj}|}{\max(F_j) - \min(F_j)}, \tag{10}$$

Then, the algorithm computes the distance function d_{rq} using the relations

$$d_{rq} = \frac{\tilde{d}_{rq}}{\sum_{l=1}^{k} \tilde{d}_{rl}}, \text{ with } \tilde{d}_{rq} = exp\left(-\left(\frac{rank(r, q)}{\sigma}\right)^2\right), \tag{11}$$

where k is number of k-nearest neighbors, rank (r, q) is the position of the qth observation among the nearest neighbors of the rth observation, and the parameter σ controls the influence of the distance [22].

2.2 Classification

Decision Trees. The decision trees are binary circular partitioning procedures with the ability of processing continuous and nominal features as targets [16]. In [8], Gepp and Kumar claim that a decision tree model is a hierarchical structure approach with nodes and direct edges and it breaks down a dataset into smaller and smaller subsets. A normal pattern of decision trees has three basic types of nodes: a root node (without coming edges or more outgoing edges), internal nodes (with one coming edge and two or more outgoing edges) and leaves representing terminal nodes (each with one coming edge and no outgoing edge) [16]. The leaf node of the decision presents the class labels of the decision trees model [8]. The method of node splitting for the decision trees is called *the classification and regression tree* (CART) [18]. The representation for the CART model is a binary tree. Each root node represents a single input variable (x) and a split point on that variable (assuming the variable is numeric). Each leaf node of the tree contains an output variable (y) which is used to make a prediction. The concept of the CART method uses the binary splitting method which means that each non-leaf node splits into two new branches. We have used the Gini's diversity index, which is similar to classical entropy (information gain) criterion. For a binary target, the Gini measure of impurity of a node t [18] is defined as

$$G(t) = 1 - p(t)^2 - (1 - p(t))^2, \tag{12}$$

where $p(t)$ is the (possible weighted) relative frequency of class 1 in the node. Using the entropy rule, we can obtained a new expression of the impurity measure, namely $G(t) = -p(t) \ln p(t) - (1 - p(t)) \ln (1 - p(t))$. The improvement information gain generated by a split of the parent node P into left and right children L and R is defined as

$$I(P) = G(P) - qG(L) - (1 - q)G(R), \tag{13}$$

where q is the possible weighted fraction of instances going left children. We have used classification learner in Matlab program to create and train three models of prediction options of decision trees as: a Fine Tree, a Medium Tree, and a Coarse Tree. Each of the above mentioned categories is detailed below.

Fine Tree: There are many small leaves for a highly flexible response function (minimum leaf size is 4, maximum number of splittings is 100 and optimization is based on setting split criterion according to Gini's diversity index).

Medium Tree: There are medium-sized leaves for a less flexible response function (minimum leaf size is 12, maximum number of splittings is 20 and optimization uses Gini's diversity index).

Coarse Tree: There are a few large leaves for a coarse response function (minimum leaf size is 36, maximum number of splittings is 4 and optimization is based on Gini's diversity index).

2.3   Our Proposed Model

Our proposed method has two steps. In the first step, after data normalization and data preparation, the dataset have been divided into a subset of 70% representing the training dataset and a subset of 30% representing the test dataset. Then, we have applied feature selection using one of the following techniques: PCA, ReliefF algorithm or a combination of PCA and ReliefF algorithm. The training dataset has been trained and tested by each of the two considered classifiers: decision trees and SVM. We have used several kinds of decision trees (fine tree, medium tree and coarse tree). As a result of the first step, we have obtained an optimum set of features. In the second step, the optimal features from the first step have been used to classify the testing dataset. Then, we have estimated the accuracy for each classifier (see Fig. 1).

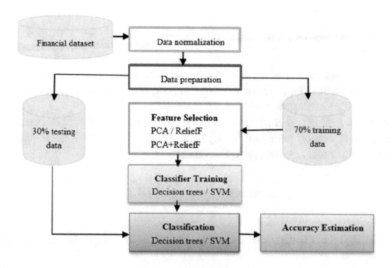

Fig. 1. Our proposed model

3   Experimental Results

3.1   Dataset Description

The German credit approval dataset consists of 1000 loan applications with 700 accepted credit cards (1 = "good credit") and 300 rejected credit cards (2 = "bad credit"). An applicant is described by 25 numeric attributes including class label in the last column of dataset [21]. The remained 24 attributes describe the status of existing checking accounts, credit history records, credit amount and purpose, an income status, an assortment of personal information such as age, sex, job, property, housing, the number of existing credits at this bank and marital status etc.

3.2   Experimental Procedure

Our model includes the following steps:

Step 1: Data collection. We have used the German credit dataset from a public benchmark (UCI machine learning repository).

Step 2: Data preparation. We have divided the dataset into a subset of 70%, representing the training dataset and a subset of 30%, representing the testing dataset. We have used data normalization according to the equation

$$\text{Normalization} = \frac{x_i - \text{mean}}{\text{std}} \tag{14}$$

Step 3: Feature selection tasks. We have performed feature selection using PCA, ReliefF and combination of PCA and ReliefF. The percentage of variance of selected components in the PCA transformed space is given by relation

$$\text{Percentage of variance in PCA transformed space} = \frac{\sum_{i=1}^{m} \lambda_i}{\sum_{i=1}^{n} \lambda_i} * 100, \qquad (15)$$

where n is the dimension of input data, and m is the number of selected components.
Step 4: Classifier training. We have trained the consider classifiers (Decision trees and SVM) using the training dataset.
Step 5: Model testing and accuracy evaluation. The proposed model has been tested for classification (Decision trees and SVM) using testing dataset. We have evaluated the accuracy of the model according to the relation

$$\text{Accuracy} = \left(\frac{TP + TN}{N}\right) * 100, \qquad (16)$$

where TP is True Positives, TN is True Negatives and N is the total number of instances.

3.3   Experimental Results

Table 1 shows the amount of variance in the PCA transformed domain associated with each component. The optimal number of components can be defined as the minimum number of components that contain the maximum variance. For our experiment, we have retained the accumulated percentage of variance in the PCA transformed space of 95.33%, corresponding to m = 20 selected components.

Table 1. Eigenvalues, percentages of variance in the transformed PCA space associated with each component and cumulated variance.

Component index	1	5	10	15	18	20	23	25
Eigenvalue	2.4602	1.6435	1.0243	0.9334	0.8002	0.4832	0.1283	0.1112
% Variance in the PCA transformed space	9.84	6.56	4.08	3.73	3.20	1.93	0.48	0.44
% Cumulated variance	9.84	37.92	61.24	80.76	90.80	95.33	99.12	100.00

Table 2 shows the results of the different classifier types consisting of SVM-RBF, decision trees-fine tree, medium tree, and coarse tree. We have considered both classification without feature selection and also with feature selection. We have found that the fine tree classifier using for feature selection a combination of PCA and ReliefF leads to the best performance. Its maximum classification accuracy is of 91.67%.

Table 2. Performance comparison of the different classifier models without feature selection and with feature selection.

Classifier model	Without feature selection	With feature selection		
		PCA	ReliefF	PCA+REliefF
SVM-RBF	75.35	76.00	86.16	85.18
Decision tree-Fine tree	74.00	72.23	88.28	**91.67**
Decision tree-Medium tree	77.33	78.00	83.00	81.67
Decision tree-Coarse tree	74.67	76.00	75.33	80.49

In Fig. 2 we have shown an example of fine decision tree with the maximum number of splits equal to 6. This cluttered-looking tree uses a series of rules of the tree form node "PCAcomponent_7 < 0.142744" to classify each good or bad credit into one of terminal nodes. To deduce the status of a credit card, we have started at the top node and have applied the algorithm rule. If a node satisfies the rule, we have taken the left path, and if not we have taken the right path. Ultimately, we reach a terminal node that assigns the observation to one of two decision regarding the card status: accepted or rejected (see Fig. 2).

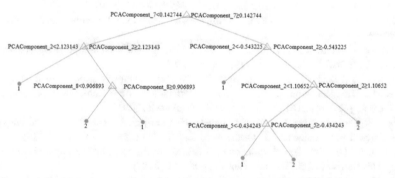

Fig. 2. Decision tree-Fine tree with 3 pruning levels (1 = "good" = acceptation; 2 = "bad" = rejection).

4   Conclusions

In this paper, we have focused on improving the accuracy of credit scoring approval systems by proposing an optimum combination cascade of feature selection followed by a corresponding binary classifier. The experiments have used the German credit dataset, available from the UCI machine learning repository. For feature selection, we have applied PCA and ReliefF algorithms. For classification, we have considered four variants, one for SVM and three ones using decision trees. The experiments have shown that for the same feature selection technique, the classification using decision tree have led to better accuracy than using SVM classifier. The best performance is obtained by combination of PCA and ReliefF for feature selection followed by fine decision tree for classification. This corresponds to an accuracy of 91.67% that is much higher when compared to the results obtained both for the case without feature selection and that using only one of the variants of PCA or ReliefF for feature selection (Table 2). The experimental results show the obvious advantage of our proposed model (feature selection with combination of PCA and ReliefF followed by decision tree-fine tree classifier) over other considered variants.

References

1. Abdelmoula, A.K.: Bank credit risk analysis with k-nearest- neighbor classifier: case of Tunisian banks. Account. Manag. Inf. Syst. **14**(1), 79–106 (2015)
2. Agre, G., Dzhondzhorov, A.: A weighted feature selection method for instance-based classification. In: 17th International Conference on Artificial Intelligence : Methodology, Systems and Applications (AIMSA), pp. 14–25. Springer, Switzerland (2016)
3. Anaei, S.M., Moradi, M.: A new method based on clustering and feature selection for credit scoring of banking customers. Int. J. Mod. Trends Eng. Res. **3**(2), 123–128 (2016)
4. Antunes, F., Ribeiro, B., Pereira, F.: Probabilistic modeling and visualization for bankruptcy prediction. Appl. Soft Comput. J. **60**, 831–843 (2017)
5. Beretta, L., Santaniello, A.: Implementing ReliefF filters to extract meaningful features from genetic lifetime datasets. J. Biomed. Inform. **44**(2), 361–369 (2011)
6. Browne, D., Prestwich, S.: Credit scoring : feature selection on machine learning algorithms (2016)
7. Damrongsakmethee, T., Neagoe, V.: Data Mining and machine learning for financial analysis. Indian J. Sci. Technol. **10**(39), 1–7 (2017)
8. Gepp, A., Kumar, K.: Predicting financial distress: a comparison of survival analysis and decision tree techniques. In: Eleventh International Multi-Conference on Information Processing-2015 (IMCIP-2015), pp. 396–404. Elsevier (2015)
9. Go, W., Lee, T., Kim, I., Lee, K.: Feature selection practice for unsupervised learning of credit card fraud detection. J. Theor. Appl. Inf. Technol. **96**(2), 408–417 (2018)
10. Gupta, A.: Classification of complex UCI datasets using machine learning and evolutionary algorithms. Int. J. Sci. Technol. Res. **4**(5), 85–94 (2015)
11. Ha, V., Nguyen, H.: Credit scoring with a feature selection approach based deep learning. In: MATEC web of Conferences (MIMT), pp. 1–5 (2016)
12. Ilgun, E., Mekic, E., Mekic, E.: Application of Ann in Australian credit card apporval. Int. Multidiscip. J. **69**(2), 334–342 (2014)
13. Louzada, F., Ara, A., Fernandes, G.B.: Surveys in operations research and management science classification methods applied to credit scoring: systematic review and overall comparison. Surv. Oper. Res. Manag. Sci. **21**(2), 117–134 (2016)
14. Neagoe, V., Ciotec, A., Cucu, S.: Deep convolutional neural networks versus multilayer perceptron for financial prediction. In: International Conference on Communications (COMM-2018), pp. 201–206. IEEE, Bucharest (2018)
15. Pendey, S., Benkatesh, N.: Analysis of German credit data using microsoft azure machine learning. J. Emerg. Technol. Innov. Res. **5**(2), 302–305 (2018)
16. Sang, H., Nam, N., Nhan, N.: A novel credit scoring prediction model based on feature selection approach and parallel random forest. Indian J. Sci. Technol. **9**(20), 1–6 (2016)
17. Shukla, A., Mishra, A., Gwalior, M.: Design of credit approval system using artificial neural network: a case study. Int. J. Eng. Res. Comput. Sci. Eng. **4**(1), 1–6 (2017)
18. Steinberg, D.: CART : Classification And Regression Trees, Chap. 10 (2009)

19. Urbanowicz, R.J., Olson, R.S., Schmitt, P., Meeker, M.: Benchmarking relief-based feature selection methods for bioinformatics data mining. J. Biomed. Inform. **4**(2), 1–21 (2018)
20. Van, S., Ha, N., Bao, H.: A hybrid feature selection method for credit scoring. EAI Endorsed Trans. Context. Syst. Appl. **4**(11), 1–6 (2017)
21. UCI. https://archive.ics.uci.edu/ml/datasets/statlog+(german+credit+data). Accessed 10 Sept 2018
22. ReliefF algorithm. https://uk.mathworks.com/help/stats/relieff.html#responsive_offcanvas. Accessed 10 Feb 2019

Online Monitoring Automation Using Anomaly Detection in IoT/IT Environment

Chul Kim[1,2], Inwhee Joe[2(✉)], Deokwon Jang[1], Eunji Kim[1], and Sanghun Nam[1]

[1] NKIA Corporation, 660 Daewangpangyoro, Seongnam, South Korea
ki4420@gmail.com, deokwon@gmail.com,
kej12516@gmail.com, smilegun@gmail.com
[2] Hanyang University,
222 Wangsimni-ro, Seongdong-gu, Seoul 04763, South Korea
iwjoe@hanyang.ac.kr

Abstract. The increase of the IoT and the cloud environment have played a significant role of making our society knowledgeable and informative.

Due to this trends the system environment gets more sophisticated and requires more system resources. In this paper, the monitoring automation without humans being involved has been proposed. It is noted that the 93.75% faults has been detected via the simulation using the proposed technique and the faults that the operators reported have been detected as well in datacenter.

Keywords: Anomaly detection · Machine learning · Monitoring automation · Time-series · IT · IoT

1 Introduction

The increase of the IoT and the cloud environment have played a significant role of making our society knowledgeable and informative.

Likewise, the use of IoT sensors and system resources becomes more pronounced and the resource environment becomes more dynamic and complicated.

Since we come to the conclusion that it might be impossible to manage the massive resources with a traditional monitoring methodology. It is suggested to utilize the monitoring automation with a machine learning.

The traditional monitoring methodologies can be categorized as follows: (1) the operator manually deal with visual monitoring the events in the chart (2) based on rulesets such as CEP and fixed threshold values, the systematic monitoring where dynamic threshold values using simple statistics are applied [6].

The manual monitoring executed by operators is good for dealing with anomaly status. Nevertheless, the number of monitoring objects is limited and the quality of monitoring varies depending on the performance of the operators.

Furthermore, it is not precise to monitor with the fixed threshold value in a systematic monitoring. The maintenance based on rulesets enables a precise monitoring, but it is expensive if the level of complication gets higher. The threshold value based on simple statistics like average and standard deviation is deficient in order to reflect the characteristics of the seasonality that each individual object has.

© Springer Nature Switzerland AG 2019
R. Silhavy (Ed.): CSOC 2019, AISC 985, pp. 96–106, 2019.
https://doi.org/10.1007/978-3-030-19810-7_10

In order to improve the traditional monitoring, the monitoring automation without humans being involved has been proposed. We propose the technique that can determine the functionality by using the statistical method with measured anomaly scores after executing to changepoints detection and anomaly detection. This is based on the model that learns unsupervised time series model online. In order to verify the technique the fault scenarios have repeated 16 times without a separate signature or rulesets, results are faults of 93.75% are successfully detected.

2 Background

2.1 Log Regression Seasonality Based Time Series Decomposition

When the data is analyzed by TSD it becomes made out of trend, seasonality, random and cyclical. This has been investigated to create the prediction data by characterizing or recombine the data. Even though the observed data severely varies the data trend, periodic seasonality can be found by analyzing factors with TSD whereas the portion that rules out the trend and seasonality can be extracted. Figure 1 shows the seasonality analysis chart of CPU usage using the decompose function of R stats package. Figure 1 exhibits the prediction method using the average and log regression seasonality, respectively. Figure 1 upper chart shows the average seasonality that reflects the average of the factors with the same seasonal index as the observed data. This has the advantage that the operation time is fast [4].

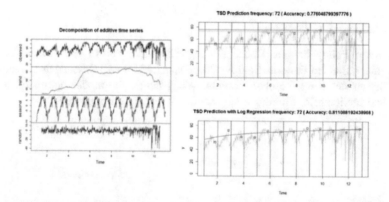

Fig. 1. Time series decomposition by R package {stats} and average seasonality of time series decomposition (hereafter TSD) vs. log regression seasonality of time series decomposition (hereafter LRSTSD) [4].

2.2 Changepoints Detection

Changepoint analysis for time series is an increasingly important aspect of statistics. A changepoint is an instance in time where the statistical properties before and after this time point differ. The first published article concerning change-points was in 1954 by E.S. Page [1, 8].

2.3   Anomaly Detection

In data mining, anomaly detection is the identification of rare items, events or observations which raise suspicions by differing significantly from the majority of the data [2]. The anomaly detection is the technique to detect the micro-cluster that is apart from mainstream for instance [3], it can recognize the labels like normal or anomaly and detect something by scoring the degree of anomaly.

3   Our Approach

3.1   Overview

We are aiming to develop the automated technique that can sustain the availability of monitoring resources by anomaly detection and adopting to the environmental changes without human being involved. First, the monitoring automation generates the time series model every 12 h by utilizing the data (4,320) collected per 10 min for one month of each resource. Second, it scores the degree of anomaly by measuring the difference between the values predicted by the time series model and observed value. Lastly, sum of anomaly scores evaluate by threshold $Q1, Q3 \pm 1.5 \cdot IQR$ is defined (hereafter $1.5 \cdot IQR$ threshold). If the values are out of the range the resource list consisting of the scores are transferred to the operators. Therefore, it enables the rapid analysis and resolution of root causes of anomaly and makes it possible to support monitoring automation.

3.2   Changepoints Detection

We develop Policy Sample Consensus algorithm that allows ones to detect a big change and corrected Random Sample Consensus (RANSAC). It leads us to realize the capability of detecting the changepoints. We generate the model optimized in each individual section after dividing each sections based on changepoints.

Fig. 2. The red vertical lines indicates the detected changepoint that implies the beginning and end of the local model.

Algorithm 1. Policy sample consensus(Changepoints detection)

1: Receive actuals a of length T
2: Initialize differences memory D
3: **for** $t = 2, T$ **do**
4:    Store $abs(a_t - a_{t-1})$ in D
5: **end for**
6: Remove $d < 2.5\ percentile$ and $d > 97.5\ percentile$ from D
7: Set distance $\alpha \leftarrow mean(D)$
8: Set number of samples $N \leftarrow 12$
9: Set step size $w \leftarrow T \cdot (N)^{-1}$
10: Sample indices γ increased by w from a
11: Accumulate all subsets ϕ that have two index from γ
12: Initialize max. number of inliers $m \leftarrow 0$
13: Initialize optimal point $p \leftarrow 0$
14: **for** $i = 1, len(\phi)$ **do**
15:    Take $l \leftarrow$ compute homography from ϕ_i
16:    Take number of inliers $c \leftarrow$ compute inliers in l under α
17:    **if** $c > m$ **then**
18:        $p \leftarrow \phi_i, m \leftarrow c$
19:    **end if**
20: **end for**

3.3 Detecting Seasonality

The detection of seasonality is executed using Autocorrelation function (ACF) or Partial Autocorrelation function (PACF) However, since this technique measures the relative analysis of entire data it is impossible to be employed in online system or massive data. Therefore, we propose the seasonality detection method that measures only the relativity within close neighborhood that consumes less computing resources than ACF does. The similar seasonality detection can be realized with the small amount of cost than ACF by using the technique studies here.

Algorithm 2. Detecting seasonality

  1: Receive actuals a
  2: Set seasonal range $T \leftarrow (len(a)) \cdot 3^{-1}$
  3: Set seasonal fitting threshold $k \leftarrow 0.6$
  4: Initialize optimal seasonal $\tau \leftarrow 0$
  5: Initialize max. fitting $f \leftarrow 0$
  6: **for** seasonal $s = 10, T$ **do**
  7:      Initialize correlation memory D
  8:      **for** $i \leftarrow 1, T$ **step** s **do**
  9:         $u \leftarrow a[i{:}i+s], v \leftarrow a[i+s{:}i+2\cdot s]$
10:         Compute $\rho \leftarrow correlation(u,v)$
11:         Store ρ in D
12:      **end for**
13:      Compute $z \leftarrow mean(D)$
14:      **if** $z > f$ **then**
15:         $f \leftarrow z, \tau \leftarrow s$
16:      **end if**
17: **end for**
18: **if** $f > k$ **then**
19:      **return** τ
20: **Else**
21:      **return** 0

3.4 Seasonal Model

When the seasonality is detected in the sections divided by changepoints the seasonality model is generated, otherwise distribution model is generated. The seasonality model is generated by applying the LRSTSD with robust regression which we previously proposed.

Algorithm 3. Create seasonal model

  1: Receive actuals a, seasonal s
  2: Initialize seasonal model π
  3: Compute robust linear regression model b as trend from a
  4: Compute log regression seasonality model g according to LRSTSD[4]
  5: Set $\pi_i \leftarrow b_i + g_i$

3.5 Distribution Model

In the section where the seasonality is not detected the trend is extracted with robust regression t. The distribution model ξ is generated with the limit value of residuals r.

$$r(x_i) = Y_i - t_i \tag{1}$$

$$\xi(upper_i, lower_i) = (t_i + Q_3 + 1.5 \cdot IQR(r(X)), t_i - Q_1 - 1.5 \cdot IQR(r(X)) \tag{2}$$

3.6   LRSTSD with Changepoints

The seasonality prediction modeling with LRSTSD has been exploited. The accuracy of the LRSTSD assumed as single model significantly reduces when the changepoint exists [7]. In order to resolve the issue it is suggested to detect the changepoints with the robust technique and generate the optimal model. This technique is called Log regression seasonality based Time Series Decomposition with ChangePoints (LTSDCP). The first step of the time series prediction modeling generation with LTSDCP is to detect the changepoints sensing the change in the model within the dataset. The second step is to detect the existence of the seasonality in separated sections divided with the changepoints when the seasonality exists the prediction model that adopts the robust model with the LRSTSD is generated. Otherwise, the distribution model is generated.

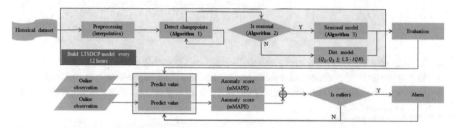

Fig. 3. Flow diagram where the LTSDCP prediction model is generated every 12 h for 1 month and anomaly detection within online data is executed.

3.7   Anomaly Detection

The separated sections divided with the changepoint generates Seasonal Model and distribution model depending on the characteristics and, in turn, the LTSDCP model is composed. The model generated in each section is called the local model. This model calculates the anomaly scores with the predicted value and the value observed online that are generated with the latest local model and, in turn, automatically executes the anomaly detection with $1.5 \cdot IQR$ threshold. The advantage of the technique is to improve the performance of prediction creating the latest model generated.

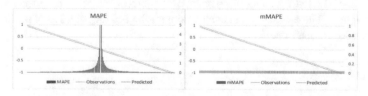

Fig. 4. In the case where the observed value is equal to the predicted one it is measured with MAPE and mMAPE as MAPE get close to zero the measured value exponentially increases. mMAPE is constant around the values around zero [5].

The detected changepoint is recorded as a label whilst it can be used in the root cause analysis. In order to calculate the anomaly score Mean Absolute Percentage Error (MAPE) and extended maximum Mean Absolute Percentage Error (mMAPE) (3) [5] are utilized. Though the observed value is close to zero mMAPE is able to measure normal scores.

$$score(X) = \begin{cases} \frac{100}{n} \sum_{i=1}^{n} |x_i - \hat{x}_i|, & if\ x_{max} < 1 \\ \frac{100}{n} \sum_{i=1}^{n} \frac{|x_i - \hat{x}_i|}{x_{max}}, & otherwise \end{cases} \qquad (3)$$

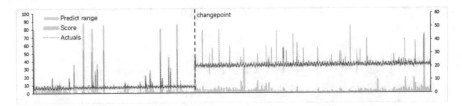

Fig. 5. Indicates the observed and predicted values and anomaly score and change point. The gray line represents the observed values while the blue line represents the predicted values calculated with LTSDCP. The orange column is the anomaly score while the red vertical line is the detected changepoints.

4 Experiments

4.1 Overview

Two experiments have been executed in order to verify the technique suggested here. First, the duration of generating model to assess the performance of LTSDCP algorithm, the number of the detected model and accuracy have been measured by mMAPE (3). Second, it has been assessed that the malfunction depending on the scenario in the practical circumstance is reproduced, it can work by itself.

4.2 Accuracy of LTSDCP vs. LSTM

In order to assess the performance of the seasonality model the performance of LTSDCP is evaluated by using the dataset of the single variable time series as can be seen in Table 1. The assessment is evaluated with the holdout validation in the ratio of train data (80) to test data (20).

Table 1. Accuracy and model time of LTSDCP vs. LSTM

Dataset	# of dataset	LTSDCP		LSTM	
		Test mMAPE	Train time	Test mMAPE	Train time
01_1.csv	1,030	19.12	124 ms	12.43	214 s
02_2.csv	8,988	1.05	6.61 s	3.52	5,365 s
03_2.csv	1,284	20.21	201 ms	7.83	147 s
04_1.csv	371	12.65	35 ms	25.95	125 s
05_1.csv	381	6.37	14 ms	18.24	125 s
06_1.csv	549	13.31	30 ms	12.19	312 s
07_1.csv	360	18.86	20 ms	7.56	145 s
08_2.csv	3,321	0.03	727 ms	11.82	2,961 s
09_1.csv	2,435	10.2	640 ms	5.21	2,229 s
10_1.csv	2,435	18.66	703 ms	20.27	1,593 s
11_1.csv	2,435	33.04	448 ms	4.78	1,423 s
12_1.csv	2,435	4.6	605 ms	10.63	3,314 s
13_2.csv	2,435	**0.01**	190 ms	6.22	1,536 s
14_1.csv	1,569	14.46	565 ms	6.00	25,976 s
15_2.csv	288	1.16	7 ms	6.35	383 s
16_1.csv	1,425	5.7	224 ms	5.79	2,648 s
17_1.csv	1,425	0.01	97 ms	7.74	236 s
18_1.csv	2,435	0.11	658 ms	19.30	1,307 s
19_1.csv	400	**0.01**	23 ms	15.32	10,479 s
20_2.csv	1,440	11.82	222 ms	4.85	1,238 s
Mean	**1,872.05**	**9.57**	**607.15 ms**	**10.60**	**3,167 s**

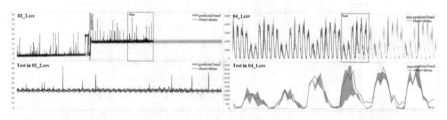

Fig. 6. Figures show some test datasets in Table 1. Furthermore, if you really want to test, you can test it via http://anny.io.

The train time is used to evaluate its feasibility to the online system. LTSDCP was evaluated with higher accuracy than LSTM because it predicts the intervals such as upper and lower. mMAPE (0.01) means that all test values are between upper and lower.

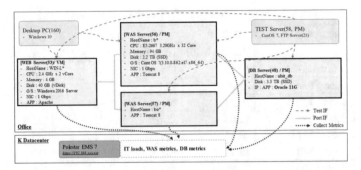

Fig. 7. Figure indicates the diagram of the established environment to test the malfunctioning scenario. The web service is established so that the scenario can be executed. In addition, two laptop computers which malfunctioning scenario can occur are connected. The system resource data measured in the web service is collected via Polestar EMS 7 system, and the anomaly score is measured.

The web service established here is used for a test environment. The test environment as the distributed system consisting of web server, WAS server, DB server gives loading to a particular system or disables the functionality. It measures the probability that can be detected the technique suggested here. The total fault number of 16 times is reproduced while the detection is tested as the automated monitoring suggested here.

Table 2. Test scenario list

No.	Owner	Scenario
1st	PC(160) → WEB(53)	HTTP Congestion
2nd	PC(160) → WEB(53)	TCP Port(80) Congestion
3rd	PC(160) → WAS(56)	Login and Http Congestion
4th	TEST(58) → WAS(56)	TCP(8009) Congestion
5th	WEB(53)	High CPU usage of other process
6th	WEB(53)	High Memory usage of other process
7th	WAS(57)	High CPU usage of other process
8th	WAS(57)	High Memory usage of other process
9th	WAS(57)	Filesystem is full
10th	WAS(57)	Disk I/O overloads
11th	DB(48)	DB Lock
12th	DB(48)	DB SQL(insert, select) Congestion
13th	DB(48)	DB SQL(insert, delete) Congestion
14th	DB(48)	exec select_nobind(5,000,000)
15th	DB(48)	Top SQL(long-running transaction)
16th	DB(48)	Session overload

The web server under the test environment operates 7 days. After that the automated detection function is demonstrated by running the malfunctioning scenario defined in Table 2.

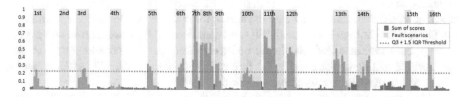

Fig. 8. Figure shows the summation of the fault scenario reproduced 16 times and anomaly score and $1.5 \cdot IQR$ *threshold* values as a horizontal line; the faults are not detected for 2nd and 4th attempts whereas all the faults is detected for the rest of attempts. Non-detected fault result from the TCP Congestion where simultaneous requests are executed. The sum of these anomaly scores keep the average values.

5   Application

5.1   Overview

The technique suggested here has been applied in the operating data center and the automated monitoring has been proven. The monitoring is executed for the environmental factors such as temperature and moisture level. Figure 9 exhibits the comparison between the fault events reported by the operator during the test period and the score collected from the monitoring automation.

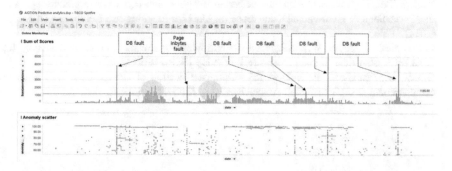

Fig. 9. Figure displays the demonstration of the monitoring automation by using Spotfire. The chart on the top indicates the summation of the score and $1.5 \cdot IQR$ *threshold* as a column chart and horizontal red dot line, respectively. The tooltips shown in the figure indicates the points that the operators physically detect the malfunction events. It is noted that all the scores exceed the threshold value. However, since the case where the anomaly score in the unknown points (two orange circle) exceeds the threshold value is detected it is about to be analyzed.

6 Conclusions

It is noted that the detection performance is 97.35% (14/16) in the scenario where the monitoring is automated without human being involved. However, subsequent research is needed to identify False Positive (FP) or True Positive (TP) for unknown alarms. If it is a True Positive the research where the descriptor is developed for additional artificial intelligence technology is executed. If it is a False Positive the resolution to prevent FP through the root cause analysis should be investigated. As mentioned above, TP (descriptor)/FP will be investigated in the future. Additionally, LTSDCP, ARIMA, Holt-winter and performance assessment will be executed.

This work was supported in part by MSIT & NIPA. (C0610-18-1017, Smart Industrial Energy ICT Convergence Consortium).

References

1. Killick, R., Eckley, I.: changepoint: an R package for changepoint analysis. J. Stat. Softw. **58** (3), 1–19 (2014)
2. Zimek, A., Schubert, E.: Outlier detection. In: Encyclopedia of Database Systems, pp. 1–5. Springer, New York (2017). https://doi.org/10.1007/978-1-4899-7993-3_80719-1. ISBN 9781489979933. Accessed 16 Aug 2018
3. Chandola, V., Banerjee, A., Kumar, V.: Anomaly detection: a survey. ACM Comput. Surv. (CSUR) **41**(3), 15 (2009)
4. Kim, C., Nam, S.-H., Joe, I.: A log regression seasonality based approach for time series decomposition prediction in system resources. In: Advances in Computer Science and Ubiquitous Computing, pp. 843–849. Springer, Singapore (2015)
5. Shin, K.-H., Kim, C.: Estimation method of predicted time series data based on absolute maximum value. J. Energy Eng. **27**(4) (2018)
6. Veasey, T.J., Dodson, S.J.: Anomaly detection in application performance monitoring data. In: Proceedings of International Conference on Machine Learning and Computing (ICMLC) (2014)
7. Carter, K.M., Streilein, W.W.: Probabilistic reasoning for streaming anomaly detection. In: 2012 IEEE Statistical Signal Processing Workshop (SSP). IEEE (2012)
8. Killick, R., Fearnhead, P., Eckley, I.A.: Optimal detection of changepoints with a linear computational cost. J. Am. Stat. Assoc. **107**(500), 1590–1598 (2012)

Cross-collection Multi-aspect Sentiment Analysis

Hemed Kaporo[✉]

Sabanci University, Orta Mahalle, 34956 Tuzla, Istanbul, Turkey
hemedkaporo@sabanciuniv.edu

Abstract. This paper proposes the use of cross-collection topic models to achieve aspect-based sentiment analysis of multiple entities simultaneously. A topic refinement algorithm that enhances semantic interpretability of topics to match that of visually identifiable aspects is presented. It is shown that, with this refinement, topics elicited from cross-collection topic models align excellently with entity aspects. Finally, the utility of opinion words returned from cross-collection topic models in investigated in the task of sentiment analysis. It is concluded that the use of such words as features for sentiment analysis yields more accurate sentiment scores than supervised counterparts.

Keywords: Cross collection topic modelling ·
Multi-entity multi-aspect sentiment analysis · Opinion mining

1   Introduction

Aspect based sentiment analysis aims at finding opinion orientation of multiple aspects of an entity(s) in a given opinionated text. Given a document collection, e.g., reviews on phoneA, aspect based opinion mining returns people's sentiments towards each individual phoneA aspect such as battery life, performance, and camera quality. This is an important research topic in practice as it helps users to evaluate how much value they ascribe to each of these distinct aspects.

Although numerous studies on multi-aspect opinion mining have been published, most of them focus on mining opinion and associated aspects within a single collection (assuming an entity per collection). Less effort has been directed to situations where a comparison of multiple entities is of essence. In [1,2], researchers attempted to compare opinion on aspects of multiple entities. The approach in both works involves determining the aspect-sentiment pairs for each entity (collection) separately and later presenting entities and their aspects for comparison. This approach has limitations on a practical level, the task becomes time consuming with the increase in number of collections to compare.

Jeong et al. [3] recently proposed a topic model named the entity sentiment topic (EST) model that uncovers emotional topics from the perspective of entities. The model was tested over a story book, where both entities and their aspects

© Springer Nature Switzerland AG 2019
R. Silhavy (Ed.): CSOC 2019, AISC 985, pp. 107–118, 2019.
https://doi.org/10.1007/978-3-030-19810-7_11

usually appear in close proximity, e.g "John's arm is effective". However, in case of review collections, reviewers usually do not explicitly mention the entity name in each review. Mixing reviews of multiple entities in a single collection makes attribution of opinions and aspects to right entities a highly complex task.

This paper proposes the use of cross-collection topic models to perform aspect-based sentiment analysis for multiple entities simultaneously. The discussion is based on the Cross-Perspective Topic Model (CPTM) proposed in [4]. A topic refinement algorithm which employs coherent cluster growth and word embeddings is presented. The algorithm produces highly semantically coherent and visually identifiable aspects. It is then asserted that, with this refinement, topics elicited from cross-collection topic models align elegantly with product aspects. Therefore, the intrinsic semantic coherence of aspects refined by the proposed algorithm is compared to topics returned directly by CPTM. We then show that, the proposed approach returns more coherent aspects than CPTM.

Traditionally, cross-collection topic models including CPTM uses the bag of words model to model opinion words from multiple collections, and usually negation indicators such as 'not' are removed during model preprocessing. Thus, no semantic relation can be inferred from elicited opinion words. We demonstrate that opinion words returned by CPTM are better features for sentiment score calculations than the use of supervised state-of-the-art approaches. To this end, we compare sentiment scores associated to aspects as measured by a sentiment lexicon over opinion words returned by CPTM to sentiment scores resulting from the approach presented in [5]. In [5], Ruder et al. present the hierarchical model of reviews for aspect based sentiment analysis. The model is considered state-of-the-art and its implementation has been used in popular applications such as Aylien [1]. We show that CPTM features outperforms the hierarchical model in matching the ground truth.

The contributions in this paper can be summarized as follows:

- We show that cross-collection topic models can be used to for aspect based sentiment analysis of multiple entities.
- A topic refinement algorithm based on coherent cluster growth and word embeddings is proposed. The algorithm to produce highly semantically coherent and visually identifiable aspects.
- We argue that, opinion words returned from cross collection topic models such as CPTM are better sentiment indicators to elicited aspects than state-of-the-art approaches.
- Extensive experiments over the hotels review dataset is conducted to evaluate both aspect quality and sentiment accuracy.

The rest of this paper is organized as follow. In Sect. 2. an overview of existing aspect-based sentiment analysis approaches is presented. In Sect. 3. the proposed aspect refinement algorithm is described. Section 4. explains the environment in which the presented claims and algorithm are tested. Evaluation is conducted in Sect. 5. and finally in Sect. 6. concluding remarks are put forth.

---

[1] Rapidminer extension for aspect based sentiment analysis.

2   Related Work

Aspect based sentiment analysis have been extensively studied in recent years. For a general survey, please refer to [6]. Generally the task is comprised of two major sub-tasks; identification of aspects of an entity(s) and mapping those aspects to their respective opinion words or phrases. Methods in literature include usage of topic models [7], hidden markov models [8], conditional random fields [9] and syntax-based methods [10].

Although these methods perform well in practise, they only appeal to the extraction of aspects and associated opinion words for a single entity at a time. Usually documents describing an entity form a collection from which aspect-based sentiment analysis is performed. In [1,2], authors attempted to compare opinion on aspects of multiple entities. The approach in both efforts involves executing aspect based opinion mining algorithms to each entity collection separately and later presenting the results for comparison. This approach has some limitations, space and time complexities increase linearly with the increase in number of collections to compare.

A family of models that has a potential to perform aspect-based sentiment analysis for multiple entities simultaneously is the class of topic models referred to as cross-collection topic models. To the best of our knowledge these models have not been used for multi-entity multi-aspect sentiment analysis of product reviews. Initial efforts such as that of Zhai et al. [11] and Paul et al. [12] focuses on information summarization mainly on cultural-political domain. In [4], Fang et al. took the problem a step further by presenting a cross-perspective topic model (CPTM). Although CPTM also focuses on political domain, it models topics and their corresponding opinion words across multiple perspectives. Based on CPTM this paper proposes a topic refinement algorithm that increases topic interpretability to match that of identifiable product aspects.

The act of equating latent topics to product aspects is not a new phenomenon. Various approaches have attempted to make extracted latent topics align well with visually identifiable product aspects. Most of these approaches use the famous Latent Dirichlet Allocation (LDA) which is the topic model proposed in [13]. While authors in [14] assumes that executing LDA over nouns of a review document is enough to return latent topics as product aspects, authors in [15] assumes product aspects are more likely to be uncovered from sentence-level word co-occurrence information than from document-level, thus they model LDA over sentences. However, since LDA uses the bag of words model to model topics, topic-words are not required to be semantically related. This makes it difficult to interpret a topic and even harder to identify it as a known aspect.

In this work, for each topic, words that are close in meaning are grouped together to form an aspect. The closeness in meaning between words is measured by using cosine similarity of embeddings of the words. Results show that, aspects found by this approach have high interpretability score as measured by normalized pairwise mutual information (nPMI) [16] than those returned by CPTM. Furthermore, we argue that opinion words returned by CPTM are better sentiment indicators of elicited aspects than state-of-the-art approaches.

3   Methodology

There are several cross-collection topic models, including the cross perspective topic model (CPTM) [4], which return topics from multiple collections. However, these models cannot be used for aspect based sentiment analysis because topics do not correspond to aspects. For instance, when CPTM is applied to a corpus of comparable review collections, the resulted output is as illustrated in Fig. 1(a).

Although a topic is comprised of related words, usually these words do not represent one distinct aspect. For example, topic 3 in Fig. 1(a) can be identified as "screen size" or "display quality" which are two different aspects. Likewise topic 2 can mean "memory capacity" or "processor speed".

Moreover, it is useful to return a numerical score indicating sentiment polarity of aspects. Traditional topic models that model opinion words return a list of opinion words which may include both positive and negative words, this makes it hard for users of such models to comprehend the overall positivity or negativity of reviewers towards an aspect. Due to these shortcomings, such models remains good models for opinion summarization but not very useful for sentiment analysis of product reviews.

This work intends to refine CPTM topics into recognizable distinct aspects as shown in Fig. 1(b). Then to utilize extracted opinion words to compute sentiment scores for each aspect as illustrated in Fig. 1(b). In this manner users can easily compare opinion on multiple aspects of competing products.

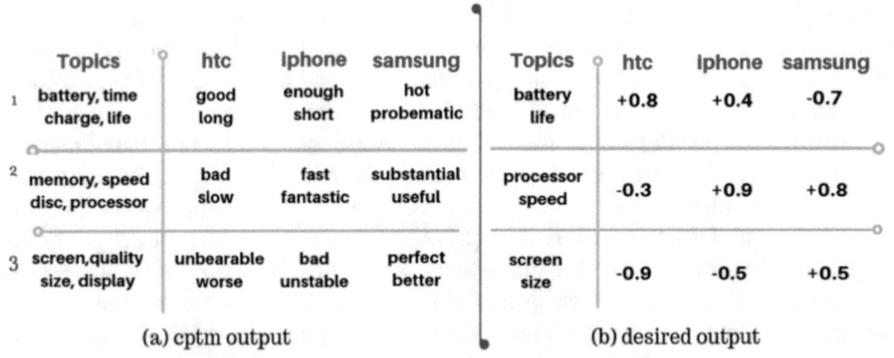

Topics	htc	iphone	samsung		Topics	htc	iphone	samsung
1 battery, time charge, life	good long	enough short	hot probematic		battery life	+0.8	+0.4	-0.7
2 memory, speed disc, processor	bad slow	fast fantastic	substantial useful		processor speed	-0.3	+0.9	+0.8
3 screen, quality size, display	unbearable worse	bad unstable	perfect better		screen size	-0.9	-0.5	+0.5

(a) cptm output	(b) desired output

Fig. 1. CPTM output vs. desired output

3.1   Topic Refinement Algorithm

Algorithm 1 is the pseudo-code of the proposed topic refinement algorithm. The algorithm takes as input a set **'T'** of k topics each containing n words i.e. $T = \{t_1, ..., t_k\}$ where $t_i = \{w_{i_1}, ..., w_{i_n}\}$. The algorithm returns a set **'A'** of k aspects each containing m words i.e. $A = \{a_1, ..., a_k\}$ where $a_i = \{w_{i_1}, ..., w_{i_m}\}$.

Algorithm 1. Topic refinement algorithm

INPUT: a set T of k topics; $T = \{t_1, ..., t_k\}$
OUTPUT: a set A of k aspects; $A = \{a_1, ..., a_k\}$

1: **for** $i \leftarrow 1, ..., k$ **do**
2:     $a_i \leftarrow (w_{i_x}, w_{i_y}) \leftarrow \text{bestPair}(t_i)$;
3:     **for** $j \leftarrow 1, ..., m - 2$ **do**
4:         $w_{i_z} \leftarrow \text{bestAddition}(a_i, t_i)$;
5:         $a_i \leftarrow a_i \cup w_{i_z}$
6:         $t_i \leftarrow t_i - w_{i_z}$
7:     **end for**
8: **end for**

The algorithm starts by finding a pair p of words that displays the highest semantic similarity in the given topic (line 2). Different semantic similarity measures between words can be used. For this case the usage of cosine similarity of words as they appear in euclidean space of propose. Representation of words in vectorial form has proven success in practice in capturing semantic relations between words. For detailed information on word embeddings please refer to [17].

Pair p then acts as initial member of the aspect cluster a_i. For every iteration, the algorithm then tries to grow the aspect cluster by finding a word within topic t_i which when added to the cluster will maximize the cluster coherence as given by the average pairwise cosine similarity (ACOSIM) of words in the cluster. See Eq. 1. The similarity between two words is denoted by $sim(w_i, w_j)$ and 'm' is the number of words in the aspect cluster.

$$ACOSIM = \frac{\sum_{i=2}^{m} \sum_{j=1}^{i-1} sim(w_i, w_j)}{\binom{m}{2}} \tag{1}$$

When the best word is found, as per line 4, it is included in the aspect and removed from the topic. Iterations continue until a desirable number 'm' of aspect representative words is reached. The procedure is then repeated to every topic in the topics set, T.

4   Experimentation

This section explains the experimental conditions set to test the performance of the topic refinement algorithm presented in Sect. 3.1. The approach used for sentiment score calculation is also elucidated.

4.1   Dataset

The data set used is a subset of three hotels; Rio mar in Puerto Rico, Iberostar and Caribe club princess in Dominican Rep from a huge hotels reviews dataset introduced in [18]. This dataset is preferred because it contains labeled aspect ratings for every review i.e reviewers are asked to provide an aspect rating ranging from 1 to 5 on seven aspects; value, room, location, cleanliness, check in/front desk, service i.e. food/drinks and business service for each review. These ratings serve as ground-truth for sentiment evaluation in Sect. 5.2. Number of documents (reviews) in each collection (hotel) is around 500, each review contains 500 words or more.

4.2 Topic Extraction

After prepossessing the documents, which includes tokenization, removing stop words and punctuations as well as converting all words to a common case (lower case). The number of topics k is set as 40, $\alpha = 50/k$, $\beta = \beta_i = 0.02$ and CPTM is run for 200 iterations over the hotels dataset described in Sect. 4.1. Findings in [19] show that these hyper parameters only affect the convergence of Gibbs sampler but not the output results. On the other hand the value of k is chosen based on prior knowledge of the dataset. It is known that number of distinct aspects to be extracted is less than 40. Thus, based on this knowledge choosing any number above forty is unrealistic. However, choosing k less than 40 results to high perplexity values, which is an indication of poor generalization.

The result of topic extraction process is a set of 40 topics each with a corresponding set of opinion words for each of the three hotels. Each topic and opinion is represented by multiple words. Please refer to [4] for an overview of CPTM's output arrangement. The cutoff point for number of topic representative words, n is crucial for aspect formulation. In the next subsection, criteria for selection of n and number of aspect representative words, m is described.

4.3 Parameter Selection for the Topic Refinement Algorithm

The topic refinement algorithm presented in Sect. 3.1. takes two hyper-parameters; number of words per topic, n and the desired number of words per aspect, m. With larger values of n the risk of including words that actually do not belong to the particular topic increases. However, larger values of n increases the space from which coherent aspect-words are derived. There is a trade-off between topic accuracy and aspect-words coherence when selecting the value of n.

Likewise, too small value of m faces the risk of concluding a wrong aspect from a given topic. Larger values of m however, reduce the aspect interpretability making an aspect nothing but a mere topic. An aspect should be emphasized by a reasonable number of words describing an identifiable item or concept.

To regulate topic accuracy we use average probability of topic words. We rely on the fundamental idea that a topic contains only the most probable words, this means, smaller value of n. On the other hand, the space for aspect extraction is regulated by using average pairwise cosine similarity (ACOSIM) of the extracted aspects. ACOSIM is computed as shown in Eq. 1. An aspect is required to have high ACOSIM, this is favoured when the value of n is large.

Figure 2 shows the variation of ACOSIM and average topic words probability against different number of topic and aspect representative words. The figure shows that, with the increase in **n**, ACOSIM increases for aspects of all sizes. However, the average probability of topic words decreases. The two graphs intersect at (n, m) and $(10, 5)$. The intersection point is the point of equilibrium, hence we fix n and m at $n = 10$ and $m = 5$.

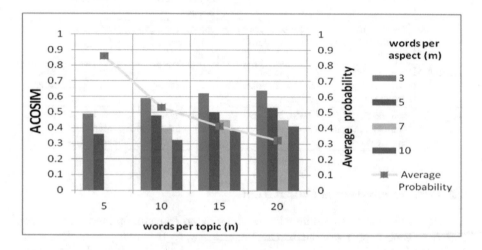

Fig. 2. Parameter selection

4.4 Topic Words Refinement

To apply the topics refinement algorithm proposed in Sect. 3.1, we set $n = 10$, $m = 5$ and run the algorithm over topic-words returned by CPTM and word vectors from glove pre-trained embeddings[2]. These embeddings are trained over common crawl (Google data) and contain 2.2 Million words. The number of dimensions in each vector is 300.

Table 1 shows five best aspects after refinement. The number of words representing an opinion is left at its initial 10 but for space considerations only 5 words per opinion are display. It can be seen that, aspect four represents *prices* while aspect fifteen is on *rooms*. While Iberostar's price is considered *high* and *expensive*, opinion words describing Caribe club indicates it to be *worth* and *inclusive*.

---

[2] https://nlp.stanford.edu/projects/glove/.

Table 1. Aspect words and their corresponding opinion words for 3 hotels

Aspects	Aspect words	Caribe club	Iberostar	Rio mar
Aspect4	prices, price, cost, pay, deal	inclusive, worth, wonderful, better, ask	expensive, less, allowed, close, funny	free, nice, little, high, take, away
Aspect15	bed, beds, room, door, reception	amazingly, clean, favorite	really, quiet, double, king, booked	king, called, told, never, given
Aspect26	dinner, lunch, breakfast, buffet, chicken	definitely, good, great, available	spanish, everyday, good, beautiful, close	best, clean, first, still, given
Aspect25	hotel, facilities, pool, beach, restaurant	great, much, better, overall, inclusive	good, great, much, fine, different	great, good, nice, enough, friendly
Aspect13	excursion, excursions, trips, trip, boat	little, fun, great, well, last	great, always, entertaining, much, snorkeling	great, good, loved, rainforest, snorkeling

4.5   Sentiment-Scores Calculation

To determine sentiment orientation of the elicited aspects, opinion words associated to each aspect are fed to the Valence Aware Dictionary and sEntiment Reasoner (VADER) [20]. VADER is a lexicon and rule based sentiment analysis tool. Although vader is primarily designed for social media text, its author argues it can be used for other domains. The output is a rating scale from -1 to 1 for each aspect. Note, in an event where two or more topics represent the same aspect, the average of their sentiment scores is return.

5   Evaluation

5.1   Aspect Quality Evaluation

To evaluate the quality of resulting aspects we compare the normalized pairwise mutual information (nPMI) [16] of aspects before and after refinement. To achieve fair comparison we concentrate only on top 5 words output by CPTM. In this manner, for each aspect, top 5 words directly returned by CPTM are compared by top 5 words refined by Algorithm 1.

Normalized pairwise mutual information is arguably the best way to evaluate topic models, please refer to [21] for detailed information on different topic modeling evaluation measures. The nPMI score ranges from -1 to 1 where -1 reflects poor topic quality and 1 represents the best topic quality in terms of semantic interpretability. nPMI is computed by using a held out wikipedia corpus[3] containing over a million articles.

---

[3] https://dumps.wikimedia.org/.

nPMI is given by Eq. 3. where $p(w_i, w_j)$ is the ratio of number of documents in the held out corpus containing both words w_i and w_j to the total number of documents. $p(w_j)$ is the ratio of number of documents in the held out corpus containing word w_j to the total number of documents. N is the total number of words in topic t.

$$npmi(t) = \sum_{i=2}^{N} \sum_{j=1}^{i-1} log \frac{\frac{p(w_i,w_j)}{p(w_i)p(w_j)}}{-p(w_i, w_j)} \tag{2}$$

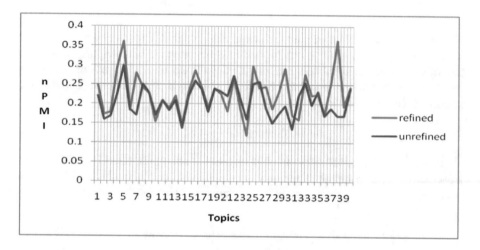

Fig. 3. Average nPMI scores for 40 topics

Figure 3 shows that, almost all 40 refined exhibit show better interpretability than unrefined aspects. Consider topic 26 which shows a spiking improvement, its initial 10 output words from CPTM are {*room, night, dinner, breakfast, nights, lunch, buffet, p, couples, chicken*} arranged in decreasing order of their probability of occurrence. When considering only top 5 most probable words; an aspect is {*room, night, dinner, breakfast, nights*} with an average pairwise cosine similarity score of 0.58. It is hard to conclude whether the described aspect is a room or food. After refinement the resulting aspect is {*dinner, lunch, breakfast, buffet, chicken*} with the average pairwise cosine similarity score of 0.67, which clearly describes food.

5.2   Sentiment Scores Evaluation

Baseline Sentiment Scores: To compute baseline sentiment scores for each aspect, the hierarchical model of reviews for aspect based sentiment analysis presented in [5] is used. The model in [5] is considered state-of-the-art as it has been used in multiple public applications including Aylien[4]. Aylien is therefore run to three hotel collections one at a time. The output of this task is a list of sentiment scores in a ternary scale i.e. positive, negative or neutral for every aspect in every review. To obtain the total sentiment score for an aspect we aggregate number of reviews that identified an aspect as positive, negative or neutral. The sentiment score for an aspect is then given by Eq. 3. Where ps, ng, nt are the number of reviews that identified an aspect '**a**' as positive, negative or neutral respectively.

$$sentiment(\mathbf{a}) = \frac{ps - ng}{ps + ng + nt} \tag{3}$$

Ground Truth: From the hotel dataset described in Sect. 4.1. which contains labeled aspect ratings for every review, we aggregate and compute average of the ratings. Since the ratings range from 1 to 5 we normalize the final average to a value between −1 to 1 using Eq. 4. Where x is a score between 1 and 5 and nScore is the normalized score.

$$nScore(\mathbf{x}) = \frac{x}{2} - 1.5 \tag{4}$$

Resulting sentiment scores for five aspects is as summarized in Table 2. The three values in each cell are the baseline scores (left), scores from CPTM features (middle) and ground truth scores in brackets (right).

Table 2. Qualitative sentiment evaluation

Hotels	Aspects (baseline, CPTM, ground truth)				
	value	rooms	cleanliness	check in	food/drinks
Caribe club	−0.09, 0.66, (0.39)	0.01, 0.44, (0.28)	0.28, 0.46, (0.42)	0.14, −0.13, (0.40)	0.24, 0.30, (0.39)
Iberostar	−0.11, 0.40, (0.62)	0.20, 0.47, (0.51)	0.40, 0.91, (0.76)	0.30, 0.80, (0.57)	0.30, 0.44, (0.72)
Rio Mar	−0.01, 0.14, (0.21)	0.04, 0.43, (0.41)	0.15, 0.93, (0.45)	0.09, 0.36, (0.41)	0.12, 0.64, (0.39)

Table 2 shows that, for almost all aspects and hotels, sentiment scores computed using CPTM features outperform the baseline in matching the ground truth. Root Mean Square Error of CPTM features versus baseline scores as measured against the ground truth is as summarized in Fig. 4.

---

[4] A rapidminer extension for aspect based opinion mining.

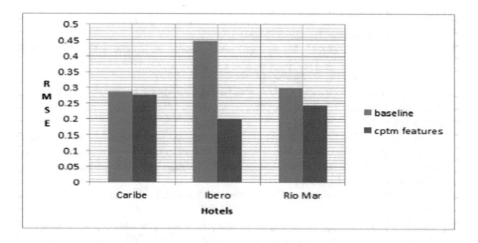

Fig. 4. RMSE of sentiment scores for baseline versus CPTM features

6   Conclusion

This paper investigates how cross collection topic models can be used to perform multi-aspect multi-entity sentiment analysis of product reviews. It is shown that by improving semantic interpretability, topics match product aspects. It is further show that the use of opinion words from these models as features for sentiment analysis outperforms state-of-the-art approaches.

For future work, it will be interesting to explore how other cross collection topic models such as cross collection LDA (ccLDA) [12] and Topic Aspect Model (TAM) [22] can be improved in this direction. Performance issues pertaining to these models can also be explored.

References

1. Liu, B., Hu, M., Cheng, J.: Opinion observer: analyzing and comparing opinions on the web. In: Proceedings of the 14th International Conference on World Wide Web, pp. 342–351. ACM (2005)
2. Jin, W., Ho, H.H., Srihari, R.K.: Opinionminer: a novel machine learning system for web opinion mining and extraction. In: Proceedings of the 15th ACM SIGKDD International Conference on Knowledge Discovery and Data Mining, pp. 1195–1204. ACM (2009)
3. Jeong, Y.-S., Choi, H.-J.: Entity sentiment analysis using topic model. In: The 18th International Symposium on Advanced Intelligent Systems (ISIS2017), KIIS (2017)
4. Fang, Y., Si, L., Somasundaram, N., Yu, Z.: Mining contrastive opinions on political texts using cross-perspective topic model. In: Proceedings of the fifth ACM International Conference on Web Search and Data Mining, pp. 63–72. ACM (2012)
5. Ruder, S., Ghaffari, P., Breslin, J.G.: A hierarchical model of reviews for aspect-based sentiment analysis, arXiv preprint arXiv:1609.02745 (2016)

6. Schouten, K., Frasincar, F.: Survey on aspect-level sentiment analysis. IEEE Trans. Knowl. Data Eng. **28**(3), 813–830 (2016)
7. Lu, B., Ott, M., Cardie, C., Tsou, B.K.: Multi-aspect sentiment analysis with topic models. In: 2011 IEEE 11th International Conference on Data Mining Workshops (ICDMW), pp. 81–88. IEEE (2011)
8. Jin, W., Lexicalized, H.H.H.A.N.: HMM-based Learning Framework for Web Opinion Mining in Proceedings of the 26th International Conference on Machine Learning. Montreal, Canada (2009)
9. Jakob, N., Gurevych, I.: Extracting opinion targets in a single-and cross-domain setting with conditional random fields. In: Proceedings of the 2010 Conference on Empirical Methods in Natural Language Processing, pp. 1035–1045. Association for Computational Linguistics (2010)
10. Hu, M., Mining, B.L.: Summarizing customer reviews KDD'04. Seattle, Washington, USA, 22–25 August 2004
11. Zhai, C., Velivelli, A., Yu, B.: A cross-collection mixture model for comparative text mining. In: Proceedings of the Tenth ACM SIGKDD International Conference on Knowledge Discovery and Data Mining, pp. 743–748. ACM (2004)
12. Paul, M., Girju, R.: Cross-cultural analysis of blogs and forums with mixed-collection topic models. In: Proceedings of the 2009 Conference on Empirical Methods in Natural Language Processing: Volume 3-Volume 3, pp. 1408–1417. Association for Computational Linguistics (2009)
13. Blei, D.M., Ng, A.Y., Jordan, M.I.: Latent dirichlet allocation. J. Mach. Learn. Res. **3**, 993–1022 (2003)
14. Naveed, N., Gottron, T., Staab, S.: Feature sentiment diversification of user generated reviews: the freud approach. In: Seventh International AAAI Conference on Weblogs and Social Media (2013)
15. Brody, S., Elhadad, N.: An unsupervised aspect-sentiment model for online reviews. In: Human Language Technologies: The 2010 Annual Conference of the North American Chapter of the Association for Computational Linguistics, pp. 804–812. Association for Computational Linguistics (2010)
16. Bouma, G.: Normalized (pointwise) mutual information in collocation extraction. In: Proceedings of GSCL, pp. 31–40 (2009)
17. Lai, S., Liu, K., He, S., Zhao, J.: How to generate a good word embedding. IEEE Intell. Syst. **31**(6), 5–14 (2016)
18. Wang, H., Lu, Y., Zhai, C.: Latent aspect rating analysis on review text data: a rating regression approach. In: Proceedings of the 16th ACM SIGKDD International Conference on Knowledge Discovery and Data Mining, pp. 783–792. ACM (2010)
19. Griffiths, T.L., Steyvers, M.: Finding scientific topics. Proc. Nat. Acad. Sci. **101**(suppl 1), 5228–5235 (2004)
20. Gilbert, C.H.E.: Vader: a parsimonious rule-based model for sentiment analysis of social media text. In: Eighth International Conference on Weblogs and Social Media (ICWSM-14) (2014). http://comp.social.gatech.edu/papers/icwsm14.vader.hutto.pdf. 20 April 2016
21. Lau, J.H., Newman, D., Baldwin, T.: Machine reading tea leaves: automatically evaluating topic coherence and topic model quality. In: Proceedings of the 14th Conference of the European Chapter of the Association for Computational Linguistics, pp. 530–539 (2014)
22. Paul, M., Girju, R.: A two-dimensional topic-aspect model for discovering multifaceted topics. Urbana, vol. 51, no. 61801, p. 36 (2010)

Information Flow Control in Interactive Analysis of Map Images with Fuzzy Elements

Stanislav Belyakov$^{(\boxtimes)}$, Marina Savelyeva, Alexander Bozhenyuk, and Andrey Glushkov

Southern Federal University, Taganrog, Russia
{lbeliacov, avb}@yandex. ru,
marina. n. savelyeva@gmail. com, andrey. drv@gmail. com

Abstract. This paper considers information flow redundancy reducement problem for decision-making followed by cartographic images dialogue process analysis. The specifics of the considered problem is displaying fuzzy cartographic objects. This includes those that have not got a preliminary cartographic processing and leads to image defects. The loss of images semantic content, however, compensated by an information about the outside world, which is carried by fuzzy objects. The paper proposes a method of managing flow based information by maximizing a workspace utility function for analysis. The authors introduced a representation of the working area by two subsets of cartographic objects: the skeleton and the environment. Representation variations with fuzzy objects that improve the quality of solving such problems as generation of decision alternatives, risk assessment of decision making and assessment of the external data sources quality are proposed. The considered case generalization can be reused by any system that provides a visual image for search and decision making to user.

Keywords: Visual analysis · Geoinformation systems · Intelligent systems · Human-machine interaction

1 Introduction

Cartographic data usage for decision-making improves the quality of results in many areas of human activity. Software tools for spatial, statistical, morphometric and topological analysis of modern geographic information systems and services contribute to a more comprehensive formation and reliable spatial image of an applied task for a user analyst [1]. From our point of view visual analytics support seems to be the most important geographic information technology. In the process of interactive search and cartographic data analysis a figurative thinking of the analyst is stimulated, his creativity increases. Visualized cartographic images enrich the mental image of the field of analysis in the user's mind.

These factors has the best effect when all layers of the geographical map have undergone professional processing. This means that all objects are consistent with each other and reliably reflect the real world. Modern geographic information systems use map data sources that do not have such consistency feature [2]. The map data sources

R. Silhavy (Ed.): CSOC 2019, AISC 985, pp. 119–127, 2019.
https://doi.org/10.1007/978-3-030-19810-7_12

are provided either by volunteer cartographers, or this data is extracted by intellectual procedures from non-cartographic sources [3].

Despite the obvious probable defects, the data under consideration has a high value and can be used within certain limits by geoinformation systems. These limits arise from a difficulty of perceiving raw data by users. The desire to obtain relevant information requires additional efforts from a user to select data and images exploration. If the amount of work is big, the visual analysis loses its meaning. Figure 1 shows an analysis topic example where pedestrians trajectories knowledge was important. You can notice that visual analysis of the current situation is messed up because of the too many GPS tracks. The perception of the main objects of the map, as buildings, structures, communications has become much more complicated and has violated the integrity of the image.

Fig. 1. Workspace with inconsistent data example.

Finding a compromise between the gains from using actual cartographic data and losses due to their lack of consistency is a well-known problem [3].

This paper proposes a method for managing the selecting cartographic objects for visual analysis process based on a specific analysis field representation. This makes it possible to construct useful visual representations for decision making.

2 Known Approaches Overview

Information flow management can be considered as a problem of providing the necessary level of perception during the mental image formation [4]. The well-known Hick's law indicates an increase in decision-making time as the number and complexity of options increase. The cognitive load on the user is increasing. Decreasing in map viewing dynamics or situation diagram could be observed during cartographic images analysis. It is known [5] that viewing from different angle, changing the scale, panning a cartographic fragment is a fundamental mental process necessary for understanding the meaning what user could see. The idea that perception plays a crucial role in the process of interactive visual analysis agrees with modern ideas about the effect of visualization on creative behavior (creativity) [6]. The results of these works indicate the need to minimize cognitive overload (cognitive overload), but do not provide a real solution for the problem.

Important patterns were revealed during a perception research and information memorization. Miller's "fragmentation" research that affects the perception and information memorization [7] is quite well known. For visual map analysis, charts and plans fragmentation means a representation structure, which is based on a targeted object group selection and establishment of these groups into the overall image. From our point of view, such objects are objects with uncertainty. Having a limited independent significance, they significantly influence an image as a whole, increasing the quality of the image perception. For sure the "fragmentation" implementation requires additional research.

Also there is a noticeable research in the field of User experience design (UX, UXD, UED) which are aimed on user behavior analysis. In particular, Jacob's law of Internet UX [8] says that professionally oriented user groups have some kind of common intuitive interfaces idea. These patterns are valuable for the field of cartographic images analysis since these images are essentially an interface for accessing a spatial information by professional-oriented user groups.

The problem of using objects with incompletely defined parameters in GIS is being investigated in the geoinformatics field [9] for quite a long time. The main aspect of the research are data models with an inaccurate, incomplete, ambiguous and contradictory information. The visual representation of such objects is realized within a cartographic visualization general principles. The disadvantage of this approach is ignoring objects interaction with each other when they are visually analyzed by the user.

Our research topic is on the edge of human-machine visual analysis systems research [10]. Human-machine visual analysis systems technical implementation and purpose are extremely diverse. GIS is one of them [11, 12]. The review [13] emphasizes the idea that visual analysis could have not a supporting role, but provide a guidance for the process of solving an applied problem. Intelligent guidance is particularly attractive. It is aimed at achieving an analysis goal, which was globally set up. This paper describes the generalized guidance systems components. This includes models for performing individual stages of tasks and knowledge types required for this. Generalization of the considered approaches does not allow directly implement the mechanism of objects visualization with uncertainty in GIS.

Intellectual management apparatus of a user dialogue with GIS are investigated in the following research [14, 15]. This work develops this research towards the use of cartographic objects with uncertainty.

In general, the analysis of publications related to the topic shows a lack of intellectual management information flow problem research during cartographic data analysis.

3   The Proposed Method of Information Flow Management

The information flow arises in the process of interaction between a client and GIS server, coming from map data execution requests. An intensity of the flow is determined by the number of cartographic objects, which are necessary for a user-analyst to form a work space area mental image in which the applied task is solved. It is known that information received from a GIS server upon client request is redundant [1]. Therefore, the flow redundancy reduction during a cartographic analysis can be considered as an improvement to the quality of solving applied problems that use spatial data.

We propose the following method for managing information flows:

- the client's dialogue with the GIS server is controlled on the client side in order to build a minimally redundant working area of a common map. The functionality is delivered by an intellectual subsystem, which uses knowledge about the properties of non-redundant cartographic images, the operations of their detailed elaboration and generalization during a visual analysis;
- the intelligent subsystem continuously analyzes client requests to the GIS server. The purpose of the analysis is to assess the work area quality which is to be generated on the client side. Insufficient quality is compensated by cartographic objects addition or removal from the workspace. These actions are performed transparently for the user, without his participation.

This method considers the formal presentation problem and workspace utility evaluation. Then it is considered in order to use cartographic objects containing uncertainty.

Let us consider a model of interactive problem solving and define an analysis field representation. Let $\Omega = \{\omega_1, \omega_2, \ldots \omega_n\}$ is a subset of objects accessible by GIS from external data sources. Due to the fact that $|\Omega|$ is very high to solve the problem the analyst constructs a local workspace $w \subset |\Omega|, w| < < |\Omega|$. Then the analyst examines the workspace by sending GIS queries that formed into set $Q = \{q_i\}$, each of them either changes the set of objects in the workspace or changes its visual representation:

$$w' = q_i(w, \Omega),$$

where w'- modified instance of the workspace. We emphasize that the workspace changes its composition even when a visual presentation changes. This is due to the need of saving a semantic content of the studied fragment. Changing a scale or angle during a cartographic analysis reflects the map detail changes. Zooming in corresponds to a "more detailed" one. Zooming out corresponds to a "more generalized" map view. At the same time, in all the cases listed above the semantic invariant of the analysis field should be preserved. These are objects and relations that are essential for analysis.

This requires an addition and removal of map objects. Through $C = \{c_k\}$ we define the set of contexts in which the sense invariants are supported. We assume that visual analysis is performed in context $c_m \in C$, if geospatial data, thematic maps and charts, rules for assessing redundancy and maintaining integrity, reflecting the professional orientation of the analysis are used. Let us define a set of objects in the workspace as a combination of two components:

$$
\begin{aligned}
&w = B \cup E \\
&B \subseteq \Omega : \forall \omega_i \in B \Rightarrow \exists q_j = true, \, j = \overline{1, Q}, \, i = \overline{1, |\Omega|}, \\
&B \cap E = \varnothing, \, E \subseteq \Omega,
\end{aligned}
\tag{1}
$$

where B is a set of map objects selected by queries Q to the database Ω. Let the set B to be a skeleton. The skeleton is formed by objects that were explicitly requested by a user through GIS dialog menus. The set E – skeleton environment designed to improve a perception of the work area. The objects of the environment were not explicitly requested by the user, but they represent important terrain features that influence the situation evaluation.

Let us define a fitness-function $I(w)$ of the workspace w. The level of work area professional perception by a user is higher when the function value is also greater. Through $R(w)$ we define computational resources of the user hardware needed to visualize a workspace w Number of cartographic elements in a workspace can be considered as a resource consumption measure.

As the workspace after executing each query $q_i \in Q$ changes its state let us define $w_j, j = 0, 1, 2 \ldots |J|$ as a sequence of these states. So index j is a discrete time in a visual analysis session. Then the sequence of states w_j is a discrete process in which the current state is the result of all previous ones. Taking into account the introduced notation, the task that the GIS should implement is formulated as follows:

$$
\begin{cases}
I(w_j) \to \max, \\
w_j = K(\Omega, c_m, w'_j, w_{j-1}, w_{j-2}, \ldots w_0), \\
w'_j = q_i(w_{j-1}, \Omega), \\
R(w_j) < R^*, \\
w_j, w'_j \subset \Omega, \\
j = \overline{0, |J|}, q_i \in Q, c_m \in C.
\end{cases}
\tag{2}
$$

Thus, GIS should ensure maximum usefulness of a workspace representation in each of its conditions. Here K – is an operator displaying the intermediate state of the workspace in a given context in the new state based on an available GIS data and the current history of visual analysis. An intermediate state is the one which the workspace enters after execution of the next user request. Formally, this state is internal to the GIS and is not available to the user for inspection. Redundancy and integrity w_j are provided by the implementation of the operator K based on the value of its argument $c_m \in C$. The current state of the workspace must satisfy the resource constraint specified by R^*.

Fitness-function is defined based on the fact that a cartographic image total elements number $N = |w|$ is the main perception factor. By increasing its value the perception becomes worse due to physiological limitations of vision. But the same effect is also observed when it was decreased due to the image sense loss. Whole perception corresponds to the subjectively determined by the user image elements number, which takes some "best" value $|w_{siml}| < N^* < |w_{complex}|$. Здесь $|w_{siml}|$ – is the number of objects in the uninformative on the subjective feeling of the user workspace, $|w_{complex}|$ – number of objects in the "too complex" image.

Then, for example, a piecewise linear fitness function for problem (2) with $j = \overline{0, |J|}$ we could define in the following way:

$$
I(w_j) = \begin{cases}
1 - (N^* - |w_j|)/N^*, & |w_j| < N^*, \\
1, & |w_j| = N^*, \\
1 - (|w_j| - N^*)/|w_{complex}|, & |w_j| > |w_{complex}|.
\end{cases}
$$

It can be seen that the maximum utility value $I(w_j) = 1$ is achieved at point N^*, which is the only maximum value. Increasing or decreasing the value of N^* leads to utility reducing and $I(w_j) \to 0$. The fitness-function parameters can be considered as a user account attributes on the GIS server.

4   Results

We will refer to the objects with uncertainty as u-objects. U-objects are not a special class of GIS objects. An object acquires uncertainty if at least one property of the object instance takes an inaccurate, indefinite, incorrect, or insufficiently reliable value [1].

Appearance of the u-objects in a workspace is possible only by an explicit user request, since the risk of their use is estimated by the analyst himself. U-objects, therefore, are always included in the skeleton. Analyzing the perception of the working area we can state the following:

- the inclusion of u-objects in the image can generate new objects, which actually display defects. A good example could be line sections that reflect the discrepancy between the track of a vehicle and the line of an underground tunnel. These objects have acquired a special meaning and should not be considered in the original context;
- U-objects can generate wrong spatial relationships. For example, exit from a parking lot directly connected to a railway means that wrong data about parking lot position. In this case, u-objects are highlighted on the map with a color, texture, or mnemonic to indicate a defect in the spatial relationship;
- map with u-objects may contain no visually detectable defects. The presence of u-objects does not change the professional perception of the image.

Thus, a workspace with u-objects can be considered as a set of objects with the following structure:

$$
w = B \cup E \cup B^+ \cup B^-, \tag{3}
$$

where B^+ - set of u-objects, that do not generate image defects, set B^- - u-objects, which are the cause of image defects.

Analysis of (3) allows us to conclude the following statements. First, the use of u-object does not change the essence of visualization control. As before, perception is limited by a number of objects to be visualized, and semantic content is provided by a proper environment. The concept of objects which are representing defects makes it possible to extend a visualization control method to maps which contain objects with uncertainty. Secondly, the representation (3) creates a new opportunity to diversify the visual analysis by changing the "semantic view" of the display. The essence of such operation is as follows: the sets (3) can be combined into a skeleton and environment in different ways, which gives leads to a different semantic images. We assume that

$$w = B \cup E \cup B^+ \cup B^- = B^* \cup E^*,$$
$$B^* = B \cup B^+ \cup B^-, E^* = E.$$

In case of including u-objects both with and without defects into the skeleton generates an image that leads to a new solution alternatives generation. This can be explained by the fact that the presence of u-objects increases the diversity of space-time and semantic situations.

If you create a visualization with a representation

$$B^* = B \cup B^+, E^* = E,$$

only properly mapped u-objects will be included to the image. Such image is useful during decision risks evaluation. Since it reflects the area of analysis as plausible as possible.

During visualization, when

$$B^* = B^+ \cup B^-, E^* = E,$$

an image of u-objects in the context of the problem to be solved is built. This allows you to evaluate the quality of data drawn from third-party sources. For example, an analyst can estimate pedestrian and cyclists traffic intensity during street illumination network planning. In this case, cycling tracking information map could be useful. As long as there is a sufficient number of observations available at the requested area.

When representation was fixated

$$B^* = B, E^* = B^- \cup B^+ \cup E,$$

the u-objects give appropriate semantic meaning to image. For example, for a house rent research, analyst could use points of interest (POI) data which was given by users of some social network who are familiar to the territory in scope. This extra information may not fundamentally affect the decision making, but may change the significance of the alternatives under consideration. Such perspective is useful for the analyst, who is looking for a more safe decision making.

Workspace representation with u-objects are summarized in the following Table 1:

Table 1. Workspace representations

Representation skeleton (B^*)	Representation environment (E^*)	Representation purpose
$B \cup B^- \cup B^+$	E	Alternative decisions generation
$B \cup B^+$	E	Risk analysis
$B^- \cup B^+$	E	Additional data quality evaluation
B	$B^- \cup B^+ \cup E$	Context semantic details strengthening

Additional positive effect presence from the application of the described representations can be justified by the following assertions.

Statement 1. A necessary condition from using a representation with u-objects additional effect appearance is to include non-empty sets in the workspace B^- and B^+.

The proof of this statement is the absence of set objects B^- and B^+ means the absence of relevant information for solving the problem. If no relevant information is received, the additional positive effect appearance is not possible.

Statement 2. A sufficient condition for additional effect occurrence using a representation with u-objects is to maximize the fitness function of the workspace $I(w)$ including non-empty sets B^- and B^+.

The proof is based on the fact that an additional positive effect arises as a result of the user-analyst understanding the visualized actual data. Reasoning is achieved only in case of the most comfortable cartographic image perception. This is the case when the fitness function is maximized.

5   Discussion

The information flow management method proposed in this paper is based on the knowledge application. The most of the practical realization difficulties are tied with the following issues. The evaluation subjectivity of a comfortable cartographic images perception level, significance evaluation specificity of object classes and relations requires flexibility in setting and modifying knowledge. This problem is a characteristic of any intelligent system and can only be successfully solved by developing automatic procedures for extracting and using knowledge.

The ability to rationally research cartographic images with relevant, but not sufficiently correct information in a dialogue with GIS is an effective approach to analyze spatial Big Data sources. With the increase in the number and volume of data sets, their professional cartographic processing capabilities will steadily decrease. From this point of view, the proposed method seems promising.

6   Conclusion

The proposed method application is not limited to GIS and geographic information technology. The considered case generalization can be reused by any system that provides a visual image for search and decision making to user. An example could be industrial technological systems, computer-aided design systems, business process management, financial monitoring and analysis.

Further research on this problem is related to the automatic search and knowledge extraction during the dialogue process, as well as an information flow control subsystem on the dialogue data machine learning.

Acknowledgment. This work has been supported by the Ministry of Education and Science of the Russian Federation under Project "Methods and means of decision making on base of dynamic geographic information models" (Project part, State task 2.918.2017).

References

1. Longley, P.A., Goodchild, M., Maguire, D.J., Rhind, D.W.: Geographic Information Systems and Sciences, 3rd edn. Wiley, New York (2011)
2. McKenzie, G., Hegarty, M., Barrett, T., Goodchild, M.: Assessing the effectiveness of different visualizations for judgments of positional uncertainty. Int. J. Geogr. Inf. Sci. **30**(2), 221–239 (2016)
3. Lee, J., Kang, M.: Geospatial big data: challenges and opportunities. Big Data Res. **2**, 74–81 (2015)
4. Hick, W.E.: On the rate of gain of information. Q. J. Exp. Psychol. **4**(1), 11–26 (1952)
5. Gibson, J.J.: A theory of direct visual perception. In: Royce, J., Rozenboom, W. (eds.) The Psychology of Knowing. Gordon & Breach, New York (1972)
6. Cybulski, J.L., Keller, S., Nguyen, L., Saundage, D.: Creative problem solving in digital space using visual analytics. Comput. Hum. Behav. **42**, 20–35 (2015)
7. Colman, A.M.: A Dictionary of Psychology, 3rd edn. Oxford University Press, Oxford (2008)
8. Jakob's Law of Internet User Experience. https://www.nngroup.com/videos/jakobs-law-internet-ux/. Accessed 02 Dec 2018
9. Schumann, H., Tominski, C.: Analytical, visual and interactive concepts for geo-visual analytics. J. Vis. Lang. Comput. **22**, 257–267 (2011)
10. Cleveland, W.S.: Visualizing Data. Hobart Press, Summit (1983)
11. Andrienko, N., Andrienko, G.: Exploratory Analysis of Spatial and Temporal Data. Springer, Berlin (2006)
12. Andrienko, N., Lammarsch, T., Andrienko, G., Fuchs, G., Keim, D., Miksch, S., Rind, A.: Viewing visual analytics as model building. Comput. Graph. Forum **37**(6), 275–299 (2018)
13. Christopher, C., Andrienko, N., Schreck, T., Yang, J., Choo, J., Engelke, U., Jena, A., Dwyer, T.: Guidance in the human–machine analytics process. Vis. Inf. **2**(3), 166–180 (2018)
14. Belyakov, S., Bozhenyuk, A., Rozenberg, I.: The intuitive cartographic representation in decision-making. In: World Scientific Proceeding Series on Computer Engineering and Information Science, vol. 10, pp. 13–18 (2016)
15. Belyakov, S., Belyakova, M., Savelyeva, M., Rozenberg, I.: The synthesis of reliable solutions of the logistics problems using geographic information systems. In: 10th International Conference on Application of Information and Communication Technologies (AICT), pp. 371–375. IEEE Press, New York (2016)

A Binary Sine-Cosine Algorithm Applied to the Knapsack Problem

Hernan Pinto, Alvaro Peña$^{(\boxtimes)}$, Matías Valenzuela, and Andrés Fernández

School of Engineering, Pontificia Universidad Católica de Valparaíso,
Valparaíso, Chile
{hernan.pinto,alvaro.pena,matias.valenzuela,andres.fernandez}@pucv.cl

Abstract. In industry, the concept of complex systems is becoming relevant due to the diverse applications in operations research. Many of these complex problems are NP-hard and it is difficult to approach them with complete optimization techniques. The use of metaheuristics has had good results and in particular, the design of binary algorithms based on continuous metaheuristics of swarm intelligence. In this article, we apply the binarization mechanism based on the percentile concept. We apply the percentile concept to the sine-cosine algorithm (SCOA) in order to solve the multidimensional backpack problem (MKP). The experiments are designed to demonstrate the usefulness of the percentile concept in binarization. In addition, we verify the efficiency of our algorithm through reference instances. The results indicate that the binary Percentile Sine-Cosine Optimization Algorithm (BPSCOA) obtains adequate results when evaluated with a combinatorial problem such as the MKP.

Keywords: Combinatorial optimization · KnapSack · Metaheuristics · Percentile

1 Introduction

The combinatorial problems have great relevance at industrial level, we find them in various areas such as Civil Engineering, Bio Informatics [1], Operational Research [2–4], resource allocation [5], scheduling problems [6,7], robust optimization [8], scheduling problems [9], covering problems [2] among others. On the other hand, in recent years, algorithms inspired by nature phenomena to solve optimization problems have been generated. As examples of these algorithms we have Cuckoo Search, Black Hole, Bat Algorithm, and sine-cosine Algorithms [10] among others. Many of these algorithms work naturally in continuous spaces and therefore must be adapted to solve combinatorial problems. In the process of adaptation, the mechanisms of exploration and exploitation of the algorithm can be altered, having consequences in the efficiency of the algorithm.

Several binarization techniques have been developed to address this situation. In a literature search, the main binarization methods used correspond to transfer

© Springer Nature Switzerland AG 2019
R. Silhavy (Ed.): CSOC 2019, AISC 985, pp. 128–138, 2019.
https://doi.org/10.1007/978-3-030-19810-7_13

functions, angle modulation and quantum approach, for more detail refer to [11]. In this article we present a new binarization method that uses the percentile concept to group the solutions and then perform the binarization process. To verify the efficiency of our method, we used the sine-cosine (SCOA) algorithm.

To check our binary percentile sine-cosine algorithm (BPSCOA), we use the well-known knapsack problem. Experiments were developed using a random operator to validate the contribution of the percentile technique in the binarization process of the sine-cosine algorithm. In addition, local search operator is used to strengthen the results. Moreover, the binary artificial algae (BAAA) and K-means transition ranking (KMTR) algorithms were used to compare our results. BAAA was developed in [12] and uses transfer functions to perform the binarization process. KMTR was developed in [13] and uses a K-means algorithm to perform the binarization. The results show that the percentile technique obtains results superior to those obtained by the random operator and that our BPSCOA algorithm shows competitive results against the BAAA and KMTR algorithms.

2 KnapSack Problem

The multidimensional knapsack problem (MKP), models resource allocation situations. The goal is to find the subset of objects that produce the greatest benefit by satisfying a set of constraints. MKP is one of the most studied NP-hard class combinatorial problems, however it remains as a challenge due to the difficulties encountered in solving medium and large size instances. A search for the last few years in the literature, we find that MKP has been solved for example in [14] using a quantum binarization technique, in [13] applied the k-means clustering technique to perform binarization, in [15] used a differential algorithm with transfer functions, and in [16] a modification of the PSO equations was used. MKP can be set as:

$$\text{maximize} \sum_{j=1}^{n} p_j x_j \tag{1}$$

$$\text{subjected to} \sum_{j=1}^{n} c_{ij} x_j \leq b_i \, , \, i \in \{1, ..., m\}. \tag{2}$$

With $x_j \in \{0,1\}$, $j \in \{1, ..., n\}$. p_j corresponds to the profit of element j. c_{ij} represents a cost associated with dimension i and element j. The constraints in each dimension i are represented by b_i. The solution can be modelled using a binary representation, in this representation a 0 means the element is not included in the knapsack.

3   Sine-Cosine Algorithm

The standard sine-cosine algorithm (SCOA) uses the sine and cosine functions to perform the exploration and exploitation of the search space. SCOA is a population-based optimization algorithm. Let X_i be a solution where $i \in \{1, ..., N\}$ and $j \in \{1, ..., D\}$ then each solution starts in as described in Eq. 3

$$x_{i,j} = x_j^{min} + rand \times (x_j^{max} - x_j^{min}) \tag{3}$$

where, N indicates the number of solutions, D indicates the size of the respective problem. x_j^{min} and x_j^{max} are the minimum and maximum value of the jth component. After initialization, the positions of each solution are updated using Eqs. 4 and 5.

$$X_i^{t+1} = X_i^t + r_1 \times sin(r_2) \times |r_3 P_i^t - X_i^t| \tag{4}$$

$$X_i^{t+1} = X_i^t + r_1 \times cos(r_2) \times |r_3 P_i^t - X_i^t| \tag{5}$$

where X_i^t is the position of current solution at t-th iteration. r_1, r_2 and r_3 are random numbers. P_i is the position of the destination point. These two equations are combined based on the random number r_4. This is shown in Eq. 6:

$$X_i^{t+1} = \begin{cases} X_i^t + r_1 \times sin(r_2) \times |r_3 P_i^t - X_i^t|, & \text{if } r_4 < 0.5 \\ X_i^t + r_1 \times cos(r_2) \times |r_3 P_i^t - X_i^t|, & \text{if } r_4, \geq 0.5 \end{cases} \tag{6}$$

In order to balance exploration and exploitation, the range of sine and cosine in Eqs. 4 and 5 is changed adaptively using the following Eq. 7

$$r_1 = a - t\frac{a}{T} \tag{7}$$

where t is the current iteration, T is the maximum number of iterations, and a is a constant.

3.1   Initialization and Element Weighting

As BPSCOA is a swarm algorithm, to begin the exploration and exploitation of the search space, the list of solutions must be initialized. For the generation of each solution, an element is randomly chosen first. The next step is to check if other elements can be incorporated, for this we must evaluate the constraints of our problem. To select the new element, we generate a list of possible elements that comply with the constraints. Each element is calculated a weight and the best weight element is selected. This procedure is repeated until no additional element can be incorporated. In the Fig. 1, the initialization algorithm is displayed.

To calculate the weight of each element, several methods have been developed. In [17] a pseudo-utility in the surrogate duality approach was proposed.

The way to calculate it is shown in Eq. 8. In this equation the variable w_j corresponds to the surrogate multiplier whose value is between 0 and 1. This multiplier can be interpreted as a shadow prices of the j-th constraint.

$$\delta_i = \frac{p_i}{\sum_{j=1}^{m} w_j c_{ij}} \quad (8)$$

A more intuitive measure focused on the average resource occupancy was proposed by [18]. It is shown in Eq. 9.

$$\delta_i = \frac{\sum_{j=1}^{m} \frac{c_{ij}}{m b_j}}{p_i} \quad (9)$$

In this article we used a variation of this last measure focused on the average occupation and proposed in [13]. this variation considers the elements that exist in knapsacks to calculate the average occupancy. In each iteration depending on the selected items in the solution the measure is calculated again. The expression of this new measure is shown in Eq. 10.

$$\delta_i = \frac{\sum_{j=1}^{m} \frac{c_{ij}}{m(b_j - \sum_{i \in S} c_{ij})}}{p_i} \quad (10)$$

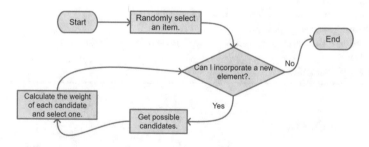

Fig. 1. Flowchart of generation of a new solution.

3.2 Percentile Binary Operator

Due to the iterative nature of swarm intelligence algorithms and considering that SCOA works in continuous space. The velocity and position of the solutions are updated in $\mathbb{R}^n$. A general way of writing the update is shown in Eq. 11. In this equation x_{t+1} represents the position of the particle x at time $t+1$. To obtain the position, we consider the function Δ, which is specific to each algorithm. For example in Black Hole $\Delta(x) = \text{rand} \times (x_{bh}(t) - x(t))$, in Cuckoo Search $\Delta(x) = \alpha \oplus Levy(\lambda)(x)$, and in PSO, Firefly, and Bat algorithms Δ can be written in simplified form as $\Delta(x) = v(x)$.

$$x_{t+1} = x_t + \Delta_{t+1}(x(t)) \quad (11)$$

The application of statistics and machine learning techniques, are used in many areas such as transports, smart cities, agriculture and computational intelligence, examples of applications are found in [4,8,13,19–22]. For the case of SCOA, it is considered a binary percentile operator to perform the passage of the continuous space to the binary space. Given the particle x, let us consider the magnitude of the displacement $Delta^i(x)$ in the i-th component and we group these magnitudes for all solutions in order to obtain the values for the percentiles {20, 40, 60, 80, 100}. At each percentile, we will assign a transition probability where the values are shown in Eq. 12. Using these transition probabilities together with the Eq. 13 binarization of the solutions is performed. The algorithm is detailed in Algorithm 1.

$$P_{tr}(x^i) = \begin{cases} 0.1, & \text{if } x^i \in \text{group } \{0,1\} \\ 0.5, & \text{if } x^i \in \text{group } \{2,3,4\} \end{cases} \tag{12}$$

$$x^i(t+1) := \begin{cases} \hat{x}^i(t), & \text{if } rand < P_{tg}(x^i) \\ x^i(t), & \text{otherwise} \end{cases} \tag{13}$$

Algorithm 1. Percentile binary operator

1: **Function** percentilebinary(vList, pList)
2: **Input** vList, pList
3: **Output** pGroupValue
4: pValue = getPercentileValue(vList, pList)
5: **for each** value in vList **do**
6:     pGroupValue = getPercentileGroupValue(pValue,vList)
7: **end for**
8: **return** pGroupValue

3.3   Repair Operator

Local search and binary percentile operators can generate infeasible solutions. There are different mechanisms to address these infeasible solutions. In this article, the repair of the solutions was considered. To perform the repair, the measure described in the Eq. 10 was used. If the solution requires repair, the element with the maximum measure is chosen and it is removed from the solution, this process is iterated until a valid solution is obtained. The solution is then improved by exploring the possibility of incorporating new elements. This stage of improvement is iterated until there are no elements that can be incorporated without violating the constraints. The pseudocode is shown in Algorithm 2.

Algorithm 2. Repair Algorithm

1: **Function** Repair(S_{in})
2: **Input** Input solution S_{in}
3: **Output** The Repair solution S_{out}
4: $S \leftarrow S_{in}$
5: **while** needRepair(S) == True **do**
6:      $s_{max} \leftarrow$ getMaxWeight(S)
7:      $S \leftarrow$ removeElement(S, s_{max})
8: **end while**
9: state $\leftarrow$ False
10: **while** state == False **do**
11:     $s_{min} \leftarrow$ getMinWeight(S)
12:     **if** $s_{min} == \emptyset$ **then**
13:         state $\leftarrow$ True
14:     **else**
15:         $S \leftarrow$ addElement(S, s_{min})
16:     **end if**
17: **end while**
18: $S_{out} \leftarrow S$
19: **return** S_{out}

---

4   Results

4.1   Insight of BPSCOA Algorithm

This section aims to find out the contribution of the percentile binary operator to the performance of the algorithm. To carry out the comparison problems cb.5.250 from the OR-library were selected. Violin charts and the Wilcoxon non-parametric signed-rank test were used to perform the statistical analyzes. In the charts, the X axis identifies the studied instances and the Y axis the % -Gap described in the Eq. 14. The Wilcoxon test is run to determine if the results obtained by BPSCOA have significant difference with respect to other algorithms. The parameter settings and browser ranges are shown in Table 1.

$$\% - Gap = 100\frac{BestKnown - SolutionValue}{BestKnown} \tag{14}$$

Table 1. Setting of parameters for Binary sine-cosine algorithm.

Parameters	Description	Value	Range
N	Number of solutions	30	[20, 25, 30]
G	Number of percentiles	5	[4,5,6]
Iteration Number	Maximum iterations	1000	[1000]

Evaluation of Percentile Binary Operator. A random operator was designed to evaluate the contribution of the percentile binary operator. This random operator has the peculiarity of performing the transitions with a fixed probability of 0.5 without considering in which percentile the variable is located. Two configurations were studied: The first one where the local search operator is included and the second where the local operator is not considered. BPSCOA corresponds to our standard algorithm. *random.ls* is the random variant that includes the local search operator. *BPSCOA.wls* corresponds to the version with percentile binary operator without local search operator. Finally *random.wls* describes the random algorithm without local search operator.

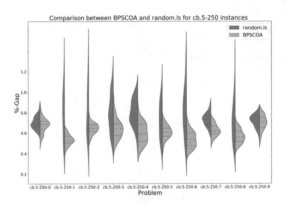

Fig. 2. Evaluation of percentile binary operator with local search operator

When we compared the Best Values between BPSCOA and *random.ls* which are shown in Table 2. BPSCOA outperforms to *random.ls*. However the Best Values between both algorithms are very close. In the Average comparison, BPSCOA outperforms *random.ls* almost in all problems. The comparison of distributions is shown in Fig. 2. We see the dispersion of the *random.ls* distributions are bigger than the dispersions of BPSCOA. In particular this can be appreciated in the problems 1, 2, 3, 5, 6 and 8. Therefore, the percentile binary operator together with local search operators, contribute to the precision of the results. Finally, the BPSCOA distributions are closer to zero than *random.ls* distributions, indicating that BPSCOA has consistently better results than *random.ls*. When we evaluate the behaviour of the algorithms through the Wilcoxon test, this indicates that there is a significant difference between the two algorithms.

Our next step is trying to separate the contribution of local search operator from the percentile binary operator. For this, we compared the algorithms *wls* and *random.wls*.

When we check the Best Values shown in Table 2, we note that *wls* not always performs better than *05.wls*. In the case of the average indicator, *wls* outperforms in all problems to *05.wls*. The Wilcoxon test indicates that the difference is significant. This suggests that *wls* is consistently better than *05.wls*.

Table 2. Evaluation of percentile binary operator

Set	Best Known	Best random.ls	Best BPSCOA	Best random.wls	Best wls	Avg random.ls	Avg BPSCOA	Avg random.wls	Avg wls
cb.5.250-0	59312	59211	59225	59158	59175	59132.1	59151.3	59071.8	59132.1
cb.5.250-1	61472	61435	61435	61409	61409	61324.6	61366.1	61288.3	61370.7
cb.5.250-2	62130	62036	62074	61969	61969	61894.4	61970.3	61801.6	61924.6
cb.5.250-3	59463	59365	59446	59365	59349	59257.8	59302.5	59136.1	59269.3
cb.5.250-4	58951	58914	58951	58883	58930	58725.6	58803.1	58693.6	58758.5
cb.5.250-5	60077	60015	60056	59990	60015	59904.6	59931.7	59837.8	59945.9
cb.5.250-6	60414	60355	60355	60348	60349	60208.2	60321.1	60230.6	60311.1
cb.5.250-7	61472	61407	61436	61407	61407	61290.8	61307.9	61233.9	61342.3
cb.5.250-8	61885	61829	61885	61790	61782	61737.1	61759.7	61644.9	61738.1
cb.5.250-9	58959	58832	58959	58822	58787	58769.1	58759.1	58653.7	58770.4
Average	60413.5	60339.9	60382.2	60314.1	60317.2	60224.43	60267.28	60159.23	60256.3
p-value							2.21 e-06		2.17 e-05

In the violin chart shown n the Fig. 3 it is further observed that the dispersion of the solutions for the case of *05.wls* is much larger than in the case of *wls*. This indicates that the percentile binary operator plays an important role in the precision of the results.

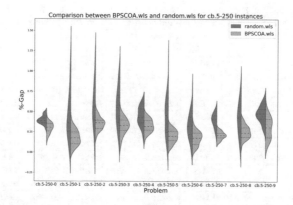

Fig. 3. Evaluation of percentile binary operator without Local Search operator

4.2 BPSCOA Comparisons

In this section we evaluate the performance of our BPSCOA with the algorithm BAAA developed in [12]. BAAA uses transfer functions as a general mechanism of binarization. In particular BAAA used the tanh $= \frac{e^{\tau|x|}-1}{e^{\tau|x|}+1}$ function to perform the transference. The parameter τ of the tanh function was set to a value 1.5. Additionally, a elite local search procedure was used by BAAA to improve solutions. As maximum number of iterations BAAA used 35000. The computer configuration used to run the BAAA algorithm was: PC Intel Core(TM) 2 dual

CPU Q9300@2.5 GHz, 4 GB RAM and 64-bit Windows 7 operating system. In our BPSCOA algorithm, the configurations are the same used in the previous experiments. These are described in the Table 1. Additionally we made the comparison with KMTR-BH and KMTR-Cuckoo binarizations. KMTR uses the unsupervised K-means learning technique to perform the binarization process. In the article [13], the Black Hole and Cuckoo Search algorithms were binarized using KMTR.

The results are shown in Table 3. The comparison was performed for the set cb.5.500 of the OR-library. The results for BPSCOA were obtained from 30 executions for each problem. In black, the best results are marked for both indicators the Best Value and the Average. For the best value indicator BAAA was higher in 3, KMTR-BH in 11, KMTR-Cuckoo in 7 and BPSCOA in 9. In the Average indicator BAAA was higher in 2 instances, KMTR-BH in 14, KMTR-Cuckoo in 15 and BPSCOA in 0.

Table 3. OR-Library benchmarks MKP cb.5.500

Instance	Best Known	BAAA Best	Avg	KMTR-BH Best	Avg	KMTR-Cuckoo Best	Avg	BPSCOA Best	Avg	Time(s)	std
0	120148	120066	120013.7	**120096**	120029.9	120082	**120036.8**	120082	120002.1	561	57.2
1	117879	117702	117560.5	**117730**	**117617.5**	117656	117570.6	117656	117527.4	575	68.2
2	121131	**120951**	120782.9	121039	120937.9	120923	**120855.1**	120782	120597.1	598	71.8
3	120804	120572	120340.6	120683	**120522.8**	120683	120455.7	**120683**	120431.6	631	68.3
4	122319	122231	122101.8	**122280**	**122165.2**	122212	122136.4	122212	122001.4	621	77.7
5	122024	121957	121741.8	**121982**	**121868.7**	121946	121824.6	121946	121623.1	701	88.3
6	119127	119070	118913.4	**119068**	**118950.0**	118956	118895.5	118956	118612.2	654	101.3
7	120568	**120472**	120331.2	120463	**120336.6**	120392	120320.4	120392	120104.1	722	96.4
8	121586	121052	120683.6	**121377**	**121161.9**	121201	121126.3	121295	121001.8	701	89.2
9	120717	120499	120296.3	**120524**	**120362.9**	120467	120335.5	120467	120287.3	731	92.1
10	218428	218185	217984.7	218296	218163.7	218291	**218208.9**	218296	217099.5	728	105.3
11	221202	220852	220527.5	220951	220813.9	220969	**220862.3**	220951	220643.1	685	126.7
12	217542	217258	217056.7	217349	217254.3	**217356**	**217293.0**	217349	217101.3	603	103.2
13	223560	223510	223450.9	**223518**	223455.2	223516	**223455.6**	223516	223256.1	695	99.5
14	218966	218811	218634.3	218848	218771.5	**218884**	**218794.0**	218848	218712.8	643	88.3
15	220530	220429	220375.9	**220441**	220342.2	220433	**220352.7**	220410	220203.1	714	146.1
16	219989	219785	219619.3	219858	219717.9	**219943**	**219732.8**	219858	219675.8	685	102.3
17	218215	218032	217813.2	218010	217890.1	**218094**	**217928.7**	218010	217799.1	602	119.7
18	216976	**216940**	**216862.0**	216866	216798.8	216873	216829.8	216866	216640.8	643	126.5
19	219719	219602	219435.1	219631	219520.0	**219693**	**219558.9**	219631	219433.9	705	132.3
20	295828	295652	295505.0	**295717**	**295628.4**	295688	295608.8	295688	295478.2	688	162.9
21	308086	307783	307577.5	307924	307860.6	**308065**	**307914.8**	307924	307699.4	717	141.4
22	299796	299727	299664.1	299796	**299717.8**	299684	299660.9	**299796**	299567.8	659	143.7
23	306480	306469	306385.0	**306480**	**306445.2**	306415	306397.3	306415	306187.3	709	156.1
24	300342	300240	300136.7	300245	**300202.5**	300207	300184.4	**300245**	299984.1	672	184.9
25	302571	302492	302376.0	302481	**302442.3**	302474	302435.6	**302481**	302187.9	587	112.2
26	301339	301272	301158.0	301284	301238.3	301284	**301239.7**	301284	301003.1	577	131.1
27	306454	306290	306138.4	306325	306264.2	306331	**306276.4**	306331	306088.9	629	181.1
28	302828	302769	302690.1	302749	302721.4	**302781**	302716.9	302771	302556.7	674	142.7
29	299910	299757	299702.3	299774	299722.7	299828	**299766.0**	299828	299582.1	627	113.4
Average	214168.8	214014.2	213861.9	**214059.5**	213964.1	214044.2	213959.1	**214030.5**	213769.6	657.9	114.3

5   Conclusions

In this article, the percentile technique is used to perform the binarization of the SCOA algorithm. It should be noted that the percentile technique can be applied in the binarization of any continuous swarm-intelligence algorithm. We used the knapsack problem to evaluate our binary algorithm. The contribution of the percentile binary operator was studied, observing that this operator contributes to the precision and quality of the solutions obtained. Additionally, we develop the comparison with the BAAA and KMTR algorithms, in this comparison, BPSCOA was able to outperform the other algorithms in the best value indicator in 9 problems. However, it was not effective in the average indicator.

As a future line of research, it is interesting to compare the performance of the percentile operator with the K-means operator used in KMTR and with the transfer functions used in BAAA, all applied in the binarization of the SCOA algorithm. Also, it is interesting to use the percentile operator to binarize other algorithms in addition to solving other NP-hard problems. From the theoretical point of view, it would also be interesting to understand how the exploration and exploitation of space are altered when we introduce the percentile operator. Another interesting research line is to explore adaptive techniques to automate the selection of parameters.

References

1. Barman, S., Kwon, Y.-K.: A novel mutual information-based Boolean network inference method from time-series gene expression data. PloS ONE **12**(2), e0171097 (2017)
2. Crawford, B., Soto, R., Monfroy, E., Astorga, G., García, J., Cortes, E.: A meta-optimization approach for covering problems in facility location. In: Workshop on Engineering Applications, pp. 565–578. Springer (2017)
3. García, J., Crawford, B., Soto, R., Astorga, G.: A percentile transition ranking algorithm applied to binarization of continuous swarm intelligence metaheuristics. In: International Conference on Soft Computing and Data Mining, pp. 3–13. Springer (2018)
4. García, J., Crawford, B., Soto, R., Astorga, G.: A clustering algorithm applied to the binarization of swarm intelligence continuous metaheuristics. Swarm Evol. Comput. (2018)
5. García, J., Crawford, B., Soto, R., Astorga, G.: A percentile transition ranking algorithm applied to knapsack problem. In: Proceedings of the Computational Methods in Systems and Software, pp. 126–138. Springer (2017)
6. García, J., Crawford, B., Soto, R., García, P.: A multi dynamic binary black hole algorithm applied to set covering problem. In: International Conference on Harmony Search Algorithm, pp. 42–51. Springer (2017)
7. Crawford, B., Soto, R., Monfroy, E., Astorga, G., García, J., Cortes, E.: A meta-optimization approach to solve the set covering problem. Ingeniería **23**(3), 274–288 (2018)
8. García, J., Peña, A.: Robust optimization: concepts and applications. In: Nature-inspired Methods for Stochastic, Robust and Dynamic Optimization. IntechOpen (2018)

9. García, J., Altimiras, F., Peña, A., Astorga, G., Peredo, O.: A binary cuckoo search big data algorithm applied to large-scale crew scheduling problems. Complexity **2018** (2018)
10. Mirjalili, S.: SCA: a sine cosine algorithm for solving optimization problems. Knowl. Based Syst. **96**, 120–133 (2016)
11. Crawford, B., Soto, R., Astorga, G., García, J., Castro, C., Paredes, F.: Putting continuous metaheuristics to work in binary search spaces. Complexity **2017** (2017)
12. Zhang, X., Wu, C., Li, J., Wang, X., Yang, Z., Lee, J.-M., Jung, K.-H.: Binary artificial algae algorithm for multidimensional knapsack problems. Appl. Soft Comput. **43**, 583–595 (2016)
13. García, J., Crawford, B., Soto, R., Castro, C., Paredes, F.: A k-means binarization framework applied to multidimensional knapsack problem. Appl. Intell. **48**(2), 357–380 (2018). Springer
14. Haddar, B., Khemakhem, M., Hanafi, S., Wilbaut, C.: A hybrid quantum particle swarm optimization for the multidimensional knapsack problem. Eng. Appl. Artif. Intell. **55**, 1–13 (2016)
15. Liu, J., Wu, C., Cao, J., Wang, X., Teo, K.L.: A binary differential search algorithm for the 0–1 multidimensional knapsack problem. Appl. Math. Model. (2016)
16. Bansal, J.C., Deep, K.: A modified binary particle swarm optimization for knapsack problems. Appl. Math. Comput. **218**(22), 11042–11061 (2012)
17. Pirkul, H.: A heuristic solution procedure for the multiconstraint zero? one knapsack problem. Naval Res. Logistics **34**(2), 161–172 (1987)
18. Kong, X., Gao, L., Ouyang, H., Li, S.: Solving large-scale multidimensional knapsack problems with a new binary harmony search algorithm. Comput. Oper. Res. **63**, 7–22 (2015)
19. García, J., Pope, C., Altimiras, F.: A distributed k-means segmentation algorithm applied to Lobesia botrana recognition. Complexity **2017** (2017)
20. Graells-Garrido, E., García, J.: Visual exploration of urban dynamics using mobile data. In: International Conference on Ubiquitous Computing and Ambient Intelligence, pp. 480–491. Springer (2015)
21. Graells-Garrido, E., Peredo, O., García, J.: Sensing urban patterns with antenna mappings: the case of Santiago, Chile. Sensors **16**(7), 1098 (2016)
22. Peredo, O.F., García, J.A., Stuven, R., Ortiz, J.M.: Urban dynamic estimation using mobile phone logs and locally varying anisotropy. In: Geostatistics Valencia 2016, pp. 949–964. Springer (2017)

Parameter Calculation in Time Analysis for the Approach of Filtering to Select IMFs of EMD in AE Sensors for Leakage Signature

Nur Syakirah Mohd Jaafar[✉], Izzatdin Abdul Aziz,
M. Hilmi B. Hasan, and Ahmad Kamil Mahmood

Centre for Research in Data Science, Universiti Teknologi PETRONAS,
Seri Iskandar, Malaysia
{nur_16001470, izzatdin, mhilmi_hasan,
kamilmh}@utp.edu.my

Abstract. The pipelines are used for transporting fluids and it is an important part of the media transportation for oil and gas. However, as pipelines are often spread across vast distances and carry certain hazardous substances, the chances for accidents such as leakage accidents in oil and gas pipelines are increased. Variety of factors lead to pipeline leakage accidents such as corrosion, vibration and other impacts affecting the safe operation of pipelines. Pipelines leakages cause both loss of product and as well as environmental damage. Acoustic emissions sensors have recently emerged as a promising tool for long distance pipeline monitoring due to the acoustic emission sensors advantages of high accuracy and low loss per distance. The signal processing is used to decompose the raw signal and the pre-processed signal will be analyzed in the time-frequency domain. Several existing signals processing methods such as Fourier Transform, Wavelet Transform can be used for extracting useful information. The parameters of Empirical Mode Decomposition [EMD] show promising results. The promising results in terms of accuracy of selections IMFs and analysis of time-frequency domain. The selected of Intrinsic Mode Functions [IMFs] IMFs are analyzed in the time domain by using two parameters which are standard deviation and variance. The selected IMFs are obtained to reveal the leakage and no leakage signatures of the pipeline.

Keywords: Signal processing method · Empirical Mode Decomposition · Intrinsic Mode Functions · Selected IMFs · Time domain analysis · Standard deviation · Variance

1 Introduction

The development of underground pipelines plays a vital role in the advancement of oil and gas in US industries as well as in Malaysia. Carbon and steel material has always been the primary choice for development of a pipeline for its long -term acceptable-safe services in the oil and gas industry [1, 2].

Majority of the existing pipelines are aged and tend to fail due to external and internal factors. The pipeline failure caused by the fatigue or the sudden change in the pipeline

© Springer Nature Switzerland AG 2019
R. Silhavy (Ed.): CSOC 2019, AISC 985, pp. 139–146, 2019.
https://doi.org/10.1007/978-3-030-19810-7_14

pressure and other critical factors that lead to the failure such as pipes and material properties, environmental conditions, and loads including internal and external [3]. The pipelines do fail when the pipeline starts to deteriorate to mechanisms such as leakage, corrosion, and suffixation [4].

The failure of the pipeline causes damage to main contributors such as multiple fatalities towards society, serious financial loss towards economic impacts and damages to the environmental [4, 5]. Incidents of the pipeline such as traffic and overstress loads leads to the occurrence of the leak regardless of the pipeline specifications, locations and fluid specifications [6]. The pipeline leak detection for the market in oil and gas industry in Asia- Oceania was forecasted that the spending ($MN) are increases of 600 to 1000 from the year of 2018 until the year 2020 [7].

The task to study the means of information from the time-frequency analysis in the signal processing studies for the leakage in the pipeline remains a challenge due to the rapid increase of costs to spend for the pipeline leak detection for the oil and gas industry. Either the signal of the pipeline leakage consists of the leakage signatures or no significant signal towards the leakage signatures and anomalies signatures such as leads to the noise signatures only resulted as the challenge in the study of the means. Thus, it will carry a significant impact on the analysis of the signatures of the leakage pipeline in the time-frequency analysis.

In the acoustic emission sensors, the time-frequency analysis is the most of well - known analysis for the signal processing. The sensor's specifications such as in acoustic emission sensors, it comes with many attributes including operating specifications and features to derive in the signal processing techniques.

To analyse the time-frequency analysis in acoustic emission sensors, signal processing techniques are used. "Signal processing is the field of study that analyses the information or attributes of the such as sound measurements through numerical analysis [8], which are done to improve detection on the components of interest in a measured signal through mathematical methods such as time series [9], spectral density estimation [10] and time-frequency analysis [11]". Some examples of signal processing techniques are Fourier transforms (FT), Wavelet transforms (WT), and Hilbert Huang transforms (HHT) [12].

This research contributes to the selection of one of the signal processing techniques in time-frequency analysis which is Empirical Mode Decomposition (EMD), where the acoustic emission sensor's data is based on the time-frequency analysis in the mathematical methods and uses the concept of spectral estimation [13, 14]. Empirical Mode Decomposition (EMD) technique is a method of decomposing down signals from high-frequency components into lower frequency components which are adaptive and highly efficient without leaving the time domain. The decomposition of the complicated data set into a finite function is done by Intrinsic Mode Function (IMF).

These methods have been discussed between researchers in many domains. The most type of discussion's domain such as in fault gear domain. These methods such as to provide feature selection on the acoustic emission sensors on the fault gear datasets [15]. One researcher focused on the Empirical Mode Decomposition (EMD) to extract the features from the measured signals to diagnose time domain features from the Intrinsic Mode Function (IMF) as the best features to increase the accuracy on gear faults dataset [15].

The parameter's analysis of the Intrinsic Mode Functions (IMFs) as the post-processing of the Empirical Mode Decomposition (EMD) however, most of the researchers did not apply deeper analysis. The existing researchers used the signal processing techniques, which often involved the adoption of Empirical Mode Decomposition (EMD) as well as Ensemble Empirical Mode Decomposition (EEMD) that do not involve deep analysis towards significant IMFs.

It has been revealed that the signal processing techniques are lacking the ability to interpret the selection of the IMF's in the EMD especially with regards to the issue of pipeline leakages detection [16–19]. There are commonly used signal processing techniques in time-frequency analysis which are Fourier Transforms [20], Wavelet Transforms [21, 22] and Hilbert Transforms [12, 19]. These techniques are still inferior in time-frequency analysis accurately. The approach of selection IMF in the signal processing techniques for the acoustic emission sensors needs to be analyzed and this approach is significant in this research.

2   Time Analysis of Acoustic Emission Sensors

2.1   Selected IMFs by Using Time Domain Analysis

In order to fulfil the second phase of selection IMFs, the algorithm, EMD will run and resulted in a series of IMFs and a residue. The selected of IMFs will be identifying based on several descriptive parameter's calculations for the time analysis. Based on Fig. 1, the IMF descriptive parameters calculations for time domain are standard deviation and variance. In order to reveal the significant IMFs leads to leakage signatures, the criteria for IMF descriptive parameter calculation are derived.

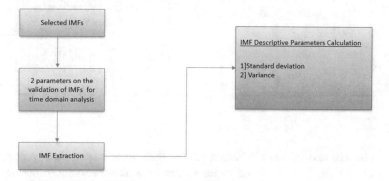

Fig. 1. IMF descriptive parameters calculation for time domain analysis

In signal processing technique, for the time and frequency analysis studies for the selected IMFs. 2 parameters which are variance and standard deviations for the time analysis are chosen for this research. 3 sets of trials been conducted which are for 2 cases, leak and no leak for both sensor 1 and sensor 2 are analysed for both parameters.

Table 1. Variance and standard deviations for the time domain analysis for leak in all cases for Sensor 1

Descriptions of Data Sets		Higher Frequencies	Lower Frequencies		Series of IMFs	Variance Sets			Standard Deviations Sets		
Conditions	Sensors	Selected of IMFs	Non Selected of IMFs	Residue	IMF's number	1	2	3	1	2	3
Leak	1	IMF 1- IMF 2	IMF 3 - IMF 15	IMF 16	1	4.4829	4.6013	5.4303	2.1173	2.1451	2.3303
					2	4.0052	4.0958	4.8165	2.0013	2.0238	2.1947
					3	0.9905	0.8902	1.1460	0.9952	0.9435	1.0705
					4	0.188	0.1766	0.1999	0.4336	0.4203	0.4471
					5	0.0655	0.0557	0.0682	0.2559	0.2361	0.2612
					6	0.0275	0.0197	0.0274	0.1659	0.1402	0.1655
					7	0.0129	0.0099	0.0126	0.1137	0.0995	0.1121
					8	0.0054	0.0065	0.0081	0.0737	0.0804	0.0901
					9	0.0038	0.0032	0.0039	0.0614	0.0567	0.0628
					10	0.0022	0.0015	0.0024	0.0474	0.0386	0.0495
					11	0.0011	0.0008	0.0010	0.0329	0.0281	0.0315
					12	0.0007	0.0005	0.0004	0.0258	0.0221	0.0190
					13	0.0003	0.0001	0.0003	0.0164	0.0115	0.0166
					14	0.0001	0.0001	0.0002	0.0113	0.0101	0.0154
					15	0.0001	0.0001	0.0001	0.0144	0.0021	0.0328
					16						

Table 2. Variance and standard deviations for the time domain analysis for no leak in all cases for Sensor 1

Descriptions of Data Sets		Higher Frequencies	Lower Frequencies		Series of IMFs	Variance Sets			Standard Deviations Sets		
Conditions	Sensors	Selected of IMFs	Non Selected of IMFs	Residue	IMF's number	1	2	3	1	2	3
No Leak	1	IMF 1- IMF 2	IMF 3 - IMF 15	IMF 16	1	1.0587	1.0587	1.0587	1.0289	1.0289	1.0289
					2	0.8355	0.8355	0.8355	0.9141	0.9141	0.9141
					3	0.3060	0.3060	0.3060	0.5531	0.5531	0.5531
					4	0.0944	0.0944	0.0944	0.3072	0.3072	0.3072
					5	0.0280	0.0280	0.0280	0.1674	0.1674	0.1674
					6	0.0081	0.0081	0.0081	0.0899	0.0899	0.0899
					7	0.0038	0.0038	0.0038	0.0618	0.0618	0.0618
					8	0.0023	0.0023	0.0023	0.0485	0.0485	0.0485
					9	0.0013	0.0013	0.0013	0.0358	0.0358	0.0358
					10	0.0008	0.0008	0.0008	0.0288	0.0288	0.0288
					11	0.0004	0.0004	0.0004	0.0199	0.0199	0.0199
					12	0.0002	0.0002	0.0002	0.0132	0.0132	0.0132
					13	0.0001	0.0001	0.0001	0.0075	0.0075	0.0075
					14	0.0001	0.0001	0.0001	0.0062	0.0062	0.0062
					15	0.0001	0.0001	0.0001	0.0024	0.0024	0.0024
					16						

The variance and standard deviation for the time domain analysis as represents in Table 1 are for the leak in all cases for sensor 1. While the variance and standard deviation for the time domain analysis as represents in Table 2 are for no leak in all cases for sensor 1. Table 1, which are for leak cases represents as IMF 1 shows the variance in set 1, set 2 and set 3 are 4.4829, 4.6013 and 5. 4303. While, for Table 2, which are for no-leak cases represents as IMF 1 is 1.0587. The IMF 2 in Table 1 for leak cases shows the variance in set 1, set 2 and set 3 are 4.0052, 4.0958 and 4. 8165. And, for the IMF 2 in Table 2 for no-leak cases shows the variance in set 1, set 2 and set 3 are 0.8355.

Table 3. Variance and standard deviations for the time domain analysis for leak in all cases for Sensor 2

Descriptions of Data Sets		Higher Frequencies	Lower Frequencies		Series of IMFs	Variance Sets			Standard Deviations Sets		
Conditions	Sensors	Selected of IMFs	Non Selected of IMFs	Residue	IMF's number	1	2	3	1	2	3
Leak	2	IMF 1 - IMF 2	IMF 3 - IMF 15	IMF 16	1	1.6445	1.7541	1.7931	1.2824	1.3244	1.3391
					2	1.6330	1.5964	1.6364	1.2779	1.2635	1.2792
					3	0.5945	0.5865	0.5995	0.7710	0.7658	0.7743
					4	0.1604	0.1618	0.1546	0.4005	0.4023	0.3931
					5	0.0520	0.0455	0.0468	0.2279	0.2134	0.2164
					6	0.0217	0.0171	0.0163	0.1475	0.1307	0.1278
					7	0.0128	0.0103	0.0099	0.1129	0.1016	0.0993
					8	0.0080	0.0058	0.0050	0.0896	0.0765	0.0709
					9	0.0047	0.0043	0.0033	0.0689	0.0658	0.0573
					10	0.0025	0.0017	0.0016	0.0501	0.0418	0.0396
					11	0.0009	0.0012	0.0009	0.0301	0.0341	0.0303
					12	0.0009	0.0007	0.0005	0.0293	0.0261	0.0223
					13	0.0008	0.0005	0.0005	0.0278	0.0130	0.0115
					14	0.0004	0.0002	0.0001	0.0201	0.0123	0.0106
					15	0.0004	0.0002	0.0001	0.0061	0.0305	0.0044
					16						

Standard deviation as the second parameter for the time domain analysis for a leak in all cases for sensor 1 represents in Table 1. While standard deviation as the second parameter for the time domain analysis for no leak in all cases for sensor 1 represents in Table 2. The standard deviation in set 1, set 2 and set 3 are 2.1173, 2.1451 and 2.3303 represented as IMF 1 in Table 1 for leak cases. While the standard deviation for no-leak cases for IMF 1 for all sets is 1.0289 as in Table 2. IMF 2 in Table 1 for leak cases shows the standard deviation in set 1, set 2 and set 3 are 2.0013, 2.0238, and 2.1947. And, for the IMF 2 in Table 2 for no-leak cases show the standard deviation in set 1, set 2 and set 3 are 0.9141.

The variance and standard deviation for the time domain analysis as represents in Table 3 are for a leak in all cases for sensor 2. While the variance and standard deviation for the time domain analysis are represented in Table 4 are for no leak in all cases for sensor 2. Table 3, which are for leak cases represents as IMF 1 shows the variance in set 1, set 2 and set 3 are 1.6445, 1.7541, and 1.7931. While, for Table 4, which are for no-leak cases represents as IMF 1 is 1. 6412. IMF 2 in Table 3 for leak cases shows the variance in set 1, set 2 and set 3 are 1.6330, 1.5964, and 1.6364. And, for the IMF 2 in Table 4 for no-leak cases shows the variance in set 1, set 2 and set 3 are 1.5092.

The second parameter which is the standard deviation for the time domain analysis for the leak in all cases for sensor 2 are in Tables 3 and 4 shows the standard deviation for the time domain analysis for no leak in all cases for sensor 2. IMF 1 in Table 3 for leak cases shows the standard deviation in set 1, set 2 and set 3 are 1.2824, 1.3244, and 1.3391. IMF 1 in Table 4 for no-leak cases shows the standard deviations in set 1, set 2 and set 3 are 1.2811. IMF 2 in Table 3 for leak cases shows the standard deviation in set 1, set 2 and set 3 are 1.2779, 1.2635, and 1.2792. And, for the IMF 2 in Table 4 for no-leak cases shows the standard deviation in set 1, set 2 and set 3 are 1.2285.

Table 4. Variance and standard deviations for the time domain analysis for no leak in all cases for Sensor 2

Descriptions of Data Sets		Higher Frequencies	Lower Frequencies		Series of IMFs	Variance Sets			Standard Deviations Sets		
Conditions	Sensors	Selected of IMFs	Non Selected of IMFs	Residue	IMF's number	1	2	3	1	2	3
					1	1.6412	1.6412	1.6412	1.2811	1.2811	1.2811
					2	1.5092	1.5092	1.5092	1.2285	1.2285	1.2285
					3	0.5763	0.5763	0.5763	0.7591	0.7591	0.7591
					4	0.1559	0.1559	0.1559	0.3949	0.3949	0.3949
					5	0.0444	0.0444	0.0444	0.2108	0.2108	0.2108
					6	0.0192	0.0192	0.0192	0.1384	0.1384	0.1384
					7	0.0112	0.0112	0.0112	0.1058	0.1058	0.1058
No Leak	2	IMF 1- IMF 2	IMF 3 - IMF 15	IMF 16	8	0.0058	0.0058	0.0058	0.0762	0.0762	0.0762
					9	0.0038	0.0038	0.0038	0.0618	0.0618	0.0618
					10	0.0018	0.0018	0.0018	0.0426	0.0426	0.0426
					11	0.0008	0.0008	0.0008	0.0288	0.0288	0.0288
					12	0.0006	0.0006	0.0006	0.0306	0.0306	0.0306
					13	0.0006	0.0006	0.0006	0.0250	0.0250	0.0250
					14	0.0004	0.0004	0.0004	0.0205	0.0205	0.0205
					15	0.0002	0.0002	0.0002	0.0136	0.0136	0.0136
					16						

3 Conclusion

Based on the result, the approach of selected IMFs from the EMD has significantly outperformed with the rest of the techniques. The proposed approach to the selection of Intrinsic Mode Functions (IMFs) is composed of two main modules. The first module is the filtration class based on the design bandpass (BP) filter and this design BP filter as the first class that the data set need to pass through before the EMD is generated. The second module applies the EMD, which is the series of IMFs results.

The series of IMFs will be analysed thoroughly before the criteria of IMFs build for the selection of IMFs. The combination of two modules which are the filtration class based on the design bandpass (BP) filter and the criteria of IMFs build for the selection of IMFs help to improve the accurate result.

The results show that the approach of selection of IMFs will lead to accurate identified filtered signatures to increase the accuracy of the selected IMF itself. The enhancement of the EMD technique by implementing together the combination of two modules can improve the approach of selection of IMFs process of AE datasets in Signal Processing Technique for Time and Frequency Analysis field. This indicates that all the combination of two modules can increase the accuracy of identified signatures for the cases by including on the data preparation and data filtering in a thorough process.

In the proposed method, it utilizes two parameters which are the variance and standard deviations for the time analysis calculated for the IMFs obtained on the decomposed of AE signals by the EMD algorithm. Most of the frequency variation is obtained through the first two IMFs which are IMF 1 and IMF 2 leads to the pertinent fundamental properties of AE signals.

The selection of the appropriate number of IMFs is of great importance in the analysis of signals using EMD and resulted as selected IMFs as post-processing. The two parameters of a combination of two IMFs which are IMF 1 and IMF 2 have been considered for producing the signatures for both sensors in both conditions.

Acknowledgements. This work was supported by the Development of Intelligent Pipeline Integrity Management System (I-PIMS) Grant Scheme from Universiti Teknologi PETRONAS.

References

1. Petro Wiki (2015). https://petrowiki.org/Pipelines
2. Iannuzzi, M., Barnoush, A., Johnsen, R.: Materials and corrosion trends in offshore and subsea oil and gas production. npj Mater. Degrad. **1**(1), 2 (2017)
3. Rezaei, H., Ryan, B., Stoianov, I.: Pipe failure analysis and impact of dynamic hydraulic conditions in water supply networks. Procedia Eng. **119**, 253–262 (2015)
4. Zardasti, L., Hanafiah, N.M., Noor, N.M., Yahaya, N., Rashid, A.S.A.: The consequence assessment of gas pipeline failure due to corrosion. Solid State Phenom. **227**, 225–228 (2015)
5. Yahaya, N., Noor, N.M., Othman, S.R., Sing, L.K., Din, M.M.: New technique for studying soil-corrosion of underground pipeline. J. Appl. Sci. **11**(9), 1510–1518 (2011)
6. Shama, A.A.M., El-Rashid, A., El-Shaib, M., Kotb, D.M.: Review of leakage detection methods for subsea pipeline
7. Pipeline leak detection market of the oil and gas industry 2018-2028, 01 May 2018. https://www.visiongain.com/report/pipeline-leak-detection-market-of-the-oil-and-gas-industry-2018-2028/
8. Signal processing [wikipedia]. https://en.wikipedia.org/wiki/Signal_processing
9. Time Series [wikipedia]. https://en.wikipedia.org/wiki/Time_series
10. Spectral density estimation [wikipedia]. https://en.wikipedia.org/wiki/Spectral_density_estimation
11. Time - frequency analysis [wikipedia]. https://en.wikipedia.org/wiki/Time%E2%80%93frequency_analysis
12. Adnan, N., Ghazali, M., Amin, M., Hamat, A.: Leak detection in gas pipeline by acoustic and signal processing-A review. In: IOP Conference Series: Materials Science and Engineering, vol. 100, no. 1, p. 012013. IOP Publishing (2015)
13. Xiao, J., Li, J., Sun, J., Feng, H., Jin, S.: Natural-gas pipeline leak location using variational mode decomposition analysis and cross-time–frequency spectrum. Measurement **124**, 163–172 (2018)
14. Mahmood, A.K.: An approach of filtering to select IMFs of EEMD in signal processing for acoustic emission [AE] sensors. In: Intelligent Systems in Cybernetics and Automation Control Theory, p. 100 (2019)
15. Park, S., Kim, S., Choi, J.-H.: Gear fault diagnosis using transmission error and ensemble empirical mode decomposition. Mech. Syst. Signal Process. **108**, 262–275 (2018)
16. Lu, C., Ding, P., Chen, Z.: Time-frequency analysis of acoustic emission signals generated by tension damage in CFRP. Procedia Eng. **23**, 210–215 (2011)
17. Li, C.J., Zhang, J.P., Zhang, Z.Q., Hu, J.L., Li, Y.: The research on leak detection technology of natural gas pipeline based on EMD. In: Applied Mechanics and Materials, vol. 494, pp. 793–796. Trans Tech Publications Ltd. (2014)
18. Ma, Y., Cui, X., Yan, Y., Ma, L., Han, X.: Leakage detection in a CO2 pipeline using Acoustic emission techniques. In: 2014 12th International Conference on Signal Processing (ICSP), pp. 78–82. IEEE (2014)
19. Hanafi, M.Y., Yusof, M.F.M., Remli, M., Kamarulzaman, M.H.: Pipe leak diagnostic using high frequency piezoelectric pressure sensor and automatic selection of intrinsic mode function (2017)

20. Mostafapour, A., Davoudi, S.: Analysis of leakage in high pressure pipe using acoustic emission method. Appl. Acoust. **74**(3), 335–342 (2013)
21. Bose, P.A., Sasikumar, T., Jose, P.A., Philip, J.: Acoustic emission signal analysis and event extraction through tuned wavelet packet transform and continuous wavelet transform while tensile testing the AA 2219 coupon. Int. J. Acoust. Vib. **23**(2), 234–239 (2018)
22. Ahadi, M., Bakhtiar, M.S.: Leak detection in water-filled plastic pipes through the application of tuned wavelet transforms to acoustic emission signals. Appl. Acoust. **71**(7), 634–639 (2010)

Model of an Intellectual Information System for Recognizing Users of a Social Network Using Bioinspired Methods

Alexey Samoylov, Margarita Kucherova$^{(\boxtimes)}$,
and Vladimir Tchumichev

Institute of Computer Technology and Information Security, Southern Federal
University, Taganrog, Russian Federation
{asamoylov,mkucherova}@sfedu.ru

Abstract. The analysis of social networks is one of the actively developing areas of research for today. In addition to methods of modeling social relations, the interest for researchers is also the methods of obtaining initial data. The analysis carried out by the authors showed that, as initial data, analysts use profile information, various types of activities. Consideration of photo albums and video files as a source of data on users of social networks was not considered. At the same time, by identifying the user's face and further searching it in photographs and video files in other profiles, an expansion of the list of basic data available to the analyst is achieved. The problems associated with the recognition and identification of the person in the photo and video are well known and not fully resolved. The greatest difficulty here is the problem of scene variation, since the angle and location of the faces in the photo and video cannot be known in advance. The article presents the results of solving the problem of scene variation, presented in the form of a model of an information system using bionspirited methods. This system and the method incorporated in it are intended for preliminary image processing in order to broaden the range of permissible scene variations.

Keywords: Pattern recognition · Object identification · Photometry ·
Intellectual system · Social networks · Analysis

1 Introduction

The analysis of social networks is one of the promising areas of research [1]. In addition to the development of a mathematical apparatus that allows the processing of extracted data, it is of interest to identify new characteristics of user interaction. One of the types of such characteristics can be the presence of shared photos and videos among users. Obtaining information about the interaction of users through photo and video analysis is based on pattern recognition technologies.

The stack of methods developed in the history of the development of image recognition [2, 3] gave a serious impetus to conducting research aimed at improving the quality and (or) speed of methods. Today, new methods for recognizing a person's face are being created [4], methods for its segmentation [5], methods for extracting attribute

© Springer Nature Switzerland AG 2019
R. Silhavy (Ed.): CSOC 2019, AISC 985, pp. 147–155, 2019.
https://doi.org/10.1007/978-3-030-19810-7_15

models and methods for identification. Newly created methods for units or fractions of a percent improve the quality and speed of recognition, but they are recognized to fight the consequence of the main problem in recognizing and identifying a person from images. This main problem lies in the unstable parameters of the scene, primarily the light and shooting angle. Developers of new methods provide some independence of recognition quality within the specified limits of scene variation [6–8]. It is for this reason that today there is no universal solution to the problem of recognition and identification of the person. However, it is worth noting that the latest results achieved mainly through the use of neural networks, as close as possible to human capabilities.

The problems of using neural networks are well known and are connected, first of all, with the procedure of their learning [9]. This makes the use of such solutions in the tasks of processing photos and videos from social networks, in particular when it is necessary to identify a previously unknown image. Alternative methods based on logical conclusions and mathematical models do not have such problems as neural networks, but are more prone to scene variations.

Managing scene parameters in the case of social network analysis is impossible because of the infinite number of variations of angles, lighting and other parameters. However, it is possible to hypothesize that adequate adjustment of image parameters may expand the limits of the work of person recognition and identification methods using images.

Such expressions as white balance, exposure, sharpness, contrast, etc., are responsible for the expressiveness of the image for the recognition and identification system. The selection of the optimal values of these parameters is a complex multi-criteria task that should be solved within a time comparable period. This task can be solved by intelligent bioinspired methods, in particular, swarm intelligence. The article proposes a model of an intellectual pattern recognition system based on bioinspired methods. The main feature of this system is the expansion of permissible for the method of boundary parameters of scene variation while maintaining the quality of recognition and identification.

2 Related Work

The problem of scene variation today is subdivided into two subcategories — a variation of the lighting and a variation of the pose. The problem of posture is beyond the scope of this article, so it will not be considered. In recent years, several approaches have been proposed to solve the problem of variable illumination in face recognition tasks [6–8, 10–16]. By their nature, these approaches can be divided into three main categories: face modeling, normalization and pre-processing of images and extraction of invariant properties.

Face modeling approaches use low-dimensional linear subspaces to simulate the variations of human faces in different lighting conditions. As an example, the method of conical lighting [11] uses the fact that a set of images of an object in a fixed pose, but under all possible lighting conditions, forms a convex cone in the image space. The lighting cones of human faces can be approximated by low-dimensional linear subspaces, the basis vectors of which are estimated using a generative model.

The generative model is constructed using a small number of training images of each person, taken using different directions of illumination. The recognition algorithm assigns the identity of the nearest light cone to the test image. Similar is the method of spherical harmonics [17], which analyzes the subspace that best approximates the image of a convex Lambert object taken from the same point, but at different distances of the light source. This algorithm uses the principal component method (PCA), due to which a low-dimensional representation of the lighting cones is obtained. One of the main drawbacks of the face modeling approach is the need for images with all types of light to build linear subspaces. This disadvantage significantly limits the practical application of this method. Despite the presence of work on the organization of optimal lighting [18], it is not possible to completely eliminate the need for the availability of source images. Also, the fact that human faces are not ideal Lambert surfaces is not taken into account [7].

Approaches to preprocessing and normalizing images do not require preparation and preliminary modeling, as well as knowledge of facial models. In [14], a private image method (QI) is proposed, which is an applied algorithm for extracting an image with normalized illumination. QI is defined as the ratio between the test image and the linear combinations of three coplanar unlit images, depending only on Albedo information, which by itself does not take into account the illumination. In [19], an image self-assessment (SQI) is introduced, which corresponds to the relationship between the test image and the smoothed version. The main advantage of SQI over QI is that only one image is required to extract the internal properties of the light. Using only one image also means that image alignment is not required. In addition, SQI works correctly in shadow regions. In [20], by analogy, a preprocessing algorithm was proposed, which consists in extracting Albedo information using the relation between the test image and the second image, which corresponds to the estimated illumination of the field. This last image is obtained directly from the test image, and, as in the SQI algorithm, image alignment is not required, and this method works correctly in shadow areas. A related approach was proposed in [6], in which the pixel intensity is normalized locally (zero mean and unit deviation) using a 7×7 window centered on a pixel. The method is based on two assumptions: (1) in a small area of the image W, the normal direction of the surface can be considered constant; (2) the light source is directional, and as follows from (1) and (2) it is conditionally constant in the region of W. Another interesting algorithm is the so-called compensation algorithm for the plane gradient or the light gradient [21], which consists in calculating the best brightness plan for the analyzed image and then exclude this plan from the image. This method allows you to compensate for heavy shadows caused by extreme light angles. In addition to the specific algorithms mentioned, general-purpose preprocessing algorithms, such as image histogram alignment, gamma correction, and logarithmic transformations, are known [8, 22, 23]. This class can include heuristic approaches based on the hypothesis that the exclusion of three of the most significant components from consideration reduces the effect of illumination changes on the quality of recognition, which was experimentally confirmed in [24]. However, in order to ensure acceptable system performance for both normally illuminated images and images with variation in illumination, it is necessary to assume that the first three main components fix variations in illumination.

The third category of methods is aimed at extracting properties that are invariant to lighting and recognition based on them. An example of such a method was proposed in [10], where edge maps, Gabor-like filters and image intensity derivatives are used as signs that are invariant to light. The main conclusion of this and similar studies is that such representations of the image are not sufficient in themselves to overcome the problem of the directionality of the light source. To solve this problem, a cosine transform (DCT) has been proposed to compensate for changes in lighting in the logarithmic domain [9]. Taking into account that changes in illumination mainly lie in the low frequency band, the main idea is to discard the DCT coefficients corresponding to the low level frequencies.

Each category of methods considered has its own advantages and disadvantages. From the point of view of research on the problem of changing the scene when working with outdoor cameras, general-purpose preprocessing methods are of most interest. The advantages of these processing methods consist in the possibility of using any applied methods of recognition and identification. Implementations known today focus on one type of transformation, since the combination of algorithms is a computationally complex task and requires new approaches and methods that are described in this paper.

3 Intellectual Information System of Pattern Recognition Based on Bioinspired Methods

The system is based on mathematical methods of recognition and identification of the individual. The detection of faces by the system is performed using the Viola-Jones method [25]. Classification is performed by the nearest neighbor method using histograms of centrally symmetric local binary patterns [26] as signs of classification. The overall algorithm of the system is shown in Fig. 1.

As can be seen from the figure, the main part of the system is based on well-known and well-established methods. For the intellectualization of the system meets a special class that performs image adjustment in accordance with the current situation. The structure of the system is represented by the following class diagram (Fig. 2).

The RecognizedFace class describes a person recognized by the classifier. Elements of this class are stored in an updated list of recognized persons, on the basis of which data is displayed on the form. The LBPTransformer class describes the CS-LBP converter. This class is a static class, has no attributes and contains only one method that performs central-symmetric LBP image transformation. The FaceDetector class describes a face detector using the Viola-Jones method in its work. The class FaceClass describes the category of persons. Each class of person corresponds to a particular person recognized. The MainWindow class describes the main application window and also implements the main application module.

The main difference from the existing pattern recognition systems is the class IntelligentImageAdjuster, which implements the principle of continuous adjustment of the parameters of the image being formed so that scene changes caused by a change in lighting do not significantly affect the ability to distinguish objects and extract their properties. Separate algorithms for adjusting the gamma, curve correction and the

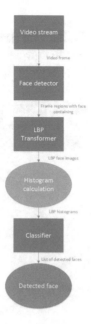

Fig. 1. Data processing algorithm.

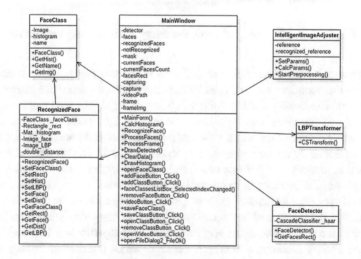

Fig. 2. Class diagram.

implementation of algorithmic transformations do not give a significant effect in this case. To adjust the image, a complex application of these algorithms is required, which in the general case is a complex multi-criteria task, which consists in finding a combination of input parameters of the image preprocessing algorithms that give the maximum effect in recognition. This task can be effectively solved by modern bioin-spired methods, such as, for example, the wolf pack method [27].

Evaluation of the ability of the system to recognize the current image parameters is carried out according to the following algorithm (Fig. 3.)

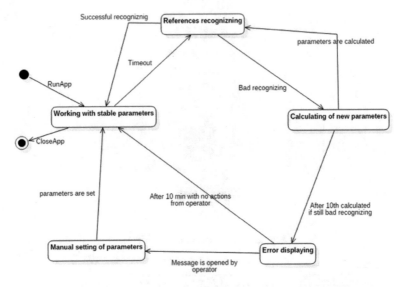

Fig. 3. Algorithm for determining the ability of the system to recognize.

The essence of the algorithm is as follows: in various places of the scene, corresponding to the boundary areas of illumination, are placed the reference images, known to the system. The locations and number of standards are determined empirically, so as to take into account variations in illumination caused by weather conditions, time of day and seasonality. At the same time, infrastructure objects and objects of natural origin that are already on the scene can be used as standards. Pattern recognition occurs at certain time intervals, and if one or several patterns were not detected as a result of the recognition, the intellectual subsystem is started, which is responsible for finding the optimal input parameters for image preprocessing algorithms. Its objective function is to minimize the number of erroneous recognition of standards located on the stage.

To adjust the image, a complex application of these algorithms is required, which in the general case is a complex multi-criteria task, which consists in finding a combination of input parameters of the image preprocessing algorithms that give the maximum effect in recognition. This task can be effectively solved by modern bioinspired methods, such as, for example, the wolf pack method [27].

4   Testing Results

The first part of the test was to choose the optimal classification method in terms of the ratio of speed and accuracy. To solve this problem, LBP histogram calculation algorithms were tested on two different data sets.

Testing was performed on the basis of facial images of the laboratory at the University of Cambridge and the database of facial images of the laboratory at Yale University. The test results showed that the classic LBP and Uniform LBP work with approximately the same accuracy. To achieve an accuracy of 90% or more, it is advisable to use image splits ranging from 4 × 4. However, it is worth noting that on the first data set, a 3 × 3 split showed itself best.

Centrally symmetric LBP when splitting an image into a small number of blocks is inferior to other local binary patterns. But when a larger number of subdomains are used, its classification accuracy rate is less than 3% from other LBPs on average. When testing on the first set of data, CS-LBP completely surpasses other patterns on a number of breaks. As a result, it can be said that centrally symmetric local binary patterns should be used in the facial recognition system in video flow, due to the high speed of work and accuracy indicators, which are almost the same as other LBPs. Comparison of the speed of the recognition system when using all three variations of LBP will be given in the section devoted to testing the application.

The best in terms of accuracy and memory cost of splitting an image into subdomains when using CS-LBP is a 4 × 4 partition, which will be used in the system being developed. This partition provides a consistently high percentage of correct classifications with not very large memory costs.

The second part of testing implied checking the speed and quality of recognition of the system itself. The video stream was captured under the same lighting conditions as the images of the training sample were captured. At the same time, different positions of the face were used when shooting to increase the percentage of correct classifications when moving people in the frame. During testing, 20 classes of individuals were loaded into the program. To form each class, 10 images were used. Of these 20 classes of individuals, 15 were created using images of people from the Internet. These classes of persons were necessary in order to evaluate the quality of the classifier's work on a large data file, as well as to test the work of the program when reading a video stream from a file.

The test results showed that the accuracy of the classifier when processing a video stream is more than 90% of correctly recognized frames. However, it is worth noting the sensitivity of the developed system to strong non-monotonous changes in lighting, as well as to changes in the position and tilt of recognizable faces that were not taken into account when taking images for shaping classes of faces. If the necessary conditions for the correct operation of the application are not observed, the classification accuracy is noticeably reduced.

5 Conclusions

The model of an intellectual information system of pattern recognition proposed in the article combines the advantages of existing approaches to the identification and identification of individuals from digital images, and also contains intelligent components that allow you to expand the possibilities of using existing methods of recognition in a changing scene. The use of bioinspired methods will allow in the mode close to real time to perform a reconfiguration of the parameters of the input image in such a way as to ensure the highest possible level of quality and speed of recognition. The resulting overhead of repeated cyclical recognition of standards is not comparable with the potential benefits of applying the proposed solution. Further studies suggest practical testing of existing bioinspired methods and algorithms, in particular the algorithm of a pack of wolves, the determination of the values of heuristic coefficients.

Acknowledgment. This work was supported by the grant of Southern Federal University, research project No. 07/2017-28.

References

1. Agiyeva, M.T.: Zadachi analiza na sotsial'nykh setyakh v marketinge. Inzhenernyj vestnik Dona (RUS) (2018). http://ivdon.ru/ru/magazine/archive/n2y2018/4889
2. Dergachev, V.V., Aleksandrov, A.A.: Metody analiza i strukturirovannogo raspoznavaniya lits v estestvennykh usloviyakh. Inzhenernyj vestnik Dona (RUS) (2017). ivdon.ru/ru/magazine/archive/n4y2017/4549
3. Zhao, W., Chellappa, R., Rosenfeld, A., Phillips, P.J.: Face recognition: a literature survey. ACM Comput. Surv. **35**, 399–458 (2003)
4. Sharif, M., Naz, F., Yasmin, M., Alyas, M., Rehman, A.: Face recognition: a survey. J. Eng. Sci. Technol. Rev. **10**, 166–177 (2017)
5. El-Khatib, S., Rodzin, S., Skobtcov, Y.: Investigation of optimal heuristical parameters for mixed ACO-k-means segmentation algorithm for MRI images. In: 2016 Conference on Information Technologies in Science, Management, Social Sphere and Medicine (issue 51), Part of the series ACSR, pp. 216–221 (2016)
6. Shin, D., Lee, H.S., Kim, D.: Illumination-robust face recognition using ridge regressive bilinear models. Pattern Recogn. Lett. **29**, 49–58 (2008)
7. Chen, W., Joo Er, M., Wu, S.: llumination compensation and normalization for robust face recognition using discrete cosine transform in logarithmic domain. IEEE Trans. Syst. Man Cybern. Syst. **36**, 458–466 (2006)
8. Ruiz-del-Solar, J., Navarrete, P.: Eigenspace-based face recognition: a comparative study of different approaches. IEEE Trans. Syst. Man Cybern. Syst. **35**, 315–325 (2005)
9. Gu, J., Wang, Z., Kuen, J., Ma, L., Shahroudy, A.: Recent advances in convolutional neural networks. Pattern Recogn. **77**, 354–377 (2017)
10. Adini, Y., Moses, Y., Ullman, S.: Face recognition: the problem of compensating for changes in illumination direction. IEEE Trans. Pattern Anal. Mach. Intell. **19**(7), 721–732 (1997)
11. Belhumeur, P., Kriegman, D.: What is the set of images of an object under all possible illumination conditions. Int. J. Comput. Vis. **28**(3), 245–260 (1998)

12. Chen, H., Belhumeur, P., Kriegman, D.: In search of illumination invariants. In: Proceedings of IEEE Conference on Computer Vision and Pattern Recognition, vol. 1, pp. 13–15 (2000)
13. Georghiades, A., Belhumeur, P., Kriegman, D.: From few to many: Illumination cone models for face recognition under variable lighting and pose. IEEE Trans. Pattern Anal. Mach. Intell. **23**(6), 643–660 (2001)
14. Shashua, A., Riklin-Raviv, T.: The quotient image: Class based re-rendering and recognition with varying illuminations. IEEE Trans. Pattern Anal. Mach. Intell. **23**(2), 129–139 (2001)
15. Xie, X., Lam, K.-M.: An efficient illumination normalization method for face recognition. Pattern Recogn. Lett. **27**, 609–617 (2006)
16. Zhang, L., Wang, S., Samaras, D.: Face synthesis and recognition from a single image under arbitrary unknown lighting using a spherical harmonic basis Morphable model. In: Proceedings of International Conference on Computer Vision and Pattern Recognition - CVPR 2005, 20–25 June San Diego, USA, vol. 2, pp. 209–216 (2005)
17. Ramamoorthi, R.: Analytic PCA construction for theoretical analysis of lighting variability in images of a Lambertian object. IEEE Trans. Pattern Anal. Mach. Intell. **24**(10), 1322–1333 (2002)
18. Lee, K.-C., Ho, J., Kriegman, D.: Acquiring linear subspaces for face recognition under variable lighting. IEEE Trans. Pattern Anal. Mach. Intell. **27**(5), 684–698 (2005)
19. Wang, H., Li, S., Wang, Y.: Face recognition under varying lighting conditions using self quotient image. In: Proceedings of 6th International Conference on Face and Gesture Recognition - FG 2004, Seoul, Korea, May 2004, pp. 819–824 (2004)
20. Gross, R., Baker, S., Matthews, I., Kanade, T.: Face recognition across pose and illumination. In: Li, S.Z., Jain, A.K. (eds.) Handbook of Face Recognition, Chap. 9. Springer (2004)
21. Sung, K., Poggio, T.: Example-based learning for viewed-based human face detection. IEEE Trans. Pattern Anal. Mach. Intell. **20**(1), 39–51 (1998)
22. Pizer, S., Amburn, E., Austin, J., et al.: Adaptive histogram equalization and its variations. Comput. Vis. Graph. Image Process. **39**(3), 355–368 (1987)
23. Shan, S., Gao, W., Cao, B., Zhao, D.: Illumination normalization for robust face recognition against varying lighting conditions. In: Proceedings IEEE Workshop on AMFG, pp. 157–164 (2003)
24. Belhumeur, P.N., Hespanha, J.P., Kriegman, D.J.: Eigenfaces vs. Fisherfaces: recognition using class specific linear projection. IEEE Trans. PAMI **19**(7), 711–720 (1997)
25. Viola, P., Jones, M.: Rapid object detection using a boosted cascade of simple features. In: IEEE Computer Society Conference on Computer Vision and Pattern Recognition (issue 1), pp. 511–518. The Institute of Electrical and Electronics Engineers, Inc (2001)
26. Ahonen, T., Hadid, A., Pietikainen, M.: Rapid object detection using a boosted cascade of simple features. IEEE Trans. Pattern Anal. Mach. Intell. **28**, 2037–2041 (1996)
27. Mirjalili, S., Lewis, A.: Grey wolf optimizer. Adv. Eng. Softw. **69**, 46–61 (2014)

A Binary Ant Lion Optimisation Algorithm Applied to the Set Covering Problem

Lorena Jorquera, Pamela Valenzuela, Matías Valenzuela[✉], and Hernan Pinto

School of Engineering, Pontificia Universidad Católica de Valparaíso,
Valparaíso, Chile
{lorena.jorquera,pamela.valenzuela,matias.valenzuela,
hernan.pinto}@pucv.cl

Abstract. The study and understanding of algorithms that solve combinatorial problems based on swarm intelligence continuous metaheuristics, is an area of interest at the level of basic and applied science. This is due to the fact that many of the problems addressed at industrial level are of a combinatorial type and a subset no less than these are of the NP-hard type. In this article, a mechanism of binarization of continuous metaheuristics that uses the concept of the percentile is proposed. This percentile concept is applied to the An Lion optimization algorithm, solving the set covering problem (SCP). Experiments were designed to demonstrate the importance of the percentile concept in the binarization process. Subsequently, the efficiency of the algorithm is verified through reference instances. The results indicate that the binary Ant Lion Algorithm (BALO) obtains adequate results when evaluated with a combinatorial problem such as the SCP.

1 Introduction

In recent years, there is a research line of algorithms inspired by natural phenomena focused on solving optimization problems. As examples of these algorithms we have Cuckoo Search [1], Black Hole [2], Bat Algorithm [3] and Sine Cosine Algorithm [4] among others. Due to the type of phenomena in which these algorithms are inspired, a large part of these algorithms operate naturally in continuous spaces. On the other hand, the combinatorial problems have a great relevance at the scientific and industrial level. We find them in different areas, such as Civil Engineering [5], Bio-Informatics [6], Operational Research [7–9], Big Data [10], resource allocation [11,12], programming problems [13,14], routing problems [15–17] among others. If we want to apply continuous algorithms to combinatorial problems, these algorithms must adapt. In the process of adaptation, the mechanisms of exploration and exploitation of the algorithm can be altered, having consequences in the efficiency of the algorithm.

Several binarization techniques have been developed to address this situation. In a literature search, the main binarization methods used correspond to transfer

© Springer Nature Switzerland AG 2019
R. Silhavy (Ed.): CSOC 2019, AISC 985, pp. 156–167, 2019.
https://doi.org/10.1007/978-3-030-19810-7_16

functions, angle modulation and quantum approach. In this article we present a binarization method that uses the percentile concept to group the solutions and then perform the binarization process. To verify the efficiency of our method, we used the BALO. This algorithm was proposed in [18] and was applied to test functions in addition to structural engineering problems.

To check our binary percentile ant lion optimisation algorithm (BALO), we use the well-known set covering problem. Experiments were developed using a random operator to validate the contribution of the percentile technique in the binarization process of the sine-cosine algorithm. In addition, to verify our results, the JPSO algorithm developed by [19] and MDBBH developed in [13] were chosen. The results show that BALO algorithm obtain competitive results.

2   Set Covering Problem

The set covering problem corresponds to a classical problem in combinatorics, and complexity theory. The problem aims to find subsets of a set. Given a set and its elements, we want to find subsets that completely cover the set at a minimum cost. The SCP is one of the oldest and most studied optimization problems. It is well-known to be NP-hard [20].

SCP is an active problem because medium and large instances often become intractable and cannot be solved any more using exact algorithms. Additionally, SCP due to its large number of instances, is used to verify the behavior of proposed new algorithms. In recent years, SCP has been approached by various continuous Swarm intelligence metaheuristics and using different methods to binarize metaheuristics. In [21], they used the teaching-learning-based optimisation with a specific binarization scheme to solve medium and large size instances. An SCP in fault diagnosis application was developed by [22]. In this article, the Gravitational Search algorithm was used and transfer functions were applied to perform the binarization. A rail scheduling problem application using SCP was developed in [13]. In [19] a Jumping PSO was used to solve SCP and in [11] CS metaheuristics were binarized using a percentile algorithm.

SCP has many practical applications in engineer, e.g., vehicle routing, facility location, railway, and airline crew scheduling [7, 23–25] problems.

The SCP can be formally defined as follows. Let $A = (a_{ij})$, be a $n \times m$ zero-one matrix, where a column j cover a row i if $a_{ij} = 1$, besides a column j is associated with a non-negative real cost c_j. Let $I = \{1, ..., n\}$ and $J = \{1, ...m\}$, be the row and column set of A, respectively. The SCP consists in searching a minimum cost subset $S \subset J$ for which every row $i \in I$ is covered by at least one column $j \in J$, i.e,:

$$\text{Minimize } f(x) = \sum_{j=1}^{m} c_j x_j \tag{1}$$

$$\text{Subject to } \sum_{j=1}^{m} a_{ij} x_j \geq 1, \forall i \in I, \text{ and } x_j \in \{0, 1\}, \forall j \in J \tag{2}$$

where $x_j = 1$ if $j \in S$, $x_j = 0$ otherwise.

3   Ant Lion Optimizer

ALO is a stochastic algorithm that is inspired by the hunting mechanism of Antlions with the goal of finding the best solutions associated with a complex problem. In this algorithm the ants are used as search mechanisms and the Antlions disturb the movement of the ant to make it more efficient in the search for solutions. The main concepts used by ALO are random walks, building tramps, trapping ants and rebuilding traps.

In the rest of the section, we detail the mathematical model that supports ALO: To model the movement of the ant, the random walk given in Eq. 3 is used.

$$X_i = [0; r(1); r(1) + r(2); ...; \sum_{j=1}^{T-1} r(j); \sum_{j=1}^{T} r(j)] \tag{3}$$

where $i = \{1, ..., dim\}$ and dim is the dimension of the search space. T is the maximum number of iterations. r is a stochastic function given by Eq. 4:

$$r = \begin{cases} 1, & \text{if rand } > 0.5 \\ -1, & \text{if rand } \leq 0.5 \end{cases} \tag{4}$$

This random walk which takes values in $\mathbb{Z}$ must be adapted to work in $\mathbb{R}$. This adaptation is executed using Equation:

$$Y_i = (\frac{x_i - a_i}{b_i - a_i}) \times (d_i - c_i) + c_i \tag{5}$$

Where a_i and b_i is the minimum and maximum respectively of X_i. On the other hand, c_i and d_i represent the minimum and maximum of Antlion trap in the ith dimension respectively. For calculating c_i y d_i in the iteration t, we use the Eqs. 6 and 7. Where $c^t = \frac{l_b}{10^w \frac{t}{T}}$ and $d^t = \frac{u_b}{10^w \frac{t}{T}}$ ($w = 2$ when $t > 0.1T$, $w = 3$ when $t > 0.5T$, $w = 4$ when $t > 0.75T$, $w = 5$ when $t > 0.9T$, and $w = 6$ when $t > 0.95T$)

$$c_i^t = AntLion_i^t + c^t \tag{6}$$

$$d_i^t = AntLion_i^t + d^t \tag{7}$$

For the update of the ants, an elitist scheme given in Equation is used 8. the antlions update occurs as follows: $Antlion = Ant$ if $f(Ant)$ better than $f(Antlion)$.

$$Ant = \frac{R_A + R_E}{2} \tag{8}$$

4   Binary Ant Lion Optimizer Algorithm

The application of statistics and machine learning techniques, are used in many areas such as transports, smart cities, agriculture and computational intelligence, examples of applications are found in [26–29, 29, 30]. In this article, we explore the percentile concept applied to binarization of continuous metaheuristics. As an algorithm to binarize, Ant Lion is used and as problem to solve, SCP. The Proposed BALO has two modules. The first module corresponds to the initialization of the feasible solutions, and is detailed in the Sect. 4.1. Once the initialization of the particles is performed, it is consulted if the maximum if iterations is satisfied. Subsequently if the criterion is not satisfied, The BALO is then run in conjunction with the percentile operator, this is detailed in Sect. 4.2). Once the transitions of the different solutions are made, we compare the resulting solutions with the best solution previously obtained. In the event that a superior solution is found, this replaces the previous one. The general algorithm scheme is detailed in Fig. 1.

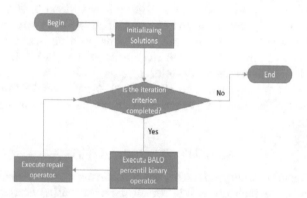

Fig. 1. Flowchart of the BALO algorithm.

4.1   Initialization

For initialization of a new solution, a column is randomly chosen. It is then queried whether the current solution covers all rows. In the case of the solution does not meet the coverage condition, the heuristic operator is called (Sect. 4.3). This operator aims to select a new column. This heuristic operation is iterated until all rows are covered. Once the coverage condition is met, the solution is optimized. The optimization consists of eliminating columns where all their rows are covered by more than one column. The detail of the procedure is shown in Algorithm 1.

Algorithm 1. Initialization Operator

1: **Function** Initialization()
2: **Input**
3: **Output** Initialized solution S_{out}
4: $S \leftarrow$ SRandomColumn()
5: **while** All row are not covered **do**
6:   S.append(Heuristic(S))
7: **end while**
8: $S \leftarrow$ dRepeatedItem(S)
9: $S_{out} \leftarrow S$
10: **return** S_{out}

4.2   Percentile Binary Operator

Since our ALO algorithm is continuous swarm intelligence metaheuristic, it works in an iterative way by updating the position and velocity of the particles in each iteration. As ALO is continuous, the update is done in $\mathbb{R}^n$. In Eq. 9, the position update is written in a general way. The $x(t + 1)$ variable represents the x position of the particle at time $t+1$. This position is obtained from the position x at time t plus a Δ function calculated at time $t+1$. The function Δ is proper to each metaheuristic and produces values in $\mathbb{R}^n$. For example in Cuckoo Search $\Delta(x) = \alpha \oplus Levy(\lambda)(x)$, in Black Hole $\Delta(x) = $ rand $\times (x_{bh}(t) - x(t))$ and in the Firefly, Bat and PSO algorithms Δ can be written in simplified form as $\Delta(x) = v(x)$.

$$x(t + 1) = x(t) + \Delta_{t+1}(x(t)) \tag{9}$$

In the percentile binary operator, we considering the movements generated by the algorithm in each dimension for all particles. $\Delta^i(x)$ corresponds to the magnitude of the displacement $\Delta(x)$ in the i-th position for the particle x. Subsequently these displacement are grouped using $\Delta^i(x)$, the magnitude of the displacement. This grouping is done using the percentile list. In our case the percentile list used the values {20, 40, 60, 80, 100}.

The percentile operator has as entry the parameters percentile list (pList) and the list of values (vList). Given an iteration, the list of values corresponds to the magnitude $\Delta^i(x)$ of the displacements of the particles in each dimension. As a first step the operator uses the vList and obtains the values of the percentiles given in the pList. Later, each value in the vList is assigned the group of the smallest percentile to which the value belongs. Finally, the list of the percentile to which each value belongs is returned (pGroupValue). The algorithm is shown in Algorithm 2.

A transition probability through the function P_{tr} is assigned to each element of the vList. This assignment is done using the percentile group assigned to each value. For the case of this study, we particularly use the Step function given in Eq. 10.

$$P_{tr}(x^i) = \begin{cases} 0.1, & \text{if } x^i \in \text{group } \{0,1\} \\ 0.5, & \text{if } x^i \in \text{group } \{2,3,4\} \end{cases} \tag{10}$$

Afterwards the transition of each particle is performed. In the case of ALO search the Eq. 11 is used to perform the transition, where $\hat{x}^i$ is the complement of x^i. Finally, each solution is repaired using the heuristic operator.

$$x^i(t+1) := \begin{cases} \hat{x}^i(t), & \text{if } rand < P_{tg}(x^i) \\ x^i(t), & \text{otherwise} \end{cases} \tag{11}$$

Algorithm 2. percentile binary operator

1: **Function** percentileBinary(vList, pList)
2: **Input** vList, pList
3: **Output** pGroupValue
4: percentileValue = getPValue(vList, pList)
5: **for each** value in vList **do**
6:    pGroupValue = getPGroupValue(pValue,vList)
7: **end for**
8: **return** pGroupValue

4.3   Heuristic Operator

The goal of the Heuristic operator is to select a new column for cases where a solution needs to be built or repaired. As input variables, the heuristic operator considers the incomplete solution S_{in} which must be completed. The operator obtains the columns that belong to S_in, then obtains the rows R that are not covered by the solution to S_in. With the set of rows not covered and using Eq. 12 we obtain in line 4 the best 5 rows to be covered. With this list of rows ($lRows$) on line 5 we obtain the list of the best columns according to the heuristic indicated in Eq. 13. Finally randomly in line 6 we obtain the column to incorporate.

$$WeightRow(i) = \frac{1}{L_i}. \tag{12}$$

Where L_i is the sum of all ones in row i

$$WeightColumn(j) = \frac{c_j}{|R \cap M_j|}. \tag{13}$$

Where M_j is the set of rows covered by Col j

Algorithm 3. Heuristic operator

1: **Function** Heuristic(S_{in})
2: **Input** Input solution S_{in}
3: **Output** The new column *colOut*
4: *lRows* ← getBestRows(S_{in}, N=10)
5: *listcolumsnOut* ← getBestCols(*LRows*, M=5)
6: *colOut* ← getCol(*lcolsOut*)
7: **return** *colOut*

5   Results

In this section we detail the behavior of BALO when it is applied to SCP. The contribution of the binary percentile operator was studied when solving the different SCP instances. Additionally, BALO was compared with other algorithms that have recently resolved SCP. To solve the different SCP instances, a PC with Windows 10, core i5 processor and 16 GB in RAM was used. The program was coded in Python 3.5. For the statistical analysis, the non-parametric Wilcoxon signed-rank test was used in addition to violin charts. The analysis of violin charts is performed by comparing the dispersion, median and the interquartile range of the distributions.

5.1   Insight of BALO Algorithm

In this module we developed experiments that allowed us to study the contribution of the binary percentile operator with respect to the quality of the solutions obtained and the iterations used when solving SCP instances. To perform this evaluation, we used the instances of Balas and Carrera. In the comparison, a random operator was designed which replaces our binary percentile operator. This random operator instead of executing transitions assigning transition probabilities per group, uses a fixed value of 0.5. To compare the distributions of the results of the different experiments we use violin Chart. The horizontal axis X corresponds to the problems, while Y axis uses the measure % - Gap defined in equation 14

$$\% - Gap = 100 \frac{BestKnown - SolutionValue}{BestKnown} \tag{14}$$

Furthermore, a non-parametric test, Wilcoxon signed-rank test is carried out to determine if the results of BALO with respect to the random algorithm have significant difference or not. The parameter settings and browser ranges are shown in Table 1.

BALO corresponds to our standard algorithm. *random.0.5* is the random variant. The results are shown in Table 2 and Fig. 2. When we compared the Best Values between BALO and *rnd.0.5* which are shown in Table 2. In the Average comparison, BALO outperforms *rnd.0.5* in all problems. The comparison of distributions is shown in Fig. 2. We see the dispersion of the *rnd.0.5*

Table 1. Setting of parameters for binary ant lion optimization search algorithm.

Parameters	Description	Value	Range
N	Number of solutions	25	[20, 25, 30]
G	Number of percentiles	5	[4,5,6]
Iteration Number	Maximum iterations	800	[700,800,900]

distributions are bigger than the dispersions of BALO. Therefore, the percentile binary operator, contribute to the precision of the results. Finally, the BALO distributions are closer to zero than *rnd.0.5* distributions, indicating that BALO has consistently better results than *random.0.5*. When we evaluate the behaviour of the algorithms through the Wilcoxon test, this test indicates that there is a significant difference between the two algorithms.

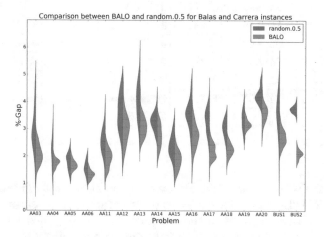

Fig. 2. Evaluation of percentile binary operator.

5.2   BALO Compared with JPSO and MDBBH

In this section, we detail the comparisons made with the objective of evaluating the performance of our BALO algorithm. For the evaluation, the larger problems of the OR-library were chosen. To develop the comparison, two algorithms were selected. The first one corresponds to a discretization Particle Swarm Optimization (PSO) technique called Jumping PSO (JPSO), [19]. The second one, is a binarization of the Black Hole technique called Multi Dynamic Binary Black Hole (MDBBH) Algorithm, [13]. JPSO uses a discrete PSO based on the frog jump. JPSO works without the concept of velocity, replacing this one by a component of random jump which allows to perform the movement in the discrete search

Table 2. Balas and carrera instances.

Instance	row	col	Density	Best Known	BALO (avg)	std	(Secs) (Secs)	rnd.0.5 (avg)	std std
AA03	106	8661	4.05%	33155	**33521.4**	124.2	207	33693.5	241.1
AA04	106	8002	4.05%	34573	**34835.2**	58.1	198	34931.0	231.2
AA05	105	7435	4.05%	31623	**31816.5**	64.8	201	31889.2	103.2
AA06	105	6951	4.11%	37464	**37577.1**	56.1	199	37651.3	97.1
AA11	271	4413	2.53%	35384	**35723.1**	131.9	187	35901.2	197.1
AA12	272	4208	2.52%	30809	**31417.4**	152.8	208	31631.3	210.2
AA13	265	4025	2.60%	33211	**33901.8**	135.1	189	34201.4	231.4
AA14	266	3868	2.50%	33219	**33787.3**	125.1	143	33967.3	121.1
AA15	267	3701	2.58%	34409	**34725.2**	112.8	141	34846.8	161.3
AA16	265	3558	2.63%	32752	**33309.6**	137.5	187	33601.3	183.9
AA17	264	3425	2.61%	31612	**31996.9**	89.1	162	32386.2	149.3
AA18	271	3314	2.55%	36782	**37233.4**	78.1	132	37417.6	148.1
AA19	263	3202	2.63%	32317	**33007.6**	77.6	146	33109.1	136.1
AA20	269	3095	2.58%	34912	**35794.1**	91.1	141	35963.1	97.8
BUS1	454	2241	1.88%	27947	**28404.6**	86.0	189	28756.1	231.6
BUS2	681	9524	0.51%	67760	**68495.3**	83.2	175	68678.9	101.2
Average				35495.56	35971.66	100.22	175.31	36164.08	165.11
p-value								1.75e-04	

space. On the other hand, MDBBH uses a binarization mechanism specific to the Black Hole algorithm. This is based on the concept of closeness to the Black Hole (BH). BH corresponds to the solution that has obtained the best value. When a solution is close to BH, the transitions are less likely to be performed. In the case of being away from the BH, the probability of transition is greater.

For instances E and F which are of medium size, the results of the MDBBH and BALO algorithms are similar when comparing their Best Value. Only in instance E.2 BALO was superior to MDBBH. When analyzing the average, JPSO obtained better results, however, the values of BALO were close, being MDBBH the one that obtained the worse performance. For the case of problems G and H, BALO was superior in problems G.1, G.2 and G.3 and MDBBH in problem H.3 when comparing their Best Value. In the case of averages, JPSO was superior in all cases. However, BALO obtained results quite close, leaving MDBBH behind. In this point we must emphasize that the percentile technique used in BALO allows binarizing any continuous swarm intelligence algorithm unlike JPSO that is specific for PSO.

Table 3. OR-Library benchmarks.

Instance	Row	Col	Density	Best Known	JPSO Avg	MDBBH Best	Avg	BALO Best	Avg	Time(s)
E.1	500	5000	10%	29	29.0	29	29.0	29	30.1	12.1
E.2	500	5000	10%	30	30.0	31	31.6	30	31.2	17.1
E.3	500	5000	10%	27	27.0	27	27.4	27	27.9	19.7
E.4	500	5000	10%	28	28.0	28	29.1	28	28.4	18.4
E.5	500	5000	10%	28	28.0	28	28.0	28	28.4	19.1
F.1	500	5000	20%	14	14.0	14	14.1	14	14.4	21.3
F.2	500	5000	20%	15	15.0	15	15.3	15	15.8	24.4
F.3	500	5000	20%	14	14.0	14	14.8	14	15.0	23.4
F.4	500	5000	20%	14	14.0	14	14.9	14	14.6	22.1
F.5	500	5000	20%	13	13.0	14	14.1	14	15.2	26.5
G.1	1000	10000	2%	176	176.0	177	178.5	177	179.2	203.5
G.2	1000	10000	2%	154	155.0	157	160.6	156	158.2	208.1
G.3	1000	10000	2%	166	167.2	168	170.4	168	171.1	235.7
G.4	1000	10000	2%	168	168.2	169	170.9	170	173.4	226.1
G.5	1000	10000	2%	168	168.0	168	169.8	168	170.2	245.8
H.1	1000	10000	5%	63	64.0	64	64.9	65	66.2	183.5
H.2	1000	10000	5%	63	63.0	64	64	65	66.6	189.3
H.3	1000	10000	5%	59	59.2	59	60	60	61.8	175.2
H.4	1000	10000	5%	58	58.3	59	60.4	59	61.4	177.3
H.5	1000	10000	5%	55	55.0	55	56.4	56	57.4	168.4
Average				67.1	67.30	67.7	68.71	67.85	69.32	110.85

6   Conclusions

In this work, the BALO algorithm based on the percentile concept was proposed to perform binarization of ALO metaheuristics. We must emphasize that this percentile concept can be applied in the binarization of any continuous metaheuristics of the swarm intelligence type. To evaluate the performance of Our percentile concept the set covering problem was used together with the ALO algorithm. Experiments were designed to evaluate the contribution of the binary operator percentile of the algorithm. It was found that the operator contributes significantly to improve the accuracy and quality of the solutions, all these visualized through violin graphics. Finally, compared to the best metaheuristic algorithm that has solved SCP, our algorithm had a lower yield of 1.74 %, which is not a big difference considering that JPSO uses a particular adaptation mechanism for PSO and the percentile concept It can be easily adapted to any technique. As future works, we believe that the autonomous search for efficient configuration is a little-explored area and it could help to improve the performance of the binarizations both in the quality of the solutions and in the times of convergence. Additionally, it is interesting to see how this percentile technique works with other metaheuristics and other NP-hard problems. Finally, we believe that using machine learning we can generate indicators or obtain correlations that allow us to understand how the percentile algorithm translates the exploration and exploitation properties from the continuous space to the binary space (Table 3).

References

1. Yang, X.-S., Deb, S.: Cuckoo search via lévy flights. In: World Congress on Nature & Biologically Inspired Computing, 2009. NaBIC 2009, pp. 210–214. IEEE (2009)
2. Hatamlou, A.: Black hole: a new heuristic optimization approach for data clustering. Inf. Sci. **222**, 175–184 (2013)
3. Yang, X.-S.: A new metaheuristic bat-inspired algorithm. In: Nature Inspired Cooperative Strategies for Optimization (NICSO 2010), pp. 65–74 (2010)
4. Mirjalili, S.: SCA: a sine cosine algorithm for solving optimization problems. Knowl. Based Syst. **96**, 120–133 (2016)
5. Khatibinia, M., Yazdani, H.: Accelerated multi-gravitational search algorithm for size optimization of truss structures. Swarm Evol. Comput. (2017)
6. Barman, S., Kwon, Y.-K.: A novel mutual information-based boolean network inference method from time-series gene expression data. PloS ONE **12**(2), e0171097 (2017)
7. Crawford, B., Soto, R., Monfroy, E., Astorga, G., García, J., Cortes, E.: A meta-optimization approach for covering problems in facility location. In: Workshop on Engineering Applications, pp. 565–578. Springer (2017)
8. Crawford, B., Soto, R., Astorga, G., García, J.: Constructive metaheuristics for the set covering problem. In: International Conference on Bioinspired Methods and Their Applications, pp. 88–99. Springer (2018)
9. García, J., Peña, A.: Robust optimization: concepts and applications. In: Nature-inspired Methods for Stochastic, Robust and Dynamic Optimization. IntechOpen (2018)
10. García, J., Altimiras, F., Peña, A., Astorga, G., Peredo, O.: A binary cuckoo search big data algorithm applied to large-scale crew scheduling problems. Complexity **2018** (2018)
11. García, J., Crawford, B., Soto, R., Astorga, G.: A percentile transition ranking algorithm applied to knapsack problem. In: Proceedings of the Computational Methods in Systems and Software, pp. 126–138. Springer (2017)
12. Astorga, G., Crawford, B., Soto, R., Monfroy, E., García, J., Cortes, E.: A meta-optimization approach to solve the set covering problem. Ingeniería, **233** (2018)
13. García, J., Crawford, B., Soto, R., Castro, C., Paredes, F.: A k-means binarization framework applied to multidimensional knapsack problem. Appl. Intell. **48**(2), 357–380 (2018). Springer
14. García, J., Crawford, B., Soto, R., Astorga, G.: A percentile transition ranking algorithm applied to binarization of continuous swarm intelligence metaheuristics. In: International Conference on Soft Computing and Data Mining, pp. 3–13. Springer (2018)
15. Franceschetti, A., Demir, E., Honhon, D., Van Woensel, T., Laporte, G., Stobbe, M.: A metaheuristic for the time-dependent pollution-routing problem. Eur. J. Oper. Res. **259**(3), 972–991 (2017)

16. Crawford, B., Soto, R., Astorga, G., García, J., Castro, C., Paredes, F.: Putting continuous metaheuristics to work in binary search spaces. Complexity **2017** (2017)
17. García, J., Crawford, B., Soto, R., Astorga, G.: A clustering algorithm applied to the binarization of swarm intelligence continuous metaheuristics. Swarm Evol. Comput. **44**, 646–664 (2019)
18. Mirjalili, S.: The ant lion optimizer. Adv. Eng. Softw. **83**, 80–98 (2015)
19. Balaji, S., Revathi, N.: A new approach for solving set covering problem using jumping particle swarm optimization method. Nat. Comput. **15**(3), 503–517 (2016)
20. Gary, M.R., Johnson, D.S.: Computers and intractability. A Guide to the Theory of NP-Completeness (1979)
21. Lu, Y., Vasko, F.J.: An or practitioner's solution approach for the set covering problem. Int. J. Appl. Metaheuristic Comput. (IJAMC) **6**(4), 1–13 (2015)
22. Li, Y., Cai, Z.: Gravity-based heuristic for set covering problems and its application in fault diagnosis. J. Syst. Eng. Electron. **23**(3), 391–398 (2012)
23. Kasirzadeh, A., Saddoune, M., Soumis, F.: Airline crew scheduling: models, algorithms, and data sets. EURO J. Transp. Logistics **6**(2), 111–137 (2017)
24. Horváth, M., Kis, T.: Computing strong lower and upper bounds for the integrated multiple-depot vehicle and crew scheduling problem with branch-and-price. Cent. Eur. J. Oper. Res. 1–29 (2017)
25. Stojković, M.: The operational flight and multi-crew scheduling problem. Yugoslav J. Oper. Res. **15**(1) (2016)
26. García, J., Crawford, B., Soto, R., Carlos, C., Paredes, F.: A k-means binarization framework applied to multidimensional knapsack problem. Appl. Intell. 1–24 (2017)
27. García, J., Pope, C., Altimiras, F.: A distributed k-means segmentation algorithm applied to Lobesia botrana recognition. Complexity **2017** (2017)
28. Graells-Garrido, E., García, J.: Visual exploration of urban dynamics using mobile data. In: International Conference on Ubiquitous Computing and Ambient Intelligence, pp. 480–491. Springer (2015)
29. Graells-Garrido, E., Peredo, O., García, J.: Sensing urban patterns with antenna mappings: the case of Santiago, Chile. Sensors **16**(7), 1098 (2016)
30. Peredo, O.F., García, J.A., Stuven, R., Ortiz, J.M.: Urban dynamic estimation using mobile phone logs and locally varying anisotropy. In: Geostatistics Valencia 2016, pp. 949–964. Springer (2017)

The Minimization of Empirical Risk Through Stochastic Gradient Descent with Momentum Algorithms

Arindam Chaudhuri$^{(\boxtimes)}$

Tata Consultancy Services, Think Campus, Bangalore 560100, India
arindam.chaudhuri3@tcs.com,
arindamphdthesis@gmail.com

Abstract. The learning problems are always affected with a certain amount of risk. This risk is measured empirically through various risk functions. The risk functional's empirical estimates consist of an average over data points' tuples. With this motivation in this work, the prima face is towards presenting any stochastic approximation method for solving problems involving minimization of risk. Considering huge datasets scenario, gradient estimates are achieved through taking samples of data points' tuples with replacement. Based on this, a mathematical proposition is presented here which account towards considerable impact for this strategy on prediction model's ability of generalization through stochastic gradient descent with momentum. The method reaches optimum trade-off with respect to accuracy and cost. The experimental results on maximization of area under the curve (AUC) and metric learning provides superior support towards this approach.

Keywords: Risk estimates · Minimization · Stochastic gradient descent · Huge datasets · AUC · Metric learning

1 Introduction

The statistical risk function with expectation for t-tuples where number of data points $t \geq 2$ plays a significant role in several machine learning scenarios. This is a representative case of supervised metric learning [1]. We try to optimize the distance function as a result of which smaller values are assigned towards point pairs with same label. Some examples worth mentioning are bipartite ranking [2], multi-partite ranking [3], pairwise clustering [4] etc. Considering a set of data points, the obvious empirical estimate for risk reaches through the average of all observation tuples [5]. It is generally of unbiased nature. Here risk estimate minimizes variance over all unbiased estimates. In past the empirical risk minimization principle has been extended towards the empirical risk of prediction rule [6]. Here the motivation is adopted from the unbiased statistics properties. These propositions do not hold for biased statistics. As the tuples grow in number, the empirical risk calculation becomes infeasible. The empirical risk function is minimized through several stochastic optimization methods. The most commonly used approach is stochastic gradient descent (SGD). Here for every step few

© Springer Nature Switzerland AG 2019
R. Silhavy (Ed.): CSOC 2019, AISC 985, pp. 168–181, 2019.
https://doi.org/10.1007/978-3-030-19810-7_17

terms are randomly towards gradient estimation [2, 7–9]. As such because of gradient estimation variance, the original SGD learning is paralyzed with slow degree of convergence [10]. Here the empirical risk function is calculated by summation of independent observations. This has led towards the growth for several variants of SGD where the reduction of variance improves the convergence factor. The variance reduction is generally achieved through exact gradient calculation [10–13] or by using non-uniform sampling schemes [14, 15]. As a result of the overwhelming possible data points, calculation for accurate gradient values or enforcing probability distribution for all data points is infeasible.

Here, empirical risk function with its structure and statistical properties are considered [16]. This helps it to take the shape of unbiased statistics through which effective SGD implementation is designed with learning process achieved using momentum. The sampling scheme's performance towards estimation of gradient estimation for SGD with momentum is analyzed. This is achieved by drawing set of data points directly with replacement and corresponding data points are formed. The challenge is towards development of estimate for unbiased statistic gradient based on data points sub-sample. Considering the maximum deviations for unbiased processes as well as the non-complete approximations [17], analysis presented here is motivated from [18]. It adopts both optimization error of SGD with momentum and estimation error for statistical finite samples. The rate bounds of non-asymptotic nature as well as convergence rates of asymptotic nature considering SGD with momentum is used towards empirical minimization of unbiased statistics. The inferences highlight conditional variance for gradient estimator impact on the convergence speed of SGD with momentum. A generalization bound is also presented which has a dependency on sampling techniques variance. Considering accuracy and cost, the hypothesis confirms estimation for incomplete unbiased statistics superiority with respect to complete variant. The results for area under the curve (AUC) maximization and metric learning considering huge data agree as per the suggested propositions. The proposed sampling strategy provides considerable performance gains towards non-uniform sampling strategies which further improves convergence. The paper is structured in the following manner. In Sect. 2 empirical risk the computational method for minimization through stochastic gradient descent is highlighted. All the experiments and results are presented in Sect. 3. The Sect. 4 gives concluding remarks.

2 Computational Method

In this section framework of proposed empirical risk minimization through stochastic gradient descent with momentum is presented. The schematic representation of empirical risk minimization is given in Fig. 1.

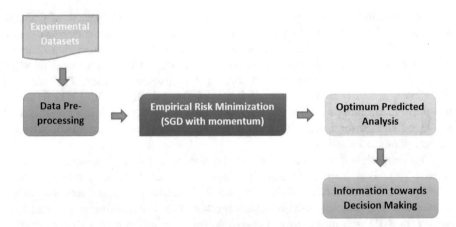

Fig. 1. The empirical risk minimization framework through stochastic gradient descent with momentum

2.1   Problem Definition

The research problem entails in minimizing the empirical risk function using stochastic gradient descent. In order to achieve this target, we setup a mathematical hypothesis which impacts the prediction model's generalization ability considering momentum based stochastic gradient descent with respect to huge datasets [19]. The challenge is to achieve optimal trade-off giving due consideration towards accuracy and cost. The various analysis results provide significant information for several decision-making activities.

2.2   Datasets

The experimental datasets for this research work are adopted and prepared from three different sources [19] viz IJCNN1, covtype and SUSY. These datasets are scaled up considerably in order to perform the experiments and analysis. IJCNN1 is binary classification neural network competition dataset hosted by Daniel Prokhorov. This dataset has around 50000 and over 90000 training and test data points respectively with 22 features. The training and test data points scaled upto 60000 and 100000 instances respectively for performing the experiments. Covtype is a 7-class classification dataset hosted by UCI repository. This dataset has over 500000 data points with 54 features. This dataset is scaled upto 600000 data points for this assignment. SUSY is binary classification dataset hosted by UCI repository. This dataset has 5000000 data points with 18 features. This dataset is scaled upto 6000000 data points for this research. The major motivation for scaling the datasets is to increase the robustness of the results.

2.3   Unbiased Statistics with Illustrations

Considering the standard version of sample mean, corresponding derivatives are calculated to reach at the generalized unbiased statistics.

Definition: Consider the following:

(a) Let $J > 1$ and $(v_1, \ldots\ldots, v_J) \in \mathbb{N}^{*J}$

(b) Let $\mathbf{Y}_{\{1,\ldots\ldots,n_j\}} = \left(Y_1^{(j)}, \ldots\ldots, Y_{n_j}^{(j)}\right), 1 \leq j \leq J$ be J independent samples of sizes $n_j \geq v_j$ and composed of independent and identically distributed random variables considering values in measurable space $\mathcal{Y}_j$ with respect to function distribution $F_j(dy)$

(c) Assume $MF : \mathcal{Y}_1^{v_1} \times \ldots\ldots \times y_J^{v_J} \to \mathbb{R}$ as measurable function which is square integrable considering the distribution $prob = MF_1^{\otimes v_1} \otimes \ldots\ldots \otimes MF_J^{\otimes v_J}$

(d) It is further assumed $MF\left(\mathbf{y}^{(1)}, \ldots\ldots, \mathbf{y}^{(J)}\right)$ takes symmetric shape considering every argument block $\mathbf{y}^{(j)}$ with $\mathcal{Y}_j^{v_j}, 1 \leq j \leq J$

Then generalized J-sample unbiased statistics of degree $(v_1, \ldots\ldots, v_J)$ with kernel MF is as follows:

$$U_{\mathbf{n}}(MF) = \frac{1}{\prod_{j=1}^{J} \binom{n_j}{v_j}} \sum_{IN_1} \cdots \sum_{IN_J} MF\left(\mathbf{Y}_{IN_1}^{(1)}; \ldots\ldots; \mathbf{Y}_{IN_J}^{(J)}\right) \tag{1}$$

Here $\mathbf{n} = (n_1, \ldots\ldots, n_J)$, $\sum_{IN_1} \cdots \sum_{IN_J} MF\left(\mathbf{Y}_{IN_1}^{(1)}; \ldots\ldots; \mathbf{Y}_{IN_J}^{(J)}\right)$ denotes summation considering every element of $\wedge$, which is the set for $\prod_{j=1}^{J} \binom{n_j}{v_j}$ index vectors $(IN_1, \ldots\ldots, IN_J)$, IN_j being a set of v_j indexes $1 \leq in_1 < \ldots < in_{v_j} \leq n_j$ and $\mathbf{Y}_{IN_J}^{(J)} = \left(Y_{in_1}^{(j)}, \ldots\ldots, Y_{in_{v_j}}^{(j)}\right) \vee 1 \leq j \leq J$.

The above definition considers the standard mean statistics to the case with $J = 1 = v_1$. When $J = 1$, $U_{\mathbf{n}}(MF)$ is the average over all v_1-tuples of observations. With $J \geq 2$, there is a multi-situation correspondence where v_j-tuple is utilized towards every sample $j \in \{1, \ldots\ldots, J\}$.

The statistics represented through Eq. (1) states that it possesses least variance considering all estimates which are of unbiased nature such that:

$$prob(MF) = \mathbb{EV}\left[MF\left(Y_1^{(1)}, \ldots\ldots, Y_{v_1}^{(1)}, \ldots\ldots, Y_1^{(J)}, \ldots\ldots, Y_{v_j}^{(J)}\right)\right] = \mathbb{EV}[U_{\mathbf{n}}(MF)] \tag{2}$$

For interested readers further discussions are available in [16]. There are certain problems in machine learning where the generalized unbiased statistics are used as performance criteria. Few such significant problems are briefly highlighted below.

2.3.1   Clustering

Consider a distance $DT : \mathcal{Y}_1 \times \mathcal{Y}_1 \to \mathbb{R}_+$, such that the partition's quality $\mathcal{PT}$ for $\mathcal{Y}_1$ towards clustering an independent and identically distributed sample $Y_1, \ldots\ldots, Y_n$ drawn from $F_1(dy)$ which is evaluated using within cluster's scatter point as:

$$\widehat{WP}_n(\mathcal{PT}) = \frac{2}{n(n-1)} \sum_{p<q} DT(Y_p, Y_q) \cdot \sum_{B \in \mathcal{PT}} \mathrm{IF}\{(Y_p, Y_q) \in \mathcal{B}^2\} \quad (3)$$

This is one sample unbiased statistics of 2nd degree having kernel as:

$$K_{\mathcal{PT}}(a, a') = DT(a, a') \cdot \sum_{B \in \mathcal{PT}} \mathrm{IF}\{(a, a') \in \mathcal{B}^2\} \quad (4)$$

Without brevity, the results can also be extended towards higher degree kernels. However, due attention needs to be considered for the kernel complexity.

2.3.2 Multi-partite Ranking

Consider J independent and identically distributed samples $Y_1^{(j)}, \ldots \ldots, Y_{n_j}^{(j)}$ with $n_j \geq 1$ and $1 \leq j \leq J$ on $\mathcal{Y}_1 \subset \mathbb{R}^s$. The scoring function $sf : \mathcal{Y}_1 \to \mathbb{R}$ has the accuracy considering J-partite ranking which is estimated empirically with respect to concordant rate for J-tuples such that:

$$\widehat{\mathrm{VROCS}}_n(sf) = \frac{1}{n_1 \times \ldots \ldots \times n_J} \sum_{j=1}^{J} \sum_{i_j=1}^{n_j} \mathrm{IF}\left\{ sf\left(Y_{in_1}^{(1)}\right) < \ldots < sf\left(Y_{in_J}^{(J)}\right) \right\} \quad (5)$$

The aforementioned quantity represents the J-sample U-statistics having degrees $v_1 = \cdots = v_J = 1$ with kernel $\overline{MF}_{sf}(y_1, \ldots \ldots, y_J) = \mathrm{IF}\{sf(y_1) < \ldots sf(y_J)\}$. It is also known as volume under the receiver operating characteristic (ROC) surface.

2.3.3 Metric Learning

For the labeled data $(A_1, B_1), \ldots \ldots, (A_n, B_n)$ based on the independent and identically distributed sample on $\mathcal{Y}_1 = \mathbb{R}^s \times \{1, \ldots \ldots, K\}$, the empirical pairwise classification performance of distance $DST : \mathcal{Y}_1 \times \mathcal{Y}_1 \to \mathbb{R}_+$ is evaluated through:

$$\widehat{R}_n(DST) = \frac{6}{v(v-1)(v-2)} \sum_{p<r<q} \mathrm{IF}\{DST(A_p, A_r) < DST(A_r, A_q), B_p = B_r \neq B_q\} \quad (6)$$

This is a single sample unbiased statistics of 3rd degree considering kernel $\overline{MF}_{DST}((a, b), (a', b'), (a'', b'')) = \mathrm{IF}\{DST(a, a') < DST(a, a''), b \neq b' \neq b''\}$.

2.4 Minimization for Unbiased Statistics with Respect to Gradients

Consider $P \subset \mathbb{R}^p, p \geq 1$ as parameter space and the risk minimization problem $\min_{\theta \in P} RM(\theta)$:

$$RM(\theta) = \mathbb{E}\left[MF\left(Y_1^{(1)}, \ldots, Y_{v_1}^{(1)}, \ldots, Y_1^{(J)}, \ldots, Y_{v_J}^{(J)}; \theta\right)\right] = p(MF(.; \theta)) \quad (7)$$

Here $MF : \prod_{q=1}^{J} \mathcal{Y}_q^{v_q} \times P \rightarrow \mathbb{R}$ is the convex loss function; $\left(Y_1^{(q)}, \ldots\ldots, Y_{v_q}^{(q)}\right)$, $1 \leq q \leq J$ represent J independent random variables with distributions $F_q^{\otimes v_q}(dy)$ on $\mathcal{Y}_q^{v_q}$ respectively such that MF has the square integrable property considering $\theta \in P$. With respect to J independent and identically distributed $Y_1^{(q)}, \ldots\ldots, Y_{n_q}^{(q)}, 1 \leq q \leq J$ samples the risk function's empirical version is $\theta \in P \vdash \widehat{RM}_{\mathbf{n}}(\theta) = U_{\mathbf{n}}(MF(.; \theta))$. This is represented as ∇_θ with respect to θ.

There are several gradient descent based learning algorithms, considering the iterations $\theta_{t+1} = \theta_t - \vartheta_t \nabla_\theta \widehat{RM}_{\mathbf{n}}(\theta_t)$ with random initiating value $\theta_o \in P$ and rate of learning with step length $\vartheta_t \geq 0$, such that $\sum_{t=1}^{+\infty} \vartheta_t = +\infty$ with $\sum_{t=1}^{+\infty} \vartheta_t^2 < +\infty$. Here we take into consideration huge datasets with sample sizes $n_1, \ldots\ldots, n_J$ of training data where calculation of empirical gradient at every iteration becomes non-manageable because of items to be averaged.

$$\widehat{eg}_{\mathbf{n}}(\theta) \stackrel{\text{def}}{=} \nabla_\theta \widehat{RM}_{\mathbf{n}}(\theta) = \left(\frac{1}{\prod_{q=1}^{J} \binom{n_q}{v_q}}\right) \sum_{IN_1} \cdots \sum_{IN_J} \nabla_\theta MF\left(\mathbf{Y}_{IN_1}^{(1)}; \ldots\ldots; \mathbf{Y}_{IN_J}^{(J)}; \theta\right)$$

(8)

The approximation done stochastically instead looks towards usage of unbiased estimator for Eq. (8) which is straightforward to calculate.

2.5   Stochastic Gradient Descent with Momentum Considering Incomplete Unbiased Statistics

A possibility lies in replacement of Eq. (8) considering complete unbiased statistics calculated through reduced size sub-samples $n_q' \ll n_q, \left\{\left(Y_1'^{(q)}, \ldots\ldots Y_{n_q'}'^{(q)}\right); j = 1, \ldots\ldots J\right\}$ drawn uniformly at random with no replacement considering actual data points. This results in following gradient estimation:

$$\widetilde{eg}_{\mathbf{n}'}(\theta) = \frac{1}{\prod_{j=1}^{J} \binom{n_q'}{v_q}} \sum_{IN_1} \cdots \sum_{IN_J} \nabla_\theta MF\left(\mathbf{Y}_{IN_1}'^{(1)}; \ldots\ldots; \mathbf{Y}_{IN_J}'^{(J)}; \theta\right) \quad (9)$$

Here summation over IN_J considers all $\binom{n_j'}{v_j}$ subsets related to set IN_j of v_j indexes $1 \leq in_1 < \ldots < in_{v_j} \leq n_j'$ and $\mathbf{n}' = \left(n_1', \ldots\ldots, n_J'\right)$. This approach is straightforward and a better estimate can be obtained considering the same cost. This fact is presented in the coming sections.

2.5.1 Empirical Gradient: Monte-Carlo Estimation

Considering the practical perspective, the alternate strategy is attributed towards simplicity. This takes help from the monte-carlo sampling which considers independent drawing with replacement from index vectors' set. This yields an incomplete unbiased statistics gradient estimator [16]:

$$\overline{eg}_A(\theta) = \frac{1}{A} \sum_{(IN_1,\ldots\ldots,IN_J)\in\mathcal{D}_A} \nabla_\theta MF\left(\mathbf{Y}_{IN_1}^{'(1)}; \ldots\ldots; \mathbf{Y}_{IN_J}^{'(J)}; \theta\right) \tag{10}$$

Here $\mathcal{D}_A$ is developed through sampling A times with replacement considering the set Λ. It may be pointed that the conditional expectation of Eq. (10) given the J observed data samples is $\widehat{eg}_n(\theta)$. The parameter A considering the terms for which average needs to be taken handles the SGD with momentum's complexity. It is observed that incomplete unbiased statistics is not unbiased statistics as such. The unbiased estimator for statistical risk's gradient $RM(\theta)$ in Eq. (10) has less accuracy than full empirical gradient in Eq. (8). Hence, it has larger variance but this increased variance results in larger reduction towards computational cost. It will be shown that considering the same computational cost ($A = \prod_{j=1}^J \binom{n'_j}{v_j}$), SGD with momentum in Eq. (10) rather than in Eq. (9) results in higher accuracy in results. The Eq. (10) has smaller variance with respect to $\nabla RM(\theta)$ except when $J = 1 = v_1$.

Proposition 1: With $A = \prod_{j=1}^J \binom{n'_j}{v_j}$ $\exists$ universal constant $ct > 0$ which yields:

$$\sigma^2(\widetilde{eg}_{\mathbf{n}'}(\theta)) \le \frac{ct \cdot \sigma_\theta^2}{\sum_{j=1}^J n'_j} \tag{11}$$

$$\sigma^2(\overline{eg}_A(\theta)) \le \frac{ct \cdot \sigma_\theta^2}{\prod_{j=1}^J \binom{n'_j}{v_j}} \tag{12}$$

Considering all $\mathbf{n} \in \mathbb{N}^{*J}$ with $\sigma_\theta^2 = \sigma^2\left(MF\left(\mathbf{Y}_{IN_1}^{'(1)}; \ldots\ldots; \mathbf{Y}_{IN_J}^{'(J)}; \theta\right)\right)$. As sampling with replacement is computationally more efficient it is considered here.

2.5.2 Condition Based Performance Analysis

Here the performance of SGD with momentum methods conditionally based upon the observed samples are discussed. For the sake of simplicity, $\mathbb{Prob}_{\mathbf{n}}(\cdot)$ represents the conditional probability measure with respect to the data and $\mathbb{Expm}_{\mathbf{n}}[\cdot]$ is the expectation of $\mathbb{Prob}_{\mathbf{n}}$. Considering matrix MT, its transpose as MT^T and $\|MT\|_{HS} := \sqrt{Tr(MTMT^T)}$ as the hilbert-schmidt norm. It is assumed that function for loss MF has l-smoothness for θ such that the gradient becomes l-lipschitz for $l > 0$.

The instance with $\widehat{RM}_n$ being α-strongly convex towards certain deterministic α is restricted here such that:

$$\widehat{RM}_n(\theta_1) - \widehat{RM}_n(\theta_2) \leq \nabla_\theta \widehat{RM}_n(\theta_1)^T (x - y) - \frac{\alpha}{2} \|\theta_1 - \theta_2\|^2 \qquad (13)$$

Here θ_n^* represents the unique minimizer. This argument may be taken towards smooth and not strong convex instance [20]. An argument considering convex analysis and stochastic optimization [20, 21] demonstrates the impact of gradient estimator's conditional variance towards empirical performance for solution developed through SGD with momentum. This supports SGD with momentum variant usage.

Proposition 2: Considering recurrence relation $\theta_{t+1} = \theta_t - \vartheta_t eg(\theta_t)$ with $\mathbb{Expm}_n[eg(\theta_t)|\theta_t] = \nabla_\theta \widehat{RM}_n(\theta_t)$ and the conditional variance of $eg(\theta)$ is $\sigma_n^2(eg(\theta))$. With the step size $\vartheta_t = \frac{\vartheta_1}{t^\alpha}$ we have the following:

(a) If $\frac{1}{2} < \alpha < 1$ then $\mathbb{Expm}_n[\widehat{RM}_n(\theta_{t+1}) - \widehat{RM}_n(\theta_n^*)] \leq \frac{\sigma_n^2(eg(\theta_n^*))}{t^\alpha} C_1$

(b) If $\alpha = 1$ and $\vartheta_1 > \frac{1}{2\beta}$ then $\mathbb{Expm}_n[\widehat{RM}_n(\theta_{t+1}) - \widehat{RM}_n(\theta_n^*)] \leq \frac{\sigma_n^2(eg(\theta_n^*))}{t+1} C_2$

with $C_1 = \vartheta_1 l 2^{\alpha-1} \left(\frac{1}{2\beta} + \frac{l\vartheta_1^2}{2\alpha-1} \right) + o\left(\frac{1}{t^\alpha}\right)$ and $C_2 = \frac{2^{\beta\vartheta_1} l \exp(2\beta l \vartheta_1^2) \vartheta_1^2}{(2\beta\vartheta_1 - 1)} + o\left(\frac{1}{t}\right)$

The SGD with momentum's convergence rate is governed considering variance term. Hence, it is required to reduce this term in order to improve its performance. The asymptotic behavior of the algorithm when $t \to +\infty$ is a source of interest with respect to considerations where:

(a) $\widehat{RM}_n(\theta)$ is two times differentiable in the neighborhood for θ_n^*
(b) $\nabla \widehat{RM}_n(\theta)$ is bounded

Theorem 1: Considering covariance matrix Σ_n^* as having unique solution towards Lyapunov equation such that:

$$\Gamma \Sigma_n^* + \Sigma_n^* \Gamma - \zeta \Sigma_n^* = \Sigma_n(\theta_n^*) \qquad (14)$$

Here $\Gamma = \nabla^2 \widehat{RM}_n(\theta_n^*)$, $\Sigma_n(\theta_n^*) = \mathbb{Expm}_n[eg(\theta_n^*)eg(\theta_n^*)^T]$ and $\zeta = \vartheta_1 > \frac{1}{2\beta}$ if $\alpha = 1$, 0 otherwise. Then with the above assumptions:

$$\frac{1}{\vartheta_t} \left(\widehat{RM}_n(\theta_t) - \widehat{RM}_n(\theta^*) \right) \Rightarrow \frac{1}{2} U^T \left(\Sigma_n^* \right)^{1/2} \Gamma \left(\Sigma_n^* \right)^{1/2} U \text{ with } U \sim \mathcal{N}(0, 1) \qquad (15)$$

With $\zeta = 0$, $\left\| (\Sigma_n^* \Gamma)^{1/2} \right\|_{HS}^2 = \mathbb{Expm}[U^T (\Sigma_n^*)^{1/2} \Gamma (\Sigma_n^*)^{1/2} U] = \frac{1}{2} \sigma_n^2(eg(\theta_n^*))$ (16)

The above theorem addresses the fact where conditional variance becomes significant in algorithm's asymptotic performance. This term dominates precision of the solution. The next sub-section highlights generalization bound [21] which explicitly depends on the gradient estimator's true variance.

2.6   Generalization Bounds

Consider $\theta^* = \text{argmin}_{\theta \in \Theta} RM(\theta)$ as the true risk minimizer. The mean for excess risk is given as:

$$\forall \mathbf{n} \in \mathbb{N}^{*J}, \text{Expm}[RM(\theta_t) - RM(\theta^*)] \leq E_1 + E_2 \tag{17}$$

With $E_1 = 2\text{Expm}\left[\sup_{\theta \in \Theta}\left|\widehat{RM}_{\mathbf{n}}(\theta) - RM(\theta)\right|\right]$ and $E_2 = \text{Expm}\left[\widehat{RM}_{\mathbf{n}}(\theta_{t+1}) - \widehat{RM}_{\mathbf{n}}(\theta_{\mathbf{n}}^*)\right]$.

The optimization error E_2 in Eq. (17), analyses learning method's generalization ability. It also requires the error of estimation E_1 control. For this we discuss the following results.

Proposition 3: Consider $\mathcal{HB}$ as collection for bounded kernels of symmetric nature on $\prod_{j=1}^{J} \mathcal{Y}_j^{y_j}$ such that $M_{\mathcal{HB}} = \sup_{(MF,y)\in\mathcal{HB}\times y}|MF(y)| < +\infty$. Also consider $\mathcal{HB}$ as vapnik-chervonenkis (VC) major functions' call with finite $VC < +\infty$. Again, $\kappa = \text{minimum}\left\{\left\lfloor\frac{n_1}{v_1}\right\rfloor, \ldots, \left\lfloor\frac{n_J}{v_J}\right\rfloor\right\}$, then for any $\mathbf{n} \in \mathbb{N}^{*J}$,

$$\text{Expm}[\sup_{MF \in \mathcal{HB}}|U_{\mathbf{n}}(MF) - p(MF)|] \leq M_{\mathcal{HB}}\left\{2\sqrt{\frac{2VC\log(1+\kappa)}{\kappa}}\right\} \tag{18}$$

Theorem 2: Consider θ_t as the generated sequence through SGD with momentum using incomplete statistic gradient estimator having $A = \prod_{j=1}^{J} \binom{n_j'}{v_j}$ terms for some $n_1', \ldots, n_J'$. Also assume that $\{RM(.; \theta) : \theta \in \Theta\}$ is a VC major class of finite VC dimension such that:

$$M_\Theta = \sup_{\theta\in\Theta,(\mathbf{Y}^{(1)},\ldots,\mathbf{Y}^{(J)})\in\prod_{j=1}^{J}\mathcal{Y}_j^{y_j}}\left|MF\left(\mathbf{Y}^{(1)},\ldots,\mathbf{Y}^{(J)};\theta\right)\right| < +\infty \tag{19}$$

with $N_\Theta = \sup_{\theta\in\Theta}\sigma_\theta^2 < +\infty$. For any $\eta \in (0,1)$ the probability is at least $1 - \eta : \forall \mathbf{n} \in \mathbb{N}^{*J}$,

$$|RM(\theta_t) - RM(\theta^*)| \leq \left(\frac{CN_\Theta}{At^\alpha} + \sqrt{\frac{D_\alpha\log\left(\frac{2}{\eta}\right)}{t^\alpha}}\right) + 2M_\Theta\left\{2\sqrt{\frac{2VC\log(1+\kappa)}{\kappa}} + \sqrt{\frac{\log\left(\frac{4}{\eta}\right)}{\kappa}}\right\} \tag{20}$$

with some constants C and D_α depending on the other parameters.

The generalization bound from the above theorem supports the usage of incomplete unbiased statistics in Eq. (10) as estimator for gradient. Generally, the results achieved are of similar nature as above theorem considering complete unbiased statistics estimator in Eq. (9) which leads towards distorted bounds. Using incomplete unbiased statistics, better performance is achieved on test dataset with reduction in iterations. This reduces number of gradient calculations which are required in order to converge towards a correct solution. In next section, experimental results verify the significance of these gains.

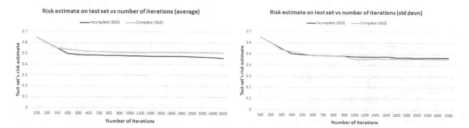

Fig. 2. The average and standard deviation for 100 runs of risk estimate with iterations (Covtype with batch size = 12, $\gamma_1 = 1$)

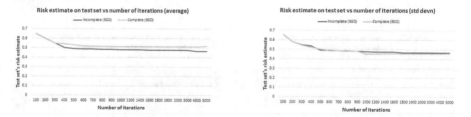

Fig. 3. The average and standard deviation for 100 runs of risk estimate with iterations (Covtype with batch size = 500, $\gamma_1 = 1$)

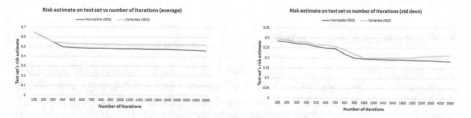

Fig. 4. The average and standard deviation for 100 runs of risk estimate with iterations (IJCNN1 with batch size = 30, $\gamma_1 = 2$)

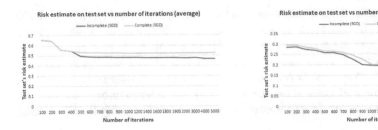

Fig. 5. The average and standard deviation for 100 runs of risk estimate with iterations (IJCNN1 with batch size = 140, $\gamma_1 = 5$)

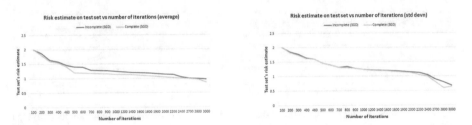

Fig. 6. The average and standard deviation for 100 runs of risk estimate with iterations (SUSY with batch size = 140, $\gamma_1 = 0.5$)

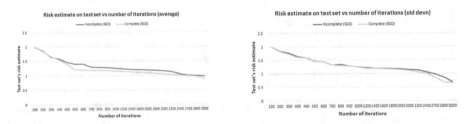

Fig. 7. The average and standard deviation for 100 runs of risk estimate with iterations (SUSY with batch size = 475, $\gamma_1 = 1$)

3   Experiments and Results

Now some experiments are provided [16] in order to have comparison of incomplete and complete unbiased statistics gradient estimators represented through Eqs. (9) and (10) in SGD with momentum. Here the dependence of these statistics is with respect to identical number of terms. The datasets are taken from [19]. For all the experiments, datasets are split randomly in ratio of 0.75:0.25 for the training and test sets respectively. This is followed by a sampling of 100K pairs considering test dataset towards test performance estimation. The step length is considered as $\vartheta_t = \frac{\vartheta_1}{t}$. The results obtained considering the number of SGD with momentum iterations.

3.1 Area Under the Curve Optimization

The binary classifier learning problem is addressed through optimization of AUC. This leads towards *VROCS* with $J = 2$. Considering independent and identically distributed observations' sequence $C_i = (A_i, B_i)$ with $A_i \in \mathbb{R}^p$ and $B_i \in \{-1, +1\}$ such that $A^+ = \{A_i : B_i = 1\}$, $A^- = \{A_i : B_i = -1\}$ and $N = |A^+||A^-|$. Here we consider a linear scoring rule with $\theta \in \mathbb{R}^p$ is the parameter which is to be learned. We use logistic loss as smooth convex function with upper bound on heavier function such that the empirical risk minimization problem takes the form:

$$\min_{\theta \in \mathbb{R}^p} \frac{1}{N} \sum_{A_i^+ \in A^+} \sum_{A_j^- \in A^-} \log\left(1 + \exp\left(s_\theta\left(X_i^-\right) - s_\theta\left(X_i^+\right)\right)\right) \qquad (21)$$

Here two datasets viz IJCNN1 and covtype are used. The variant values towards initial step length ϑ_1 and batch length A are used. The results are averaged considering 100 runs for SGD with momentum and shown in Figs. 2, 3, 4 and 5. It is observed that incomplete unbiased statistic estimator is always superior to its variant which is of complete nature. There is a small performance gap between both strategies. However, for parameters which become relevant with respect to practical situations, the difference becomes substantial. A small variance for SGD with momentum runs with incompleteness is also observed.

3.2 Metric Learning

Now considering the metric learning case where given a sample of N independent and identically distributed observations $C_i = (A_i, B_i)$ with $A_i \in \mathbb{R}^p$ and $B_i \in \{1, \ldots \ldots, c\}$. The pseudo distance $DT_M(a, a')$ are utilized here. The logistic loss I used towards convex and smooth surrogate for Eq. (6) such that the empirical risk minimization problem takes the form:

$$\min_M \frac{6}{N(N-1)(N-2)} \sum_{i<j<k} \text{IF}\{B_i = B_j \neq B_k\} \log\left(1 + \exp\left(DT_M\left(A_i, A_j\right) - DT_M(A_i, A_k)\right)\right)$$

$$(22)$$

The binary classification dataset SUSY is used here. The Figs. 6 and 7 highlights the fact that performance gap between two strategies is quite large here. This is in confirmation with gap for variance considering incomplete and complete approximations. It is much greater considering one-sample unbiased statistic for 3rd degree involving metric learning than for two-sample unbiased statistic for 1st degree involving AUC optimization.

4 Conclusion

In this research work an analysis of SGD with momentum is presented with empirical estimates for objective function which take the shape towards generalized unbiased statistics. The thoughts presented here are not applicable for biased statistics. This scenario is available in several statistical learning problem scenarios. The estimation of gradient is dependent with respect to incomplete unbiased statistics which is achieved through sampling data points considering replacement. The analysis highlights the predictive rules' generalization ability involving optimization and estimation error. The SGD with momentum surpasses the trivial form with respect to the same cost of computation considering sub-sampled data points with no replacement. The performance gains are highly appreciable considering huge data towards non-uniform sampling strategies which further improves convergence. The solution tries to approximate distribution in order to reach good trade-off considering performance as well as cost. The future work looks towards performing the risk minimization experiments with other commonly available large datasets such as criteo, criteo_tb, HIGGS, splice-site, mnist8m etc.

References

1. Bellet, A., Habrard, A., Sebban, M.: Metric Learning. Morgan and Claypool Publishers, San Rafael (2015)
2. Zhao, P., Hoi, S., Jin, R., Yang, T.: AUC maximization. In: Proceedings of 28th International Conference on Machine Learning, pp. 233–240 (2011)
3. Fürnkranz, J., Hüllermeier, E., Vanderlooy, S.: Binary decomposition methods for multipartite ranking. In: Proceedings of Joint European Conference on Machine Learning and Knowledge Discovery in Databases, pp. 359–374 (2009)
4. Clémençon, S.: On U-processes and clustering performance. In: Proceedings of 24th International Conference on Neural Information Processing Systems, pp. 37–45 (2011)
5. Lee, A.J.: U-Statistics: Theory and Practice. Marcel Dekker, New York (1990)
6. Clémençon, S., Lugosi, G., Vayatis, N.: Ranking and empirical risk minimization of U-Statistics. Ann. Stat. **36**(2), 844–874 (2008)
7. Norouzi, M., Fleet, D.J., Salakhutdinov, R.: Hamming distance metric learning. In: Proceedings of 25th International Conference on Neural Information Processing Systems, pp. 1070–1078 (2012)
8. Kar, P., Sriperumbudur, B., Jain, P., Karnick, H.: On the generalization ability of online learning algorithms for pairwise loss functions. In: Proceedings of 30th International Conference on Machine Learning, pp. III-441–III-449 (2013)
9. Qian, Q., Jin, R., Yi, J., Zhang, L., Zhu, S.: Efficient distance metric learning by adaptive sampling and mini-batch stochastic gradient descent. Mach. Learn. **99**(3), 353–372 (2015)
10. Johnson, R., Zhang, T.: Accelerating stochastic gradient descent using predictive variance reduction. In: Proceedings of 26th International Conference on Neural Information Processing Systems, pp. 315–323 (2013)
11. Le Roux, N., Schmidt, M.W., Bach, F.: A stochastic gradient method with an exponential convergence rate for finite training sets. In: Proceedings of 25th International Conference on Neural Information Processing Systems, pp. 2663–2671 (2012)

12. Mairal, J.: Incremental majorization-minimization optimization with application to large-scale machine learning. arXiv:1402.4419 (2014)
13. Defazio, A., Bach, F., Lacoste-Julien, S.: SAGA: a fast-incremental gradient method with support for non-strongly convex composite objectives. In: Proceedings of 27th International Conference on Neural Information Processing Systems, pp. 1646–1654 (2014)
14. Needell, D., Ward, R., Srebro, N.: Stochastic gradient descent, weighted sampling and the randomized Kaczmarz algorithm. In: Proceedings of 27th International Conference on Neural Information Processing Systems, pp. 1017–1025 (2014)
15. Zhao, P., Zhang, T.: Stochastic optimization with importance sampling for regularized loss minimization. In: Proceedings of 32nd International Conference on Machine Learning, pp. 1–9 (2015)
16. Chaudhuri, A.: Some investigations on empirical risk minimization through stochastic gradient with momentum algorithms. Technical report, TR–9818, Samsung R&D Institute Delhi India (2018)
17. Clémençon, S., Robbiano, S., Tressou, J.: Maximal deviations of incomplete U-processes with applications to empirical risk sampling. In: Proceedings of 13th SIAM International Conference on Data Mining, pp. 19–27 (2013)
18. Bottou, L., Bousquet, O.: The tradeoffs of large-scale learning. In: Proceedings of 20th International Conference on Neural Information Processing Systems, pp. 161–168 (2007)
19. https://www.csie.ntu.edu.tw/~cjlin/libsvmtools/datasets/
20. Bach, F.R., Moulines, E.: Non-asymptotic analysis of stochastic approximation algorithms for machine learning. In: Proceedings of 24th International Conference on Neural Information Processing Systems, pp. 451–459 (2011)
21. Nemirovski, A., Juditsky, A., Lan, G., Shapiro, A.: Robust stochastic approximation approach to stochastic programming. SIAM J. Optim. **19**(4), 1574–1609 (2009)

Variable Step Size Least Mean Square Optimization for Motion Artifact Reduction: A Review

Khalida Adeeba Mohd Zailan$^{(\boxtimes)}$, Mohd Hilmi Hasan,
and Gunawan Witjaksono

Centre for Research in Data Science, Universiti Teknologi PETRONAS,
32610 Seri Iskandar, Perak, Malaysia
{khalida_18000732, mhilmi_hasan,
gunawan.witjaksono}@utp.edu.my

Abstract. Many algorithms have been developed to reduce the motion artifact effect on Photoplethysmograph (PPG) technology and to increase the accuracy of the health monitoring device reading. It is found that existing solutions are still lacking in getting high accuracy of heart rate reading. Therefore, we propose a research to formulate an improved motion artifact reduction approach using variable step-size least mean square (VSSLMS). The objective of this paper is to review VSSLMS for motion artifact reduction. A total of eight manuscripts, collected from ISI, Scopus and Google Scholar indexing databases, were critically reviewed. The review revealed that VSSLMS is better than LMS in reducing the motion artifact in slow motion and high-speed motion. For future work, the VSSLMS results will be formulated with regression machine learning.

Keywords: Variable Step-size Least Mean Square · Photoplethysmograph · Accuracy · Health monitoring system

1 Introduction

Nowadays, mobile device has benefited human life in many ways, which one of it is in medical sector that the user could monitor their health through wearable device to detect their heart rate. Advanced technology in wearable sensors causing a revolutionary in medical sector. The existence of wearable sensors could help in monitoring health through monitoring the user's health condition and diagnosing a possible disease. Moreover, by having a monitoring health system, it could always remind the user to always aware of their health conditions and being in a good state. Wide accessibility of this technology also enhances the demand of having health monitoring system regardless age and gender to have a better quality of life. Health monitoring system also could facilitate the users to get a treatment or go to a check-up earlier and will give the patients some information about their health condition [1].

Optical sensors play a significant role in today's world by advancing smart technology and wearable devices that one of it could detect the pulse to calculate heart rate. As a result, we can witness an increase in demand for wearable heart rate monitoring

© Springer Nature Switzerland AG 2019
R. Silhavy (Ed.): CSOC 2019, AISC 985, pp. 182–190, 2019.
https://doi.org/10.1007/978-3-030-19810-7_18

devices which is now become a growing trend globally. Optical sensor is good in generating real time information like current heart rate, which could be a good sign of real time health condition [5]. Photoplethysmographs (PPG) are a popular, non-invasive health monitoring device for measuring heart rate, blood oxygen saturation, blood pressure, cardiac output, and respiration because of low cost [3] and simple hardware implementation [4]. PPG is a technology that uses optic sensor technology to detect blood volume changes within the tissue.

Technology of Photoplethysmographs (PPG) is the current optical sensor technology for wearable health monitoring system [3, 4]. Numerous prototypes and commercial health monitoring systems [1] had been developed using this PPG technology. Unfortunately, the implementation of wearable optical sensor is mostly targeted for static use [2]. This means that, the target is when the person who uses it is not moving. A static use does not consider the changes in the measurement conditions and noise sources from the tissue and blood as person moves around. Meanwhile, for the case of big impact movement, an impact error in the reading of signal occurs because of the existence of motion artefact that resulted from a movement between the skin and sensor [6]. Previous works on PPG stated that the disadvantages of using this technology is that when a motion exists on the sensor, the reading will also capture the motion artefact reading [2, 7, 8]. Motion artefact will lead to less accurate health reading, which can lead to false alarm especially when there is a big motion impact on the sensor [1, 4]. Dr. Gregory Marcus, director of clinical research for University of California, San Francisco's Division of Cardiology stated that it is not unusual for a patient goes to see a doctor because of false alarm by health monitoring system. This false alarm case is due to inaccurate system, which will cost the patient money [1].

Many algorithms have been developed to reduce the motion artifact effect on PPG technology and increase the accuracy of the health monitoring device reading. In this paper, we focus on calculating the step-size value which affect the heart rate reading. LMS is basically an adaptive filtering algorithm which focuses on the step size that represents the adjustment between the speed of adaption and the noise in steady state. Unfortunately, LMS algorithm may not work well in high motion activity [2, 7–9].

The objective of this paper is to review Variable Step-size Least Mean Square (VSSLMS), an enhanced type of LMS, for motion artifact reduction. This paper is written as follows; next section will present literature review, the it will be followed by results section. This paper will end with conclusion section, containing summary of the work as well as some future works.

2 Literature Review

Motion artifact is a motion that creates error in information or readings. Any part of body movement could cause sensor to lose contact with the skin by the deformation of tissues occurred. This resulted from the escaped or disturbed by the ambient light of transmitted light from the optical sensor since when there is a movement in human body, biologically, the tissues will also move [1, 10–12]. Different types of motion could create different effect on the PPG signal [3, 11]. High speed movement activity such as exercise or run normally affects the performance of PPG sensor in detecting pulse peak and estimating pulse rate [13]. The artifact is quite difficult to be filtered out than other type of noises since it does not have any fixed frequency spectrum and it is fluctuated by the intensity of the motion [2].

Motion artifact (MA) can be minimized by employing flexible electrodes or high input impedance front end amplifiers [14]. Placing the sensor closely to the skin and allowing it moves and flex with the skin, the noise from ambient light can be reduced as well as the impact of motion artifacts [15]. One study proposed motion artefact reduction by the correlation between six typical daily motions and patterns of the artifact were analyzed, and an artifact reduction algorithm was designed using the motion data [2, 16].

2.1 Existing Least Mean Square

Many algorithms have been developed to reduce motion artifact effect on PPG technology and increase the accuracy of the health monitoring device reading. The research in [2], implemented Least Mean Square based Active Noise Cancellation method to reduce the motion artifacts. The algorithm was applied to the accelerometer data to detect the motion reading of the user which could lead to easiness to detect the motion artifact. Adaptive filter in the research consists of two parts which are digital filter and adaptive filter algorithm [2]. Digital filter is for using fixed bandwidth meanwhile adaptive filter algorithm is to determine the proper filter coefficient for extracting the information either information from input, environment or output characteristics by changing the filter bandwidth [2]. Despite the satisfactory performance of the results from the algorithms, there are a few drawbacks occurred which are poor performance during high speed running because of the algorithm is more effected on daily life motions; and individual variations in term of physical properties especially in skin tissue and blood pressure.

This paper by Chan et al. [17] implemented Least Mean Square Variable Step-Size but they paper used time varying step size parameter LMS algorithm. This algorithm analyzes the step-size with various parameter by adjusting the step size and rely on the correlation of error at a time. As the result, the algorithm managed to overcome the shortage in regulating the step size which in the end could produce better output in term of misadjustment, reduce the noise and increase convergence speed because of the sensitivity of the time varying step size parameter.

Meanwhile paper by Li et al. [18], also implemented a new Least Mean-Square (LMS) adaptive filtering algorithm with variable step size. They proposed a new LMS adaptive filtering algorithm based on the correlation of error at any particular time and it also adjusts the step-size. Instead of using fixed step-size, they proposed the algorithm with variable step-size since the motion of the user differs among each another and the former cannot estimate at how distance the motion of the user. Thereby the sensitivity of the time-varying step-size parameter for the convergence speed will increase and reduce the noise.

The research by Zhang et al. [7] explained that this research wants to improve the corrupted asystole and ventricular fabrication frequency from chest compression (CC) using electrocardiogram (ECG) technology, which is also categorized under non-intrusive method to calculate heart rate. This research proposed a novel method combining an amplitude spectrum area (AMSA) and least mean square (LMS) filter. The LMS filter used in the research is to reconstruct CC artifact by filtering cardiopulmonary resuscitation (CPR) artifact adequately using only the CC frequency and to detailed artifact component losses. Meanwhile as for the AMSA analysis, to classify either ventricular fibrillation or asystole. The output from this research stated that LMS filter brought significant effect in increasing the performance of the frequency from chest compression.

A combination of constrained independent component analysis (cICA) and adaptive filter, proposed by Peng et al. [9] to get the clear PPG signals which are free from motion artifact. The cICA filter used to extract the PPG signal from corrupted PPG signal by the motion artefact. Meanwhile the adaptive filter was applied to recover the amplitude information of the PPG signal. LMS filter was used as the adaptive algorithm in this paper. Step size is very critical in adjusting the convergence speed and stability of the algorithm which could cause degradation of the performance of the algorithm because of improper step size. cICA algorithm could fail in producing correct output if motion artefact controls the same period with the PPG signal. The research also stated that other algorithms may be tested such as recursive least square (RLS), normalized least mean square (NLMS) or others to improve the characteristics of the LMS-based adaptive filter.

Yousefi et al. [12] suggested of using a novel algorithm which consists of five steps which are normalized least mean square (NMLS), fundamental period extraction, update the rule of Beta, continue with adaptive noise canceler and lastly SpO2 and heart rate extraction. This paper used NLMS filter to remove the noise from red and infrared signal which are the tissues effect because of the simplicity compared to other inconsistency signal energy filter. Unfortunately, because of the complex mathematical filters used in this algorithm, it resulted in poor performance due to the rapid heart rate changes regardless the threshold is adaptively adjusted to have the best result.

Wijshoff et al. [8] implemented reduction of periodic motion artefact using measurement of step rate, which can continued with estimation and reduction of motion artefact. Motion reference signal via a second-order generalized integrator (SOGI) with a frequency-locked loop (FLL) was used as the calculation of step rate process, which is an essential motion frequency that could describe the motion artifact using harmonic model of quadrature component. Quadrature component could avoid the artifact measurement from undesired frequency-shifted with only two coefficients need to be estimated per frequency component piloting to a short filter. LMS filter was used in this paper to estimate the coefficient which could construct the motion artifact estimation. The algorithm could decrease the motion artifact, but the algorithm can only deal with slowly-varying periodic motion artifact.

As per discussed above were existed research on LMS adaption in PPG technology to reduce the motion artifact. LMS is basically an adaptive filtering algorithm which focuses on the step size that represents the adjustment between the speed of adaption and the noise in steady state. Since motion artefact is random, selection of step size is very critical in controlling the stability and convergence speed of the algorithm; and it requires careful adjustment [17, 18]. Small step size, is required for small excess mean square error, hence resulted in slow convergence. Large step-size is needed for fast adaption but also gives large excess mean square. Meanwhile a too large step-size may result in loss of stability. Mean Squared Error (MSE) is a measure of how close a fitted line is to data points. The smaller the means squared error, the closer the measurement to finding the line of best fit. Hence, the objective of LMS filter is to minimize MSE [19].

$$e(n) = d(n) - X^T(n)W(n) \tag{1}$$

$$W(n+1) = W(n) + \mu e(n)X(n) \tag{2}$$

Formulas (1) and (2) are the basic formula for fixed step-size LMS algorithm. Formula (1) is used to get the $e(n)$ which is the estimation error by subtracting between desire signal $d(n)$ with; vector of input signal at time n $(X^T(n))$ and estimated tap-weight vector at time n $(W(n))$. Continue with formula (2) that calculate weight of the filter $(W(n+1))$ by adding weight of the filter $W(n)$ with total of step size measurement (μ) times estimation error $(e(n))$ and vector of input signal $X(n)$.

The step size must be selected to balance the inconsistent goals of fast convergence and small steady-state error. However, under different noise environments, step-size may be set as very small to ensure the algorithm stability. In addition, standard LMS algorithm may not work well in special environment, such as impulsive noise. Normalized Least Mean Square (NLMS) may be considered as the first variable step-size modification of LMS but it focuses only on adjusting one parameter only [20], which usually adjusting the step size on the input signal power. Unfortunately, the NLMS algorithm may be time-consuming, hence need to be avoided using this algorithm in some time-critical applications [20].

$$\mu(n) = \beta(1/(1 + exp(-\alpha|e(n)e(n-1)|)) - 0.5)) \tag{3}$$

Variable step size methods are more effective than the LMS algorithm at tracking in nonstationary environments. The step sizes are adjusted individually as adaptation progresses. If the step size is satisfied the necessary condition, then the algorithm will be stable. The variable step size LMS was proposed to increase the value of step size when it is far from optimum and decrease the value of the step size when it is near to the optimum [1]. Variable step size LMS focus on calculate the step size $(\mu(n))$ as in formula (3), instead of using fixed step size value as in LMS formula. As per shown in the Table 1 below, the comparison study on existing LMS algorithms in PPG technology.

Table 1. Comparison study for existing least mean square algorithm usage in PPG technology

Algorithm	Paper	Limitations	Algorithms	Field gap
LMS	Artifacts in wearable photoplethysmographs during daily life motions and their reduction with least mean square based active noise cancellation method, 2012	– The artifact analysis suggests that body motions directly distort the PPG signal, making it difficult to distinguish the pulse peaks – poor performance of the algorithm during high speed running – individual variations	– Butterworth – LMS	Enhancing the algorithm during high speed running
	A method to differentiate between ventricular fibrillation and asystole during chest compressions using artifact-corrupted ECG alone, 2017	The test done for ECG technology only and not PPG	– LMS – AMSA	
	Reduction of periodic motion artifacts in photoplethysmography, 2016	Instead, the most prominent spectral component varied between the step rate and its (sub)harmonic, or the spectral activity was unstructured. The algorithm can only deal with slowly-varying periodic motion artifacts. When the motion frequency and PR coincide, no improvement can be obtained	– Butterworth – second-order generalized integrator (SOGI) and frequency-locked loop (FLL) – LMS	
	Motion artifact removal from photoplethysmographic signals by combining temporally constrained independent component analysis and adaptive filter, 2014	For LMS algorithm, the step size is very critical in controlling the stability and convergence speed of the algorithm. Improper step size may degrade the performance of our cICA-LMS algorithm	– low pass filter – cICA – adaptive filter (LMS)	
NLMS	A Motion-tolerant adaptive algorithm for wearable photoplethysmographic biosensors, 2014	– The problem was mathematically formulated, and at each stage of enhancement, a reference signal was generated and utilized in NLMS adaptive filter	– NLMS – period extraction – update rule of Beta – adaptive noise canceler – SpO2 & HR extraction	Time-consuming, hence not suitable for time critical applications
VSS LMS	Adaptive reduction of motion artifact from photoplethysmographic recordings using a variable step-size LMS filter, 2002	– The reference signal reflects the motion artifact, but it does not show the extent of the variation of the motion artefact – The algorithm does not test on high speed motion	Variable step-size LMS	
	New LMS adaptive filtering algorithm with variable step size, 2017	The test only compared between LMS algorithm and SVSLMS algorithm	Variable step-size LMS	

Table 2. Comparison study between LMS,NLMS and VSSLMS

Features		Least mean square		
		LMS	NLMS	VSSLMS
Efficiency	Accuracy	Yes [2, 7–9]	Yes [20]	Yes [17, 18]
	Time	Fast [2]	Slow [12]	Fast [12]
High speed motion		–	–	Yes [17, 18]
Computational complexity		Low [2, 7]	Low [12]	Low [17, 18]

As stated in the Table 2, comparison between existing LMS algorithms which are LMS itself, normalized LMS and variable step-size LMS. Variable step-size LMS is chosen as the most suitable algorithm based on convincing discussion as per above.

In term of application, this research aims to remove motion artifact which cover in slow motion until high speed motion or variable activity to get accurate reading of heart beat since the existence of motion artifact cause the inaccurate reading. LMS algorithm is a simple algorithm which could reduce the motion artifact in daily activity which only low motion artifact occurs. Which is why VSSLMS is the selected algorithm for this research that could reduce motion artifact in both situations either slow motion or high-speed motion [17, 18].

All LMS family could get accurate reading, just the different is the time taken to execute the algorithm and get the result since fast in getting the heart beat result also needed because heart rate could stop at any time and if something happens to the heart, the smart band could send a notification to the emergency contact number provided by the user. This is why real-time result also needed in this research even though the main objective in this research is to get high accuracy in heart beat reading.

The VSSLMS algorithm also could cater for high speed motion with low computational complexity which could lead to fast convergence time. As per illustrated in Table 2, most of the spectrum algorithm using more than two algorithms which in the end will cause high complexity in computational process and increase the processing time for the computational process [3].

3   Conclusion

The existing LMS algorithm have a problem in high motion activities since the algorithm is based on fixed step size, while every motion has different step-size. In this review paper, the focus is to compare between the existing LMS algorithms and compare which LMS-based heart rate reading is the most suitable and accurate for motion artifact reduction which could cater slow and high-speed motions. The outcomes from this paper have become part of the motivation of our current research work, which is to optimize VSSLMS through the implementation of regression machine learning, to reduce motion artifact.

References

1. Caitlin, M.: The Apple Watch 4 is giving some cardiologists pause. https://www.tomsguide.com/us/apple-watch-series-4-ekg-sensor,news-28081.html. Accessed 28 Sept 2018
2. Han, H., Kim, J.: Artifacts in wearable photoplethysmographs during daily life motions and their reduction with least mean square based active noise cancellation method. Comput. Biol. Med. **42**(4), 387–393 (2012)
3. Sun, B., Zhang, Z.: Photoplethysmography-based heart rate monitoring using asymmetric least squares spectrum subtraction and bayesian decision theory. IEEE Sens. J. **15**(12), 7161–7168 (2015)
4. Ye, Y., He, W., Cheng, Y., Huang, W., Zhang, Z.: A robust random forest-based approach for heart rate monitoring using photoplethysmography signal contaminated by intense motion artifacts. Sensors **17**(2), 385 (2017)
5. Sawh, M.: ECG explained: why the HR tech from the Apple Watch Series 4 is a big deal. https://www.wareable.com/health-and-wellbeing/ecg-heart-rate-monitor-watch-guide-6508. Accessed 28 Sept 2018
6. Redesigned Apple Watch Series 4 revolutionizes communication, fitness and health. https://www.apple.com/newsroom/2018/09/redesigned-apple-watch-series-4-revolutionizes-communication-fitness-and-health/. Accessed 24 Sept 2018
7. Zhang, G., Wu, T., Wan, Z., Song, Z., Yu, M., Wang, D., Li, L., Chen, F., Xu, X.: A method to differentiate between ventricular fibrillation and asystole during chest compressions using artifact-corrupted ECG alone. Comput. Methods Programs Biomed. **141**, 111–117 (2017)
8. Wijshoff, R., Mischi, M., Aarts, R.: Reduction of periodic motion artifacts in photoplethysmography. IEEE Trans. Biomed. Eng. **64**(1), 196–207 (2017)
9. Peng, F., Zhang, Z., Gou, X., Liu, H., Wang, W.: Motion artifact removal from photoplethysmographic signals by combining temporally constrained independent component analysis and adaptive filter. BioMed. Eng. OnLine **13**(1), 50 (2014)
10. Warren, K., Harvey, J., Chon, K., Mendelson, Y.: Improving pulse rate measurements during random motion using a wearable multichannel reflectance photoplethysmograph. Sensors **16**(3), 342 (2016)
11. Tautan, A.-M., Young, A., Wentink, E., Wieringa, F.: Characterization and reduction of motion artifacts in photoplethysmographic signals from a wrist-worn device. In: 2015 37th Annual International Conference of the IEEE Engineering in Medicine and Biology Society (EMBC) (2015)
12. Yousefi, R., Nourani, M., Ostadabbas, S., Panahi, I.: A motion-tolerant adaptive algorithm for wearable photoplethysmographic biosensors. IEEE J. Biomed. Health Inform. **18**(2), 670–681 (2014)
13. Lo, F.P.-W., Meng, M.Q.-H.: Double sensor complementary placement method to reduce motion artifacts in PPG using fast independent component analysis. In: 2016 38th Annual International Conference of the IEEE Engineering in Medicine and Biology Society (EMBC) (2016)
14. Majumder, S., Mondal, T., Deen, M.: Wearable sensors for remote health monitoring. Sensors **17**(12), 130 (2017)
15. Khan, Y., Ostfeld, A.E., Lochner, C.M., Pierre, A., Arias, A.C.: Monitoring of vital signs with flexible and wearable medical devices. Adv. Mater. **28**(22), 4373–4395 (2016)
16. Clarke, G.W.J., Chan, A.D.C., Adler, A.: Effects of motion artifact on the blood oxygen saturation estimate in pulse oximetry. In: 2014 IEEE International Symposium on Medical Measurements and Applications (MeMeA) (2014)

17. Chan, K., Zhang, Y.: Adaptive reduction of motion artifact from photoplethysmographic recordings using a variable step-size LMS filter. In: Proceedings of IEEE Sensors (2002)
18. Li, H., Tong, N., Liu, N., Jiang, J.: A new variable-step-size LMS adaptive filtering algorithm. In: 2008 9th International Conference on Signal Processing (2008)
19. Schack, T., Sledz, C., Muma, M., Zoubir A.M.: A new method for heart rate monitoring during physical exercise using photoplethysmographic signals. In: 2015 23rd European Signal Processing Conference (EUSIPCO) (2015)
20. Bismor, D., Czyz, K., Ogonowski, Z.: Review and comparison of variable step-size LMS algorithms. Int. J. Acoust. Vibr. **21**(1), 24–39 (2016)

Identification of KDD Problems
from Medical Data

Andrea Nemethova[1(✉)], Martin Nemeth[1], German Michalconok[1],
and Allan Bohm[2,3]

[1] Faculty of Materials Science and Technology in Trnava,
Institute of Applied Informatics, Automation and Mechatronics,
Slovak University of Technology in Bratislava, Bratislava, Slovakia
{andrea.peterkova, martin.nemeth,
german.michalconok}@stuba.sk
[2] Faculty of Medicine, Comenius University in Bratislava, Bratislava, Slovakia
allan.bohm@gmail.com
[3] Research Institute of Academy, Bratislava, Slovakia

Abstract. In our paper, we continue with our research and in cooperation with Faculty of Medicine and Research Institute called Academy. In this paper, we would like to present our knowledge about the identification of KDD (Knowledge discovery in database) problems from medical data. Our research is focused on medical data from the field of cardiology. We are drawing attention to the main issues like identifying the diagnosis or multiple diagnoses of the patient, identification of the influence of the medical parameters on the patient's prognosis and identifying the parameters of a particular diagnosis. We have used various data mining method to achieve the proper results. In our paper, we present the best results that are representing our research.

Keywords: KDD · Data mining · Medical data

1 Introduction

The usage of data mining methods can be normally seen in the field of industry, banking, and marketing [1]. In these areas, data is mostly collected for further processing and evaluation to improve the production process, assess product demand, or create a tailor-made product for the customer. However, there is a growing interest in applying data mining methods to the field of medicine as well. Although this area typically does not store data primarily for such processing, but rather for archiving, the uncontrolled area offers the opportunity to explore unresolved problems that can be mastered by data mining methods.

The knowledge or rather called information pyramid divides terms into 4 groups where data is at the lowest pyramid level. Every other term that is placed in the pyramid above includes basics from the lower group. E.g. each acquired knowledge consists of at least one information and each information contains at least one data item. It works the other way around. E.g. we can convert any information to some extent to knowledge or even wisdom (Fig. 1).

© Springer Nature Switzerland AG 2019
R. Silhavy (Ed.): CSOC 2019, AISC 985, pp. 191–199, 2019.
https://doi.org/10.1007/978-3-030-19810-7_19

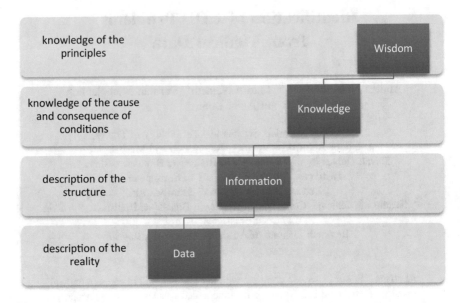

Fig. 1. The information pyramid that divides term into four groups: data, information, knowledge and wisdom [2]

Data are all numbers, facts, symbols and characters that represent a description of reality [2]. They make a basic description of matter, event, or phenomenon without meaning. The data itself does not mean anything. They have no noticeable value. Information is already data with some meaning. When we add value to the data, we will go up in the pyramid and we get the information. Knowledge is defined as information acquired theoretically or practically. If information is available to us to make a decision, the knowledge has become familiar to us. Wisdom is at the very top of the pyramid. Wisdom is the ability to effectively use knowledge for the benefit or knowledge of growth. Using knowledge, we are gradually acquiring experiences that increase the level of wisdom in the area.

An example may be entry 11,9. This entry does not mean anything by itself. However, if MPV is attributed to the number 11.9, it is likely that this number is expressed as mean of platelet volume (MPV). If we know what are the reference values for this parameter, we can conclude that the patient has a mean platelet volume elevated. If the patient has an increased mean platelet volume, it may (but may not) be an ischemic heart disease or a myocardial infarction (MI) (Table 1).

Table 1. The example of entry in the level information pyramid with explanation of what is the value to us

Level of information pyramid	What is the value to us?	Example
Entry	Nothing	11, 9
Information	Answer to the question what	MPV 11,9
Knowledge	Answer to the question how	Elevated mean platelet volume
Wisdom	Answer to the question why	If this value is increased, it may be an MI

2   Background and Specification of the Field of Solution

Cardiovascular diseases are one of the most serious problems of our population. In Slovakia, cardiovascular diseases account for 53% of total mortality and are one of the main causes of shorter life expectancy. Of cardiovascular disease, ischemic heart disease is the most common cause of mortality. Of these, the most common ischemic heart disease is responsible for 19% of all deaths. One of the most serious complications of ischemic heart disease is acute coronary syndrome (ACS), which is most commonly caused by the rupture of atherosclerotic (AS) plaque [3, 4].

Atherosclerotic plaques can be divided into stable and vulnerable, the first type of which continues in the sense of gradual narrowing of the vessel flow, unlike the second type, which may be clinically ill for a long time, but at a certain point there is a cracking with the endothelium of the vessel and thrombus - ACS. Early identification of the vulnerable plaque could lead to a better stratification of the risk of ischemic heart disease as well as to more effective therapeutic prophylaxis in the sense of percutaneous coronary intervention [4, 5].

The multifactorial nature of the AS prediction risk makes it difficult to investigate this phenomenon by traditional statistical methods. Because many types of ischemic heart disease have long been developing and ultimately irreversible, it is essential to make the most of the effort of prevention and the early recognition of pathological processes in the early stages [4, 5]. The methods of data mining and knowledge discovery from databases could significantly improve the ability of early recognition unhealthy states described above.

The process of data mining is important in acquiring knowledge because only one parameter is not enough to determine the diagnostic conclusion. In medicine, consideration should be given to the combination of several medical parameters. However, the analysis of the impacts of a combination of multiple parameters is very challenging, because a large amount of data needs to be analyzed to obtain the relevant conclusion. The manual equivalent of such analysis is time-consuming, inefficient, and often unattainable.

The problems that can be addressed in the field of medicine through the data mining methods can be divided into the following groups:

- Identifying the diagnosis or multiple diagnoses of the patient,
- Identification of the influence of medical parameters on the patient's prognosis,
- Identifying the parameters of a particular diagnosis,
- Detection of significant values of certain parameters of a given diagnosis,
- Determining the patient's diagnosis,
- Prediction of the development and course of biomarkers and health status,
- Prediction of health status and diagnoses on the population sample,
- Determination of procedures and the associated prediction of material consumption or financial costs.

In our research we decided to use the data mining methods to identify the effect of medical parameters on the patient's prognosis, resulting in the area of decision support in determining the appropriate therapy for the patient. Nowadays, the current problem area is in practice, because in determining the therapy, the physician can only rely on the results of the examinations, which he assesses individually and estimates their mutual influence based on their experience. This assessment is based on the specific experience and knowledge of each physician, as the experience and knowledge of each expert is different.

By using data mining methods, it is possible to obtain new knowledge from the data, for example in the form of relationships between individual parameters, their impact on monitored parameters, or in the form of predictions of future development of individual thresholds. Addressing this issue contributed to designing a patient diagnosis model based on the results of the examinations, or other additional data that are proven to be relevant to the correct diagnosis in the review process.

Data we used was divided into 4 classes of patients. Each patient overcame the ischemic heart disease. Each class represents the final outcome of the patients after the underwent treatment. First class consists of patients with the chronic chest pain (which is also called angina pectoris). Second class consists of patients that have encountered non-fatal myocardial infraction. Next class consists of patients that have encountered fatal myocardial infraction because of the ischemic heart disease and the last class of patients consists of patients with successful treatment (healthy patients).

Every patient was represented by 56 parameters and there were 354 patients in our data set. These parameters were mixture of laboratory blood tests, screening examinations and clinical parameters.

2.1   Identifying the Diagnosis or Multiple Diagnoses of the Patient

The identification of the diagnosis of the patient is the classification problem. To solve such a classification problem, it is possible to use the SVM method or neural networks. In our research, we use both of these methods to compare the results. Each of these methods can be used to solve the same type of task, but it differs with the principle of their operation. Both of these methods are applied to the training data set for the purpose of verifying the suitability of their use. When assessing the training algorithm we were mainly concerned with parameters like learning error and test error. These parameters tell us about the success with which these algorithms were able to learn on a given set of data. Both methods are implemented in STATISTICA 13, where we performed the testing itself. At first we have tested the SVM algorithm on the training data set.

In the first type of chosen SVM method, the training process involves minimizing the error function, the function of learning error. This minimization for this type is mathematically expressed as follows [8, 9]:

$$\frac{1}{2}w^T w + C \sum_{i=1}^{N} \xi_i,$$ (1)

subject to restrictions:

$$y_i\left(w^T \varnothing(x_i) + b\right) \geq 1 - \xi_i \, a \, \xi_i \geq 0, i = 1, \ldots, N$$ (2)

Where C represents the constant of the capacity, w is the coefficient vector, b is the constant, and ξ_i represents the parameters for manipulating the nonparametric data (inputs). Index i indicates N training cases. Kernel $\varnothing$ is used to transform data from inputs to function spaces. It should be remembered that the higher the value of C, the more the error of learning is penalized.

We have set the number of iterations to 1000 iterations and the learning error value for the learning process was set to 0,001. It means that the learning process does not have to go through all the iterations. If the learning error gets to 0,001 before the 1000 iterations are reached, the learning process is immediately terminated.

Another method, which we have used for identifying the diagnosis were neural networks. As a dependent variable of classification process, we chose the parameter "outcome", which expresses the resulting prognosis of the patient's health. As continuous inputs, we have selected all the parameters that belonged to this category, and we selected all the categorical parameters from the categorical inputs except for the "outcome" parameter, as we defined it as a categorical target (Table 2).

Table 2. The result of SVM method used for identification of patient's diagnosis with the division of data and accuracy of classification

SVM method results		
Division of data	Total number	449
	Training set	335
	Testing set	114
Accuracy of classification	Training set	100,00%
	Testing set	99,12%

For the training set, we chose 70% of the data that was randomly selected from the data file. For the test set, 15% of data from the dataset fell, and for the validation set it was also 15%. A validation set of data is used to minimize neuronal network syndrome. Using this set of data, the training process verifies whether the training process leads to an increase of the network accuracy. This verification is achieved by using a validation set of data to which the neural network has not yet been exposed. If a situation arises that the accuracy of the training data set increases, but the accuracy on the validation set remains unchanged or decreases, the learning process must be terminated. The following table shows the precision of the neural network classification in the training data set.

2.2   Identification of the Influence of Medical Parameters on the Patient's Prognosis

The identification of the influence of medical parameters on the patient's prognosis is an important issue, because the mortality of patients with the ischemic heart disease can be also caused by not considering the influence of one or combination of medical parameters of patients. In our data set, we were first evaluating the impact of individual parameters that affect the parameters surveyed. In our case, we have verified the effect of individual parameters on the patient's outcomes. The output of this is so-called "Importance plot," a graph or table showing the selection of parameters from a data file that are ranked according to the importance with respect to the effect on the monitored parameter of the resulting prognosis. This order is determined by the Chi-square test method. This output can be considered informative and serves as an initial view of the data file parameters (Tables 3 and 4).

Table 3. The result of neural networks method used for identification of patient's diagnosis with the classified outcome

Neural network results				
Outcome	Fatal IM	Nonfatal IM	Recurrent angina pektoris	Healthy
Total classified	58,00	55,00	74,00	128,00
Correct classified	55,00	55,00	73,00	128,00
Uncorrect classified	3,00	0,00	1,00	0,00
Correct (%)	94,82	100,00	98,64	100,00
Uncorrect (%)	5,17	0,00	1,35	0,000

Table 4. The degree of impact of selected individual parameters on the resulting prognosis of the patient's outcome

Medical parameter	Impact the outcome	Degree of impact
LVEF	Yes	92
Number of stents	Yes	68
Lesion <50	Yes	55
Diastolic	Yes	54
HDL-C	Yes	48
Laterrarwall	Yes	45
Inferiorwall	Yes	45
Anteriorwall	Yes	45
KDIGO	Yes	45
Nitrat	Yes	45
Mg	No	8
UREA	No	7

2.3   Identifying the Parameters of a Particular Diagnosis

To obtain relationships between individual medical parameters we have decided to use the decision tree method. To create a decision tree, we decided to use the C&RT algorithm (Classification and Regression trees). This algorithm creates a tree in such a way that it passes through all the possible divisions of data at each step using all the values of all the predictors, and at the same time looks for the best of these divisions. These divisions are searched by a parameter that is also referred to as data purity.

This means that dividing is good if we get two more homogeneous data sets than we do by using another data partition. The Gini index is the most widely used for solving tasks of the classification type. The value of this index gets zero if the division includes cases that belong to one and the same class. The highest value is reached if the size of the classes after applying the division is the same. The Gini index is mathematically defined as follows [6, 7]:

$$g(t) = \sum_{j \neq i} p(j/t)p(i/t), \tag{3}$$

where p (j/t) is the probability of category j in the given node and p (i/t) is the likelihood of a poor classification of category j as category i.

The rules resulting from the tree must be represented correctly. These rules are represented as a path from the initial node to the end node. The end node in our case expresses the resulting prognosis of the patient, and tree branches leading to this end node represent dividing parameters. Parameters from the created tree are represented in the Table 5.

Table 5. The relationships between selected individual medical parameters influencing the final outcome of the patient obtained by the C&RT method

Parameter	Value of parameter	Outcome
Simvastatin	<= 0,5	Patient is healthy
Simvastatin LVEF	>0,5 <= 34,97	Fatal IM
Simvastatin LVEF	>0,5 <= 40,25 at the same > 34,97	Nonfatal IM
Simvastatin LVEF	>0,5 >40,25	Recurrent angina pectoris

3   Results

In the data mining process, we had a set of comprehensive medical parameters that described the patient's health. Among these parameters were parameters obtained from laboratory assays, such as biochemical and haematological examinations. For identification of the influence of medical parameters on the patient's prognosis, the impact of individual parameters on the monitored parameter was calculated. We determined the impact quotient based on the Chi square test for each parameter separately, finishing the parameters from the most influential parameter to the parameter with the least impact on the resulting prognosis. The Fig. 2 shows the degree of impact of the parameters that have an impact on the final prognosis (Table 6).

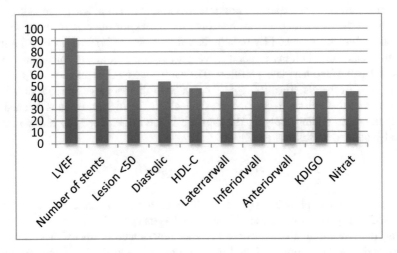

Fig. 2. The degree of impact of chosen parameters that have the influence on the final outcome of the patients.

Table 6. The summary of the KDD problems solved in our research and the used methods to achieve the best results

KDD problems from medical data	Used methods
Identifying the diagnosis or multiple diagnoses of the patient	SVM (Support vector machines) NN (Neural networks)
Identification of the influence of medical parameters on the patient's prognosis	Chi-square test method
Identifying the parameters of a particular diagnosis	C&RT algorithm (Classification and Regression trees)

The next step in the data mining process was to identify significant relationships between parameters influencing the patient's prognosis. For this purpose, we used the C&RT algorithms for decision trees. For identifying the diagnosis or multiple diagnoses of the patient were used the SVM method and neural networks

4   Conclusion

Our paper was devoted to the main KDD problems in the data mining in the field of medicine. We have described the main problems and recommended methods manage these problems. We also present our best results, which we obtained by application these methods on our medical data set. In future we aim to narrow our research to one of mentioned problems and to investigate it at deeper levels. However the main issue with KDD and data mining in medicine in general is the insufficient amount of data. To be able to perform deeper research it is needed to apply information systems to healthcare facilities to collect real structural data. Many examination results are even at this stage of technology stored in a paper form which makes it difficult to find, collect and process to a structural form into database systems.

Acknowledgments. This publication is the result of implementation of the project: "UNIVERSITY SCIENTIFIC PARK: CAMPUS MTF STU - CAMBO" (ITMS: 26220220179) supported by the Research & Development Operational Program funded by the EFRR.

This publication is the result of implementation of the project VEGA 1/0272/18: "Holistic approach of knowledge discovery from production data in compliance with Industry 4.0 concept" supported by the VEGA.

References

1. Ngai, E.W.T., Xiu, L., Chau, D.C.K.: Application of data mining techniques in customer relationship management: a literature review and classification. Expert Syst. Appl. **36**(2), 2592–2602 (2009)
2. Frické, M.: The knowledge pyramid: a critique of the DIKW hierarchy. J. Inf. Sci. **35**(2), 131–142 (2009)
3. Statistics of death. Eurostat, Eurostat explained
4. Gofman, J.W., Young, W., Tandy, R.: Ischemic heart disease, atherosclerosis, and longevity. Circulation **34**(4), 679–697 (1966)
5. Sudhakar, K., Manimekalai, M.: Study of heart disease prediction using data mining. Int. J. Adv. Res. Comput. Sci. Softw. Eng. **4**(1) (2014)
6. Breiman, L.: Classification and Regression Trees. Routledge, New York (2017)
7. Loh, W.-Y.: Classification and regression trees. Wiley Interdiscip. Rev. Data Min. Knowl. Discov. **1**(1), 14–23 (2011)
8. Ma, Y., Guo, G. (eds.): Support vector Machines Applications. Springer, New York (2014)
9. Cai, Y.-D., et al.: Support vector machines for predicting protein structural class. BMC Bioinform. **2**(1), 3 (2001)

Determination Issues of Data Mining Process of Failures in the Production Systems

Martin Nemeth$^{(\boxtimes)}$, Andrea Nemethova, and German Michalconok

Faculty of Materials Science and Technology in Trnava,
Institute of Applied Informatics, Automation and Mechatronics,
Slovak University of Technology in Bratislava, Bratislava, Slovakia
{martin.nemeth, andrea.peterkova,
german.michalconok}@stuba.sk

Abstract. The implementation of data mining methods is equally important in the industrial area as well as their deployment in other areas. Among the benefits that new knowledge can bring include, for example, optimization of production and other processes, fault prediction, or maintenance prediction. In our research we are focused on production processes and obtaining new knowledge from them. This paper deals with the data from automotive industry and with predicting and identifying failures in such production process in automotive industry. The industrial area is appropriate for KDD research because the processes are fully describable and they have finite number of states. Even possible failures can be described before the process is alive. Even though the number of possible states of the system is finite, it is hard to make predictions by hand. Therefor the methods of data mining are appropriate.

Keywords: Production systems · Failures · Data mining

1 Introduction

At present, most production systems have technical means to monitor the whole or the part of the production process, the state of the machines and equipment on a process-level. Storing and further analyzing these data can bring a new knowledge to optimize and streamline the process.

Process control systems with a hierarchical structure consist of analytical data from the process control level and from the data from the analytical level. The level of data analysis is focused on the accumulation of data from the process control level to the data warehouses where the OLAP [4] analyzes are performed.

Knowledge discovery level needs to be added as a next level of data analysis. Figure 1 illustrates a conceptual scheme of the knowledge acquisition system, which also takes into account the level of knowledge acquisition.

For data analysis using the data mining methods, it is necessary to know how to collect required data [5]. In our research, we have focused on the method of collecting data about emerging failures in chosen production process. The following Fig. 2 shows the hierarchical structure of the elements of the selected part of our selected assembly production line.

© Springer Nature Switzerland AG 2019
R. Silhavy (Ed.): CSOC 2019, AISC 985, pp. 200–207, 2019.
https://doi.org/10.1007/978-3-030-19810-7_20

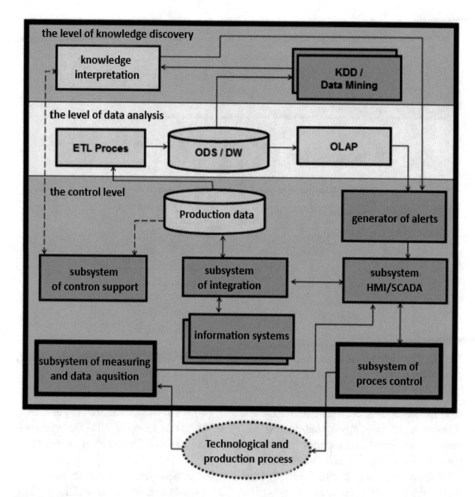

Fig. 1. The conceptual scheme of the knowledge acquisition system, also taking account of the level of knowledge acquisition [6]

As can be seen from the Fig. 2, at the lowest level there are individual working positions that consist of machines and equipment. These devices are equipped with various sensors that monitor the assembly process. These work positions are grouped into logical units that have an internal GMM number with the corresponding number in the system.

2 Background of the Data and Methods

The entire assembly line is connected to a PROFIBUS industrial network. The entire assembly is controlled by a programmable logic controller (PLC) manufactured by Siemens. The type of PLC's are S7 - 400. A SCADA system build in Simatic WINCC,

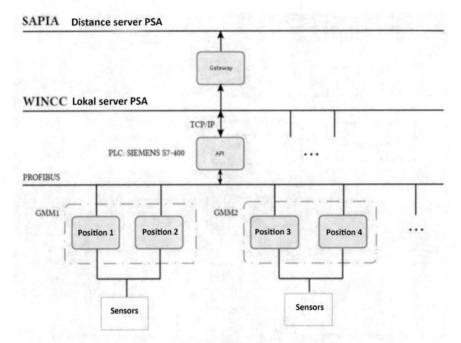

Fig. 2. The hierarchical structure of the elements of the selected part of our selected assembly production line

also from Siemens, is located above the PLC level. This system visualizes the assembly process in real time by communicating with the PLC. Data from this SCADA system is further transferred to a superior SAPIA system, from which reference data on emerging malfunctions were directly exported.

Failure data was in the form of records with multiple describing parameters like name, the time of occurrence of the failure. The records also included additional data, such as the duration of the failure and the location of the failure. A control program generated these records. This program includes data blocks that contain information about positions with respect to the GMM (signature of the station), the identity of the individual sensors on the positions and the predefined reports.

If one or more sensors detect an undesired state or event, the control program generates a record about the occurrence of that particular event (failure) based on the data from the data blocks. This entry will be displayed on a workplace after a query from the WINCC system and stored in the database. This database is located on servers that are physically located in the enterprise. These records are available from this local location for one month. After this time, the records are rewritten. However, the logs are transferred from the local server and are stored on a remote server with the old system called SAPIA. At this point, the data is archived for six months prior to their phased disposal.

The programmable logic controller also receives process data obtained directly from the sensors on the assembly line. These sensors capture various variables such as power, speed, or moments. Also the cycle times of individual workplaces are measured. These data are stored in the PLC internal memory, but they do not have the capacity to store such data for long periods of time. An example of a state belonging to the fault category is the absence of a semi-finished product at any one-job post after a certain period of time. Once this state has been registered, a record of this fact is generated.

2.1 Identification of the Influence of Available Parameters on the Emerging Failures in the Production Process

The first step in the data analysis process using selected data mining methods was the application of interactive drill-down analysis on the data. Drill-down analysis was performed in STATISTICA 13 from StatSoft. This analysis passes through selected parameters and examines the effect of the selected parameters on each other. After selecting the first partitioning parameter, the histogram of the first selected parameter is displayed first. Subsequently, based on the selection of the category from the first histogram of the selected first parameter, the histogram of the next parameter is displayed, but this is already related to the previous selection. Before applying the Drill-down analyses have made the necessary selection of data that we subjected to the analysis.

From the primary data file, we excluded all records that belonged to the status or alarms category by filtering these records from the selection. Therefore, the analysis was performed only for records that belonged to the fault category (Fig. 3 and Table 1).

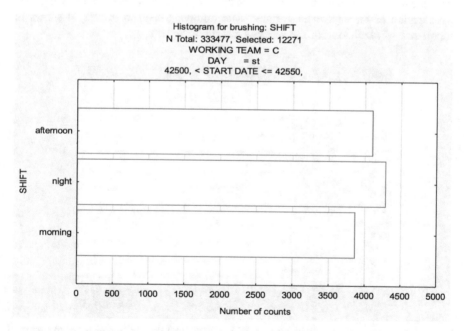

Fig. 3. The histogram which is showing the distribution of the calculations in the case of the *work shift*

Table 1. The number of failures according to the day in the week and shift who was responsible for the failure

Shift	Number of failures	Day (most occurring)
Afternoon	4123	Wednesday
Night	4287	Wednesday
Morning	3786	Monday

2.2 Analysis of the Emerging Failures Using Classification Trees

We have decided to use Classification trees to analyse the failure of the sensor. To analyse this failure type we first had to edit the data file appropriately. From the original data file, we have selected a subset of the data by filtering the data. This subset corresponded to a record of the occurrence of a failure and consisted of 236,116 records.

The selected subset of the data file thus served as the input data set for the classification tree algorithm. To create a tree, it was necessary to specify the parameters within each category. As a dependent variable, we chose the *Workgroup* parameter. Another category of variables was categorical predictors, where we selected parameters *Day* and *Work Shift* as variables. The third category was a continuous predictor. In this group, we have selected the *Duration of a failure* parameter.

After specifying the necessary parameters and creating the classification tree, we have previewed the importance of the individual predictors selected for the correct classification of the dependent variable. The following chart in the Fig. 4 shows the importance of predictors for the selected classification task.

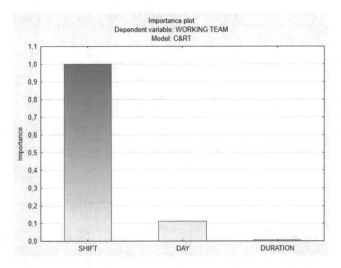

Fig. 4. The chart showing the importance of predictors and depend variables for the selected classification task

As shown in the Fig. 4, the classification tree algorithm evaluated input data as the most important parameter for classifying the parameter *Workgroup* and the parameter *Working shift*. The least important parameter in this task was the *Duration of Failure* parameter, which in the resulting classification tree, whose structure is shown in next the Fig. 5, and it did not even match as a partitioning parameter.

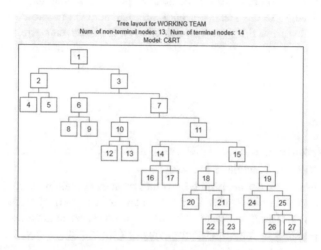

Fig. 5. The structure of the classification tree gained by the C&RT method with the number of terminal nodes for the parameter *Working team*

The structure of the classification tree itself consists of 6 levels, including 13 internal nodes and 14 end nodes. The path from the initial node to each of these end nodes can be represented as one rule. This means that the classification tree found 14 rules for this task.

2.3   Identify Recurring Sequence of Failures in the Data File

Another possibility of failure data analysis is the identification of frequently recurring failure sequences. The sequence represents recurrent occurrences of at least two failures of the same or different type. In order for two or more failures to occur to be considered as a recurring sequence of failures, their recurring occurrences must be within approximately the same time interval and should occur at least three times in the data file.

Finding recurring sequences of failures in the manufacturing system can help to better understand the mechanism of failures in the system itself. Although the failure data we have obtained from the real production system does not contain detailed technical data describing the causes of individual failures in the system, identification of such failure sequences can help to identify the relationships affecting the failures in the system.

To find repetitive failure sequences in the system, we decided to use the cluster analysis method. To find repetitive failure sequences in the dataset, we needed to perform cluster analysis at multiple hierarchical levels, requiring clusters to be defined at each level to achieve the required degree of similarity in clusters. For this purpose, we have decided to use the k-means algorithm because the hierarchical clustering algorithm did not offer this option. K-Means clustering algorithm partitions n objects into k clusters in which each object belongs to the cluster with the nearest mean. This method produces k different clusters of greatest possible distinction. The best number of clusters k leading to the greatest separation (distance) is not known as a priori and must be computed from the data. The objective of K-Means clustering is to minimize total intra-cluster variance, or, the squared error function, which is defined as shown in the Fig. 6 [1–3]:

$$J = \sum_{j=1}^{k} \sum_{i=1}^{n} \left\| x_i^j - c_j \right\|^2, \tag{1}$$

where J is objective function, k represents number of clusters, n is number of classes, $\left\| x_i^j - c_j \right\|$ is s distance function, c is a centroid.

The purpose of cluster analysis was to find clusters containing records of a failure that occurred at similar or at the same time, had similar or equal duration times, but occurred within different weeks. Based on such clustered aggregates of data, it is furthermore possible to track the surroundings of these failures and thus to find the possible recurring sequences of arising failures.

3   Conclusion

The manufacturing industry is generating more and more data that has the knowledge potential to streamline and optimize processes. The presence of such data is important for the data analysis using data mining tools. In our research we have focused on data about emerging failures that do not contain detailed technical information describing the causes of the failure.

To collect detailed data about the process, technical changes in the process are needed. These changes could include the addition of sensors or creation of database or change in the control program. In our research we show, that even with just logs generated by the control system it is possible to successfully perform data mining analysis and gain new knowledge.

In future we will be focused on creating data mining models based on physical model of a production system. This model consists of several PLC's of similar type as the one mentioned in this paper. These PLC's are controlling several machines, which are together forming a complex model of a production system.

Acknowledgments. This publication is the result of implementation of the project: "UNIVERSITY SCIENTIFIC PARK: CAMPUS MTF STU - CAMBO" (ITMS: 26220220179) supported by the Research & Development Operational Program funded by the EFRR.

This publication is the result of implementation of the project VEGA 1/0272/18: "Holistic approach of knowledge discovery from production data in compliance with Industry 4.0 concept" supported by the VEGA.

References

1. Sayad, S.: An introduction to data science, K-Means clustering, 2010-2019. https://www.saedsayad.com/clustering_kmeans.htm
2. Ankerst, M., et al.: OPTICS: ordering points to identify the clustering structure. In: ACM SIGMOD record, pp. 49–60. ACM (1999)
3. Wagstaff, K., et al.: Constrained k-means clustering with background knowledge. In: ICML, pp. 577–584 (2001)
4. Govrin, D., et al.: System and method for analyzing and utilizing data, by executing complex analytical models in real time. U.S. Patent No 6,965,886 (2005)
5. Fayyad, U., Piatetsky-shapiro, G., Smyth, P.: From data mining to knowledge discovery in databases. AI Mag. **17**(3), 37 (1996)
6. Tanuska, P.: Thesis of inauguration presentation, MTF STU (2013)

Indonesian Food Image Recognition Using Convolutional Neural Network

Stanley Giovany, Andre Putra, Agus S. Hariawan, Lili A. Wulandhari[✉],
and Edy Irwansyah

Computer Science Department, School of Computer Science,
Bina Nusantara University, Jakarta 11480, Indonesia
lwulandhari@binus.edu

Abstract. Food image recognition becomes more interesting because it can be useful in health industry to obtain many information from a food image such as calorie, nutrition, carbohydrates, fats and protein. The challenging part of food image recognition is to recognize a food image with different background, intensity, and perspective. Convolutional Neural Network (CNN) seems to be the right choice to build a powerful model that able to recognize food image accurately. Current researches in food recognition use American fast food and Japanese food as the dataset. Different dataset has different treatment to obtain good result. Color and presentation has important role as features. Therefore, this research proposes to recognize five kinds of popular Indonesian food such as meatball (bakso), grilled chicken (ayam bakar), satay (sate), gado-gado (mixed vegetables with peanut sauce), and rendang using Convolutional Neural Network approach. Convolutional Neural Network with standard architecture and inception-v4 model are chosen as techniques with Adam as the optimizer function. Experimental results show the implementation of CNN for image recognition can achieve the top-1 testing accuracy around 76.3% for standard network and 95.2% for inception-v4 network.

1 Introduction

After the succession of AlexNet [7] in ILSVRC (ImageNet Large Scale Visual Recognition Competition) 2012 [11], convolutional neural network has been adopted and improved to produce a better result of image recognition especially for large image dataset such as ImageNet [1]. Previous researches had developed food images recognition using CNN with various datasets. In research of [4], CNN used to recognize 10 food categories on food logging system and achieve 93.8% accuracy. In another research [14], deep convolutional neural network used to recognize 100 food categories in UEC-FOOD100 dataset and achieve 72.26% accuracy. Food image recognition becomes more interesting because it can be useful in health industry to obtain many information from an image such

The authors thank to Bina Nusantara University for the research grant and supporting this research.

© Springer Nature Switzerland AG 2019
R. Silhavy (Ed.): CSOC 2019, AISC 985, pp. 208–217, 2019.
https://doi.org/10.1007/978-3-030-19810-7_21

as calorie, nutrition, etc. The challenging part of food image recognition is to recognize a food image with different background, intensity, and perspective. Current researches use American fast food [15] and Japanese food as the dataset [5]. Different dataset has different treatment to obtain good result. Color and presentation has important role as features. Therefore, this research proposes to recognize five kinds of popular Indonesian food such as meatball (bakso), grilled chicken (ayam bakar), satay (sate), gado-gado, and rendang using Convolutional Neural Network and deep learning approach, also analytical result for adam optimizer which is giving an impact to the training process.

Convolutional neural networks is popularly used for recognition task especially in image recognition after the success of [8] on handwritten digit recognition. Later on, deep convolutional neural networks are invented for the larger image dataset such as ImageNet [1]. The AlexNet as introduced in [7] is one of deep convolutional neural networks architecture that winning the 2012 ImageNet competition [11]. Later on, very deep convolutional neural networks are invented. The GoogLeNet as introduced in [13] is one of very deep convolutional neural networks architecture. Recently, Inception-v4 architecture is introduced [12] as the refinement from GoogLeNet (Inception-v1) that achieve very good performance at relatively low computational cost [12].

In this research, standard CNN which is constructed from the beginning is compared to established Inception-v4 in food recognition. Inception-v4 applied transfer learning which has been trained in Imagenet dataset. This paper is arranged into five section. First section describes background of research and related work of CNN in food recognition. Section 2 presents dataset preparation which is followed by Algorithm development and optimizer and loss function in Sects. 3 and 4 respectively. Experimental results and analysis of performance comparison from two model of CNN is presented in Sect. 5. Last, this paper explains conclusion of the development algorithm, experimental result and analysis in Sect. 6.

2   Dataset Preparation

This research used Indonesian Food Image dataset. It consists of 5 food categories that are meatball (bakso), grilled chicken (ayam bakar), satay (sate), gado-gado, and rendang. The images were collected from the web which can be uploaded widely and open. Google captcha feature is added into the web to avoid spam from the uploaders. Besides the system, data collection is also conducted by crawling food image from the Google using *Bulk Media Downloader* at the Firefox to record url of the image which will be downloaded through *wget* in Linux. The data is labeled by human manually into 5 categories which is shown in Fig. 1. In order to exclude duplicate image, the system checks new uploaded image with existing image using hashing md5 approach. Admin of the system validates images data periodically according following criteria:

- It contains a whole food in an image
- It has minimum other objects beside the food.

Fig. 1. Indonesian food image dataset

Total images in the dataset are 11000 for 5 categories, with detail meatball (bakso): 2144, grilled chicken (ayam bakar): 2451, satay (sate): 2136, gado-gado: 2088, and rendang: 2181 data sample. The collected images have various size with some images are too wide or too long. These images are cropped to eliminate parts of the image that are too wide or long and take only the image with minor objects in the background. In the next step, all images are resized to 2 different sizes, namely 128 × 128 and 299 × 299.

3   Indonesian Food Image Recognition Algorithm

In this research, two models of CNN are implemented to be compared to observes which model gives the highest accuracy for food image recognition. Detail of the models are presented in Subsects. 3.1 and 3.2 respectively.

3.1   Standard Architecture

Images with size 128 × 128 with 3 RGB channels are used as input. The next layer consists of three convolutional layers, three pooling layers, two dropout layers, and three fully-connected layers. Each convolutional layer and the second fully-connected layer used ReLU [9] as the activation function. Dropout [2] is adopted in this architecture to reduce the overfitting. The third max pooling layer is dropped out by 25% and the second fully-connected layer is dropped out by 50%. The output layer used softmax as the activation function. The detail of convolution process and architecture of standard CNN is shown in Fig. 2 and Table 1 respectively.

In the third convolution process, from 72 filters, The 25% drop out process is conducted randomly and obtained 16 × 16 × 54 output size. This results are flatten into 13824 neuron to fully connected step. This fully connected applies 1 hidden layer with 128 neurons and RELU as activation function. Hidden layer to output operation applies 50% drop out randomly, and generate 64 neuron.

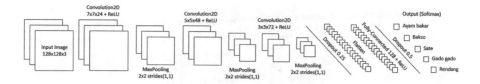

Fig. 2. Standard architecture

Table 1. Convolution process for standard CNN

	Convolution 1	Convolution 2	Convolution 3
Filter size	7×7	5×5	3×3
Number of filter	24	48	72
Max pooling	2×2	2×2	2×2
Stride	1	1	1
Output size	$64 \times 64 \times 24$	$32 \times 32 \times 48$	$16 \times 16 \times 72$

Output neurons are set into five neurons which represents five types of Indonesian food. Standard CNN uses cross- entropy as loss function and ADAM as the oprimizer. Experiment results will be shown in Sect. 5.

3.2 Inception-v4 Architecture

Inception-v4 architecture is one of the very deep convolutional networks introduced in 2016. In this research, Inception-v4 architecture is implemented to this dataset with different optimizer from the original paper [12]. The overall scheme of Inception-v4 is shown in Fig. 3.

Images with size 299×299 with 3 RGB channels are used as input. Batch normalization [3] is applied after each convolution before the activation. The activation function used for each convolution is ReLU [9]. In this research, the output of the Inception-v4 model only has 5 neurons (each neuron indicate each food category). Softmax is used as the activation function for the output layer.

Inception - v4 model consist of several modul namely stem, inception, reduction, average pooling, drop out and fully connected. Each modul has their own architecture with variation of layer and number of filters. In the implementation, training is carried out using mini batch learning, with 32 data sample in each batch. Cross entropy and Adam optimizer is also applied as loss function and learning function respectively. Initial kernel uses pre-trained model Inception - v4 where stem, inception-A, reduction-A, inception-B and reduction-B modul are frozen such that there are not weight updated. It uses weights from trained model using ImageNet data. Result of inception-v4 is described in Sect. 5.

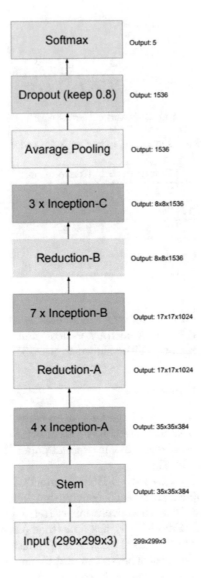

Fig. 3. Inception-v4 scheme

4   Optimizer and Loss Function

Adam optimizer is used to train the model. Adam is an algorithm for first-order gradient-based optimization of stochastic objective functions, based on adaptive estimates of lower-order moments [6]. In order to adjust the weights after the forward pass, Adam calculates the new weight with formula as shown in Eq. 1

$$\theta_t = \theta_{t-1} - \alpha \frac{\hat{m}_t}{\hat{v}_t + \epsilon} \tag{1}$$

where, α indicates the learning rate and θ_t is weights at timestep t. The ϵ is a small value that used to avoid division by zero. $\hat{m}_t$ and $\hat{v}_t$ can be calculated with formula as shown in Eqs. 2 and 3.

$$\hat{m}_t = \frac{m_t}{1 - \beta_1^t} \tag{2}$$

$$\hat{v}_t = \frac{v_t}{1 - \beta_2^t} \tag{3}$$

β_1 and β_2 are exponential decay rates, where $0 \le \beta_1 < 1$ and $0 \le \beta_2 < 1$. The β_1^t and β_2^t denote β_1 and β_2 to the power t. The m_t and v_t can be calculated with formula as shown in Eqs. 4 and 5.

$$m_t = \beta_1 \cdot m_{t-1} + (1 - \beta_1) \cdot g_t \tag{4}$$

$$v_t = \beta_2 \cdot v_{t-1} + (1 - \beta_2) \cdot g_t^2 \tag{5}$$

The g_t^2 indicates the elementwise square $g_t \odot g_t$. The g_t is gradients w.r.t. stochastic objective at timestep t, it can be calculated with formula as shown in Eq. 6.

$$g_t = \nabla_\theta f_t(\theta_{t-1}) \tag{6}$$

The $f(\theta)$ is the stochastic objective function with parameters θ. In this research, cross-entropy is used as the objective function. The cross-entropy is a function that measure how close the predicted output of the model matches the desired output, if it exactly matches, it will equals to zero, otherwise it will be a positive number. The cross-entropy function is shown in Eq. 7.

$$H(L, O) = -\sum_i L_i \log O_i \tag{7}$$

where L denotes the desired output and O denotes predicted output.

4.1   Transfer Learning

When implementing an Inception-v4 model to this dataset, transfer learning is adopted [10]. The initial weight of the model is loaded from the pre-trained Inception-v4 model. Some layers are frozen before the model is trained using this dataset. During the training, the weight of the frozen layers is not updated. Some layers which being freeze are layers from the Input until the Reduction-B modules.

The model is trained using Adam optimizer with cross-entropy as the loss function. The learning rate is set to 0.001 with a batch size of 32 examples, the value of β_1 is set to 0.9, the value of β_2 is set to 0.999, the value of ϵ is set to $10 - 8$, and the number of epochs is 100.

5 Experimental Results and Analysis

In the first experiments, we use only 5588 data sample which is divided randomly into 4588 data for training, 500 data for validation and 500 data for testing. During training, we observe the top-1 training accuracy and validation accuracy of Inception-v4 model and Standard model for the same dataset. The results are presented in Figs. 4 and 5.

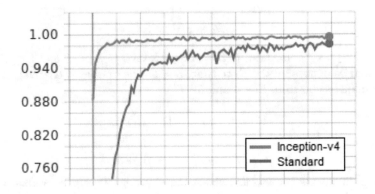

Fig. 4. Top-1 training accuracy during training of Inception-v4 network and Standard network

From the results above, each model is trained for 100 epochs. The model used for testing is taken from epoch where the validation accuracy is the highest. The parameter used and the testing accuracy is presented in Table 2.

Table 2. Top-1 testing accuracy of Inception-v4 network and Standard network.

Model	Top-1 testing accuracy (%)
Standard	76.3
Inception - V4	95.2

Figure 4 shows that inception-v4 has better accuracy compare to standard CNN, which is also confirmed by testing accuracy with 18.9% higher. However, in validation process (Fig. 5), it is shown that inception v-4 accuracy fluctuates in the range of 20% to 96%. While standard CNN tends stable, yet trapped in accuracy around 70%. This condition describes that on average inception-v4 gives higher accuracy compare to standard CNN, except in stability of validation, inception-v4 has not fulfilled it. To overcome this condition, we add number of data sample into 11000 which is divided randomly into 8800 data for training, 1100 data for validation and 1100 data for testing.

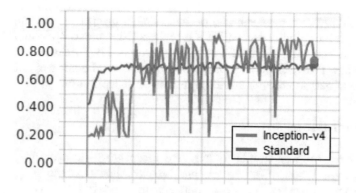

Fig. 5. Top-1 validation accuracy during training of Inception-v4 network and Standard network

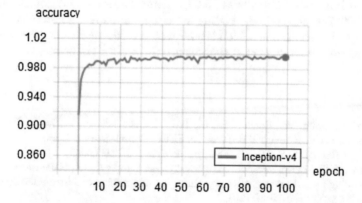

Fig. 6. Top-1 training accuracy of inception-v4 using 11000 data

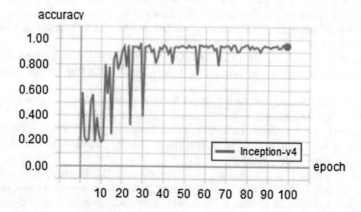

Fig. 7. Top-1 validation accuracy of inception-v4 using 11000 data

Using this 11000 data sample, performance of inception-v4 is improved and stable, both for training dan validation process (Figs. 6 and 7). While for testing accuracy, it improves into 96.8% or in other words, 1.6% higher than using 5588 data sample. This achievement shows that quantity and quality of data has large influence to performance of learning process on inception-v4.

6   Conclusion

This research propose standard CNN and inception-v4 to classify five types of Indonesian food image. The experimental results show that the inception-v4 are capable to recognize the food image better than standard CNN even with various shapes and appearances. Quantity and quality of training data also play an important role to obtain good performance of the model. It is shown that using 5588 data inception-v4 achieves 95.2% testing accuracy and not stable, while using 11000, it achieves 96.8% testing accuracy and more stable and convergence after 70epoch. Additional of food type will provide more interesting research scope and can be implemented significantly in various applications.

References

1. Deng, J., Dong, W., Socher, R., Li, L.-J., Li, K., Fei-Fei, L.: ImageNet: a large-scale hierarchical image database. In: IEEE Conference on Computer Vision and Pattern Recognition, CVPR 2009, pp. 248–255. IEEE (2009)
2. Hinton, G.E., Srivastava, N., Krizhevsky, A., Sutskever, I., Salakhutdinov, R.R.: Improving neural networks by preventing co-adaptation of feature detectors. arXiv preprint arXiv:1207.0580 (2012)
3. Ioffe, S., Szegedy, C.: Batch normalization: accelerating deep network training by reducing internal covariate shift. arXiv preprint arXiv:1502.03167 (2015)
4. Kagaya, H., Aizawa, K., Ogawa, M.: Food detection and recognition using convolutional neural network. In: Proceedings of the 22nd ACM International Conference on Multimedia, pp. 1085–1088. ACM (2014)
5. Kawano, Y., Yanai, K.: Real-time mobile food recognition system. In: Proceedings of the IEEE Conference on Computer Vision and Pattern Recognition Workshops, pp. 1–7 (2013)
6. Kingma, D.P., Ba, J.: Adam: a method for stochastic optimization. arXiv preprint arXiv:1412.6980 (2014)
7. Krizhevsky, A., Sutskever, I., Hinton, G.E.: Imagenet classification with deep convolutional neural networks. In: Advances in Neural Information Processing Systems, pp. 1097–1105 (2012)
8. LeCun, Y., Boser, B.E., Denker, J.S., Henderson, D., Howard, R.E., Hubbard, W.E., Jackel, L.D.: Handwritten digit recognition with a back-propagation network. In: Advances in Neural Information Processing Systems, pp. 396–404 (1990)
9. Nair, V., Hinton, G.E.: Rectified linear units improve restricted boltzmann machines. In: Proceedings of the 27th International Conference on Machine Learning (ICML 2010)
10. Pan, S.J., Yang, Q., et al.: A survey on transfer learning. IEEE Trans. Knowl. Data Eng. **22**(10), 1345–1359 (2010)

11. Russakovsky, O., Deng, J., Hao, S., Krause, J., Satheesh, S., Ma, S., Huang, Z., Karpathy, A., Khosla, A., Bernstein, M., et al.: Imagenet large scale visual recognition challenge. Int. J. Comput. Vis. **115**(3), 211–252 (2015)
12. Szegedy, C., Ioffe, S., Vanhoucke, V., Alemi, A.A.: Inception-v4, inception-resnet and the impact of residual connections on learning. In: AAAI, vol. 4, p. 12 (2017)
13. Szegedy, C., Vanhoucke, V., Ioffe, S., Shlens, J., Wojna, Z.: Rethinking the inception architecture for computer vision. In: Proceedings of the IEEE Conference on Computer Vision And Pattern Recognition, pp. 2818–2826 (2016)
14. Yanai, K., Kawano, Y.: Food image recognition using deep convolutional network with pre-training and fine-tuning. In: 2015 IEEE International Conference on Multimedia & Expo Workshops (ICMEW), pp. 1–6. IEEE (2015)
15. Yang, S., Chen, M., Pomerleau, D., Sukthankar, R.: Food recognition using statistics of pairwise local features. In: 2010 IEEE Conference on Computer Vision and Pattern Recognition (CVPR), pp. 2249–2256. IEEE (2010)

Prevention of Local Emergencies in the Mechanical Transport Systems

Stanislav Belyakov[✉], Marina Savelyeva, Alexander Bozhenyuk,
and Andrey Glushkov

Southern Federal University, Taganrog, Russia
{lbeliacov, avb}@yandex.ru,
marina.n.savelyeva@gmail.com, andrey.drv@gmail.com

Abstract. The paper presents with the problem of detecting emergency situations in a mechanical transport system and taking measures to prevent them. The mechanical transport system is described by a multi-product network model whose flows have variable intensity. The cause of the emergency is the interaction of the flows on the conveyor. The condition for the occurrence of a local accident situation is given by a logical expression linking the intensities of individual product flows. The condition is formulated by experts. Emergency prevention strategies are selected based on the condition of occurrence. The strategy is to choose a backup path for transmitting a flows or reduce its intensity at the input. The ambiguity and uncertainty of the input parameters and the solutions being formed is overcome by the introduction of a special way of representing knowledge - an image of the threat of an accident. The image includes the previously recorded accident precedent and a set of its permissible transformations. The permissible transformations reflect the analytical experience of the expert, obtained as a result of a posteriori analysis of the precedent. The concept of permissible transformations makes it possible to increase the reliability of decisions made. Algorithms for transforming the experience of analyzing and preventing threats are given. The limits of application of the proposed method are discussed.

Keywords: Mechanical transport system · Reconfiguration ·
Intelligent system · Image analysis of precedents

1   Introduction

A distinctive feature of the mechanical transport systems (MTS) is the presence of a conveyor network. They are intended to carry transported objects. The nodes of the conveyor network are mechanical switches for the direction of transport of an individual object. All information about the object and transportation parameters is contained in a label attached to it. MTS has an arbitrary number of input and output conveyors. This makes it possible to move cargo between different points in space. A typical example of MTS are baggage handling systems at airports. In general, the MTS should be attributed to the IoT (Internet of Things) systems due to the fact that the transportation process is closely related to the network information exchange.

© Springer Nature Switzerland AG 2019
R. Silhavy (Ed.): CSOC 2019, AISC 985, pp. 218–228, 2019.
https://doi.org/10.1007/978-3-030-19810-7_22

MTS nodes are equipped with controllers. They perform cargo flow control in real time. This case may have distinctive implementation complexity. In the simplest case, the objects are redirected in a fixed manner according to the address of the destination node. In other cases the direction of transmission in the node is selected and based on various parameters. These can be the state of the MTS, the state of flows of different directions, restrictions on the cost and time of transportation, the risk of accidents and the consequences of transport network failures that have already occurred. Routing of the transported objects is the main tool of adaptation of the MTS to the external environment.

Emergency situations during transportation is an important problem of analysis and synthesis of MTS. One aspect of the problem is identifying hazardous situations that could potentially lead to an accident. Possible losses can be minimized if it is possible to identify the danger and find a way to eliminate the threat. Such a strategy for the operation of MTS is promising but has not been studied enough.

This paper analyzes the model of protection against emergency situations. The model is based on expert data of the risk of interaction of heterogeneous flows. For example, an increase in the intensity of two flow, whose objects are light but different in large dimensions, can lead to traffic jam or loss of cargo units. In another case, quite heavy but compact in size elements of several high-intensity streams can cause a slowdown in the speed of a separate conveyor or its temporary stoppage. In any case, it should be assumed that experts are able to formulate the logical conditions in which the corresponding threat arises. Subsequent actions of MTS should allow it to be eliminated.

2   Known Methods

In general, MTS solves the following problem:

$$\begin{cases} E_{fare}(t) + E_{loss}(f) \to \min, \\ t \in [T_1, T_2], \\ f \subseteq F, \end{cases} \tag{1}$$

where $E_{fare}(t)$ is the cost of transportation during the time interval $[T_1, T_2]$ of the multicommodity flow, $E_{loss}(f)$ are the losses caused by emergency situations of a set of potentially possible emergency situations F. In this paper, it is assumed that situations that arise due to the interaction of multicommodity flows are included in F. Each MTS conveyor carries a set of dissimilar objects. At certain levels of intensity of their arrival on the conveyor, emergency situations may arise. For instance, large items can push each other off the conveyor, small heavy objects can cause congestions in front of the direction switches, the speed pulsations of one of the streams can damage the packaging of other current elements, etc. Incidents relate to both transported objects and the transport system itself. In any case this increases the value $E_{loss}(f)$ in (1), it can be reduced by increasing the cost of organizing transportation. The essence of this approach lies in the leading of control of multi-product flows (multicommodity flows) in the local areas of possible accidents. If such a threat is real, MTS controllers must consistently redirect one or more threads to prevent the threat. This mechanism is the subject of research in this paper.

Network flow analysis methods have been in-depth theoretically investigated [1, 2]. There are known methods for solving problems of minimizing the cost of multicommodity flows with various kinds of constraints. In the paper [3] the theorem of a capacity of network for multicommodity flows was proved. However, this result determines the sign of the accident situation for any subnetwork of the MTS in the absence of mutual influence of flows.

There are known works in which the change in the network capacity is studied under various constraints. For example, models of amplification (attenuation) of fluxes are known [4]. Their application in the concerned problem is difficult, since the gain function is not continuous and monotonous.

The impact of flows on each other on the conveyor can be formalized by predicate. This predicate depends on the assignment of flow intensity values to specified intervals. In this case, to detect an emergency, need to solve the continuous knapsack problem. The role of the backpack is played by possible ways in the network [5]. The flow distribution found is a solution in case of the accident. The application of these results is difficult because of the fact that the predicate may include complex conditions. For instance, such conditions as "a sharp increase in flow A", "smooth fluctuations in flow A", "synchronous changes in flows A and B". The presence of dynamics complicates the search for solutions.

The common problem with the use of network analysis methods is the inaccuracy, ambiguity and uncertainty of the description of the mutual influence of flows. Experts are able to describe the logic of the effects of flows on each other in a qualitative rather than quantitative way. Because of this description the constructed solutions cannot be exact. The use of non-clear logic and linguistic variables [6, 7] does not give the expected result. The reason is that the membership functions of the parameters of situations differ significantly. Therefore, for each local threat it is necessary to define its own linguistic variables. They cannot be reused later. As a result, the experience is unique and cannot be extended to the new cases.

It can be use other intelligent control methods, if it is considered the uncertainty and incompleteness of the original data to solve the problem (1). In particular, the case based reasoning (CBR) [8]. The presence of expert knowledge and experience in implementing solutions is the serious argument in favor of applying CBR. It was used in the works [9, 10], that investigated the conceptual model of precedent images. This model requires a further analysis of the mechanism for transferring experience to new precedents of local threats.

Recently, there has been noticed an interest in the Disaster Area Network (DAN) [11]. The subject of research in these papers is network protocols. They provide information exchange for the implementation of decisions to eliminate disasters. The principles of protocol exchange can be used to solve the problem (1) in the MTS. This requires the analysis of necessary information links.

The analysis of the well-known works indicates the need for further study of the way to solve the problem (1) based on the use of experience in preventing local threats of emergency situations.

3   Method of Detecting Local Threats and Their Prevention

The proposed method includes the following steps.

The first step is to describe the image of a local threat. It was observed by experts in practice and was analyzed in order to identify the underlying links between the elements of the precedent. If this was achieved then the meaning of the situation and the decision taken is considered to be revealed. In the work [9] it was proposed to reflect the meaning of a set of permissible transformations of the situation and decisions. Valid transformations are the way of representing knowledge. This knowledge is much closer to figurative thinking and intuition than traditionally used [12].

The situation was observed and analyzed by the expert then its image was described by the expression

$$I = <c, H(c)>, c \subseteq H(c), H(c) \subseteq J \tag{2}$$

where c is the center of the image corresponding to a precedent with actually fixed values of the observed and measured parameters of the situation. $H(c)$ is admissible transformations of the center which preserve the meaning of the situation as a whole. The $J = J_1 \times J_2 \times \ldots \times J_n$ is figurative space, in which the centers and transformations of the images are defined, is described according to the problem to be solved. In this case the axes of measurement of space $(J_i, i = \overline{1, n})$ are:

- network diagram. For example, the diagram may show the local conveyer exposed to the threat, as well as ways to ensure the redirection of flows to counter the threat;
- the time axis. It displays events and time intervals describing situations;
- the axis of values of the intensities of multicommodity flows through which the interaction of flows is displayed;
- a set of axes showing the mass, dimensions of the flow elements, their resistance to mechanical stress, elasticity of shape, etc.

In this case we can see that the axes are either continuous or discrete. Some object in the image space is linked by a hyperlink with a point on some of the listed axes. Note that the same principle is implemented by geographic information systems. In these systems the continuous plane of a geographical map is linked by references to discrete elements—records in databases, files, and Internet sites.

As an example, Fig. 1 is shown by a thick solid line a fragment of the network with a conveyor. The paths leading the flows to the fragment in question are shown by dotted lines. Allowable fragment transformations (that is conveyors, the addition of which to the original network fragment does not change the meaning of the situation and the solution applied) are also shown in dotted lines.

In Fig. 2 a possible solution is illustrated by a double dashed line. The solution is to redirect part of the stream, bypassing the dangerous MTS fragment. The permissible transformation of the solution, which does not change the meaning of the situation as a whole is shown by a gray dashed line. This solution uses a different path to redirect flow.

The described image is useful because it reflects not a single precedent but a family of common scenarios for the sequence of events. Note that the inaccuracy or ambiguity

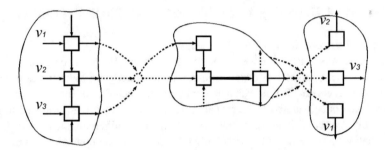

Fig. 1. The local area of the MTS included in the image of an emergency

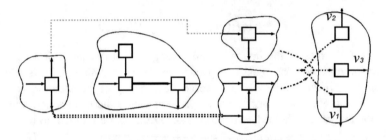

Fig. 2. Valid transformations of the solution

of permissible transformations naturally complement the description of the image not excluding and not compensating each other.

The second step consists of determining the areas of application of the synthesized image. This is an attempt to apply experience to other situations. The experience of counteracting the threat in a given area of network requires making sure that the necessary conditions for an emergency exist. Next it is necessary to make sure that in fact the hypothetical situation is equivalent to that studied earlier. If this is so, the decision of opposing is based on the deductive conclusion that the new situation is a special case studied earlier. Formally, the mechanism of transfer of experience in this case can be described as follows:

$$<\bar{c}, H(\bar{c}) > \ = F_{TR}(<c, H(c) >$$

(3)

Here $\bar{c}$ is the transformed center of the image. $H(\bar{c})$ is its valid transformations; F_{TR} is transformation function. The transformation invariant should be the meaning of situations and solutions. The meaning is represented by valid conversions as indicated above. Therefore, the conversion of an admissible transformation is possible if it does not violate the topology of the target domain. The function F_{TR} uses the topology of the space J, i.e. all relationships specific in this case for MTS and its elements, multi-commodity flows, time chart of events, etc. The result of applying the transformation function is a new image according to which it is planned to counter the threat of the accident. Transformed images of given situation form the area of application of the experience of its prevention:

$$J_c = F_{TR} <c, H(c)>, J_c \subseteq J.$$

The third step is to develop a protocol for implementing protection against threats in the MTS. The technical implementation of the threat protection is based on the software configuration of the network controllers in order to fix the measured flow parameters and the state of the MTS as well as the coordinated implementation of accident prevention actions. Consistency is achieved by implementing a protocol that fixes the onset of a pre-emergency condition, the implementation of protective actions and the disappearance of an emergency. New parameter values are loaded into the controllers when the characteristics of the multiproduct flow change at the MTS inputs.

4   Transformation Function for Local Threats

Let us consider the features of the implementation of the function transformation. The implementation consists in development of an algorithm that constructs a transformed object at a given point in the space of images. Formally, the transformation function is partial i.e. it is not defined for all possible argument values. Accordingly, the algorithm can normally terminate without constructing a transformed object.

As it follows from expression (3): F_{TR} represents the set $H(c)$ to the set $H(\bar{c})$. The image is described by heterogeneous data, so the function F_{TR} should be considered as a set of transformation functions for each of the coordinate axes of the image space:

$$F_{TR} = \{f_{J_i}\} i = \overline{1, n},$$

where n is the dimension of the space of images. The condition for the existence of the transformed image is

$$f_{J_i}(<c, H(c)>) \neq \varnothing, i = \overline{1, n}.$$

As $H(\bar{c}) \neq \varnothing$, any of the existing transformations can be chosen as the center of the transformed image. This corresponds to the definition of the image (2).

The algorithm should be built according to the specifics of a particular dimension of the space of images. The main requirement for the algorithm of the function's implementation is the maintenance the topological limitations of the space of images.

To transform local threats, the following transformation algorithms should be synthesized in the MTS:

- for sections of the transportation network;
- for the relationship between the moments of the occurrence of events, their duration;
- for the conditions of the mutual influence of multi-product flows on each other;
- for the distribution of flows in the MTS.

Further, let us consider examples of algorithms for implementing transformation functions for analyzing the factors of their complexity.

5   Examples of Transformation Functions

Let us consider the algorithm for transforming the MTS subnet. The input data for the algorithm is a set of image subnets, one of which is included in the center of the image. Figure 3 is shows an example of a subnet center and its admissible transformations. In Fig. 3a shows the center in the form of a graph, arcs of permissible transformations are depicted. In Fig. 3b, c, d are illustrated those variants of the subnet structure that do not change the meaning of the image.

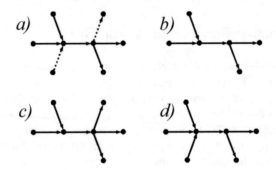

Fig. 3. Set of subnet centers and its valid transformations

The transformation invariant is the topology of the transportation network. The transformation algorithm is described as follows:

1. To extract the description of the set of admissible center transformations from the database of the image and create the collection *AdmissibleTransformations*.
2. To select the *AnalyzedRegion* on the network graph.
3. *TransformationNumberCount* = 0, to create an empty *ExecutedTransformation* collection.
4. For each *Element* of the collection *AdmissibleTransformations* to execute:
   a. if the *Element* and *AnalyzedRegion* are isomorphic, then to increase by 1 the *TransformationNumberCount* and to add the *Element* to the *ExecutedTransformation* collection.
   b. otherwise to continue the cycle.
5. The end of the algorithm.

The nonzero value of the *TransformationNumberCount* indicates the existence of a transform function value. The key action of the algorithm is to check isomorphism of graphs. This task is solved by procedures from modern programming libraries [13]. The time complexity of the isomorphism check procedure determines the complexity of the algorithm as a whole.

Another example would be the function of transforming the conditions of the interinfluence of multiproduct flows. The transformation invariant is the network topology and the intensity of product flows at the network inputs. Algorithm of transformation can be built as follows:

1. To perform a parsing of the logical threat condition and to form a collection of *AdmissibleTransformations* of possible values of intensity of flows that can lead to the accident.
2. To set the network inputs and the intensity of the input flows of each product as an object of the *AnalyzedSituation*.
3. *TransformationNumberCount* $= 0$, to create an empty *ExecutedTransformation* collection.
4. For each *Element* of the collection *ExecutedTransformation* to execute:
   a. if the *Element* and the *AnalyzedSituation* generate an admissible solution to the problem of distributing input streams in the network, then to increase by 1 the *TransformationNumberCount* and to add the *Element* to the *ExecutedTransformation* collection.
   b. otherwise to continue the cycle.
5. The end of the algorithm.

The value of the ExecutedTransformation indicates the existence of the value of the transformation function, as it does in the previous algorithm. Finding an admissible solution to the problem of stream distribution is also solved in the standard way [13].

As a numerical example, let us consider the situation when three input flows have intensities $v_1 = 10$, $v_2 = 15$, $v_3 = 12$ and the accident is detected on the certain conveyor. Expert analysis showed that the logical condition for the accident is

$$A = ((v_1 = 3)or(v_1 = 5))and(v_2 \approx 7)and(v_3 > 2)and(v_3 < 6).$$

Possible conversions for flux intensities are values (v1 = 6 or v1 = 4), v2 = 8, (v3 = 2 or v3 = 6).

Then the set of permissible transformations is represented by the set of tuples

$$v_1' \times v_2' \times v_3' = \{3,4,5,6\} \times \{7,8\} \times \{2,3,4,5,6\}.$$

Through v_1', v_2', v_3' the set of intensity values of each of the flows is denoted. The set is the result of analysis of the logical expression.

The validation of the solution of the distribution problem is implemented basing on the network topology. For example, the nominal throughput of the conveyor is 8 units. In this case, none of the tuples will give a valid solution. Consequently, the threat condition does not transform into a given conveyor.

6 Discussion

The advantage of the proposed method is the use of deep expert knowledge. This knowledge is extremely difficult to mining. The concept of representation of situations and decisions by images makes it possible to manifest deep knowledge in the form of a description of various possible transformations. The expert does not receive an explanation of how the transformation is constructed. He receives the reliably constructed instance of the transformation. An instance is more credible than an explanation of how it was designed. The reason is that the expert may not be aware of all the causes of the situation and acts instinctively. Intuition is based on deep knowledge, and it increases the importance of using possible transformations [14].

Representation by images is intended for accumulating the experience of analyzing situations, but not the experience of observing them. The principal difference consists of identifying not only the essential features of situations, but also the essential elements of their behavior. Behavior is a reaction to changes in external conditions. Exact repetition of precedents of experience in the real world is impossible; therefore, the lack of knowledge about behavior gives rise to unreliable hypotheses.

The proposed method explicitly uses the topology of the space of the problem being solved. The classic CBR topology is implicitly reflected by the situation proximity metric. Any metric is given by the expression of a high level abstraction [8]. However, broad generalizations are dangerous because of their unreliability. As the above examples of transformation algorithms show, the price for taking into account the topology and increasing the reliability is a sharp increase in algorithmic complexity. It happens due to the use of combinatorial search.

To assess the effectiveness of the proposed method, we compare it with adaptive routing based on finding the shortest path [2]. We assume that the weights of the network segments correspond to the cost of transportation. Then we can assume that when the threat is detected for the local section of the MTS, it is possible to increase its weight and reduce the flow. This is the reaction to the occurrence of the threat. The analysis shows the following disadvantages of this method:

- redistributing flows within the network can only be achieved by the change of segment weights. Total flow through the network remains unchanged. Accordingly, the threat is not fundamentally eliminated. We can only state the global increase in the cost of transportation. A centralized analysis of the situation is necessary for developing a counteraction decision. This is contrary to the mechanism of local adaptation of the network;
- since there is no specific decision to reduce the intensity of the flow of a single product at the time of the threat of the accident, it can be assumed that it can be built adaptively by gradually increasing the weight of the network segment. If such an approach is used, in order to eliminate the threat in principle, it is necessary to ensure an internal reserve of network transmission capacity, which can be used to redistribute flows. The technical implementation of such a "buffer" is more expensive than the effect on the intensity of input flows.

7 Conclusion

The attractiveness of using intellectual mechanisms to counter threats to MTS lies in the possibility of accumulating and applying experience. The change in operating conditions over time increases the significance of "sensible" decisions that are made in emergency situations. This experience is highly appreciated, despite the presence of difficult-to-explain causal relationships in the decisions made. The mechanism for preventing local threats described in this paper is based on the special presentation of experience, which increases the possibility of its use in other situations.

Further investigation of the problem is supposed to be carried out in the direction of improving the algorithms for transferring experience both to different networks and to different threats.

Acknowledgment. This work has been supported by the Ministry of Education and Science of the Russian Federation under Project part, State task 2.918.2017.

References

1. Medhi, D., Ramasamy, K.: Network Routing: Algorithms, Protocols, and Architectures. The Morgan Kaufmann Series in Networking, New York (2018)
2. Ford, L.R., Fulkerson, D.R.: Constructing maximal dynamic flows from static flows. Oper. Res. **6**, 419–433 (1958)
3. Frank, H., Frisch, I.T.: Communication, Transmission, and Transportation Networks. Addison-Wesley, New York (1971)
4. McBride, R.D., Carrizosa, E., Conde, E., Munoz-Marquez, M.: Advances in solving the multi-commodity flow problem. Interfaces **28**(2), 32–41 (1998)
5. Cormen, T.H., Leiserson, C.E., Rivest, R.L., Stein, C.: Introduction to Algorithms, 3rd edn. 1312 p. MIT Press, Cambridge (2009)
6. Bozhenyuk, A., Gerasimenko, E., Kacprzyk, J., Rozenberg, I.: Flows in networks under fuzzy conditions. In: Studies in Fuzziness and Soft Computing, vol. 346. Springer, Heidelberg (2017)
7. Chanas, S.: Fuzzy optimization in networks. In: Kacprzyk, J., Orlovski, S.A. (eds.) Optimization Models Using Fuzzy Sets and Possibility Theory. Theory and Decision Library (Series B: Mathematical and Statistical Methods), vol. 4. Springer, Dordrecht (1987)
8. Lenz, M., Bartsch-Spörl, B., Burkhard, H.-D.: Case-Based Reasoning Technology: From Foundations to Applications. Springer, Heidelberg (2003)
9. Belyakov, S., Belyakova, M., Savelyeva, M., Rozenberg, I.: The synthesis of reliable solutions of the logistics problems using geographic information systems. In: 10th International Conference on Application of Information and Communication Technologies (AICT), pp. 371–375. IEEE Press, New York (2016)
10. Belyakov, S., Savelyeva, M.: Protective correction of the flow in mechanical transport system. In: Computer Science On-line Conference. Springer (2017)
11. Miranda, K., Molinaro, A., Razafindralambo, T.: A survey on rapidly deployable solutions for post-disaster networks. IEEE Commun. Mag. **54**(4), 117–123 (2016)
12. Shapiro, S.C.: Artificial Intelligence. Encyclopedia of Artificial Intelligence, 2nd edn. Wiley, New York (1992)

13. Karlsson, B.: Beyond the C++ Standard Library: An Introduction to Boost. Addison-Wesley, New York (2005)
14. Kuznetsov, O.P.: Kognitivnaya semantika i iskusstvennyy intellekt. Iskusstvennyy intellekt i prinyatie resheniy **4**, 32–42 (2012)

Image Augmentation Techniques for Road Sign Detection in Automotive Vision System

Paulina Bugiel[(✉)], Jacek Izydorczyk, and Tomasz Sułkowski

Institute of Electronics, Silesian University of Technology, Gliwice, Poland
bugiel.paulina@gmail.com

Abstract. Well-prepared image data set for any automatic object detector is a crucial step in training a convolutional neural network. A vast collection of diverse images with corresponding labels is the ultimate goal for any researcher. Data acquisition, however, is the most time-consuming and tedious task in building an object detector. Therefore, various methods of generating new data from existing one have been developed and used. When facing the task of building an object detector for automotive purposes, a question appeared whether all commonly used augmentation methods are suitable in this context. This led to an investigation of different techniques as well as proposing and testing a new one, which is simple in principle but gives very promising results.

Keywords: Automotive · Object detection · Machine vision ·
Image augmentation · Convolutional neural networks ·
Sample placement

1 Introduction

As more vehicles appear on roads, being a road user is becoming more dangerous. Traffic, insufficient space for safe maneuvering, as well as overall hurry and irritation, contribute to life-threatening situations. The use of modern technology to aid drivers in these difficult conditions is vital, hence the increasing interest in advanced driver assist systems (ADAS) in the automotive industry. Features like lane keeping or obstacle detection are available on the market but there is still a lot to improve. City traffic is a complicated case not only for human drivers but for ADAS algorithms as well. Thankfully, recent advancements in deep learning, especially for visual data, contribute greatly to ADAS capabilities. Great emphasis has been put on visual sensors in recent years. Currently developed systems use up to 12 cameras and the goal is not only to support the driver but to enable safe autonomous driving in city traffic.

Majority of the state of the art visual object detectors use Region-based Convolutional Neural Networks (R-CNN) [1] and supervised learning which gives very good results but requires a vast amount of labeled training samples. Gathering and labeling images are both time-consuming and expensive, thus various

© Springer Nature Switzerland AG 2019
R. Silhavy (Ed.): CSOC 2019, AISC 985, pp. 229–242, 2019.
https://doi.org/10.1007/978-3-030-19810-7_23

methods of data augmentation have been proposed and used [2]. These methods, however, were developed mostly for general purpose object detectors, and may not be suitable for all types of data.

This paper will focus on comparing different available augmentation techniques as well as introduce a new technique of data augmentation.

2   Background and Related Research

2.1   Region-Based Convolutional Neural Networks

R-CNN and its variations eg. Fast [3] or Faster R-CNN [4] is currently the most widely used algorithm for object detection in images. It consists of two steps: region proposal (eg. [6]) and object classification, both being relatively computationally inexpensive compared to performing object detection in a single step. The first step produces a series of regions in the image that have a high probability of holding an object of interest. The method used for region proposals does not have to be accurate as the proposals are verified in the classification step. Classification consists of producing a feature vector with convolutional neural network for each region proposal and computing probability of the object belonging to a given class [5]. An up to date review on different network architectures can be found in [7].

2.2   Darknet and YOLO

Darknet [8] is an open source neural network framework written in C and CUDA that supports CPU and GPU computation. It is easily scalable as it compiles to one executable file that can be run on any computer with Linux operating system. This makes it a good fit for computing clusters. Network architecture and input data are fed into the framework via text files. YOLO [9] is a deep convolutional network architecture designed by Darknet developers. It has several different base versions (YOLOv2, YOLO9000 [10] or the most recent YOLOv3 [11]) as well as reduced versions requiring less computing power. YOLO is competitively accurate and fast, making it possible to work in real time. Its methodology shows a great resemblance to Faster R-CNN.

2.3   Cameras and Object Detection in Automotive Industry

A typical ADAS [12] system is based on camera and radar output as well as lidar for development purposes [13]. Active safety is a branch of automotive industry that uses camera most often for applications like e.g. collision detection or lane keeping as well as monitoring the driver's eyes for a sign of fatigue. Cameras play likewise an important role in autonomous driving development being a natural next step of ADAS. Virtual driving simulators, which aid the development of both ADAS and autonomous driving [14], also utilize object detection, the same as the real systems.

3   Motivation

Labeling ground truth manually is becoming too expensive due to the number and size of video streams as well as a number labeling rules. As the human factor is an obligatory component in providing ground truth for machine learning, it cannot be ruled out completely. It can, however, be enhanced with automatic object detection to provide proposals to the person performing the labeling. Instead of marking all objects in each frame of the video, it would be required to only accept or reject proposed labels. Moreover, rejected frames could be corrected and used to improve systems accuracy.Automatic detection will allow decreasing the amount of disk space required to store the data. In most cases, the requirement is to record a fixed number of events in given conditions (eg. a number of vehicles encountered in snow, rain, at night etc.). If the developed system could predict environmental conditions it would be easy to estimate the contents of the recordings and choose the most information-dense set. Vehicle drive recordings, however, differ from general benchmark data used in machine vision. For instance, the camera is mounted in a fixed position, the resolution never varies and color space is rarely RGB. Hence it is possible that not all commonly used image augmentation techniques would be applicable for the data. Also, new techniques may be beneficial when used for car cameras and not for benchmark data. This fact was a motivation for performing analysis of augmentation technique and their influence on automatic object detector accuracy.

4   Environment

4.1   Darknet Modification

In Darknet, image augmentations are already implemented. However, as these were the test subjects, they were disabled in the framework. Augmented images were fed into the network as part of training data sets.

4.2   Hardware and System

All computations presented in this article were performed on a desktop computer with quad-core Intel® Core™ i5-7400 3.00 GHz CPU, GeForce GTX 1060 3 GB graphics and Ubuntu 16.04 LTS operating system. Darknet with GPU acceleration requires CUDA support (version 9.2 was used), where video memory is the most important factor in terms of neural network training speed.

4.3   Detected Objects

Available labeled video data include objects suitable for automatic object detection e.g. traffic lights, pedestrians (including bike riders), vehicles (including motorcycles), traffic signs etc. Traffic signs were chosen as a base for training object detector as they have fairly uncomplicated shapes and colors, can be easily categorized into classes based on visual appearance. Working on all available objects would be much more time and storage space consuming (building and managing and manipulating multiple data sets). Traffic signs are thus a good fit for performing research.

Table 1. Object classes examples

ID	Name	Content	Examples
1	info_dark	information signs with dark background	
2	info_white	information signs with white background	
3	no_entry	no entry signs	
4	order_dark	order signs on dark background	
5	order_white	order signs on white background	
6	priority	priority / end of priority	
7	speed_lim	white circular speed limit	
8	speed_lim_usa	white rectangular speed limit	
9	stop	stop signs	
10	warning_diamond	diamond shaped warning signs	
11	warning_triangle	triangle shaped warning signs	
12	yield	yield signs	

4.4   R-CNN Architectures

Two neural network architectures were used for training in order to achieve an objective comparison of results. The first architecture had 23 convolution layers, the second one only 9. The latter showed much-decreased accuracy and training time compared to the deeper network. Both architectures are similar to YOLOv2 and Tiny YOLOv2 with a change in input size so that it corresponds to training images 4:3 aspect ratio (Fig. 1).

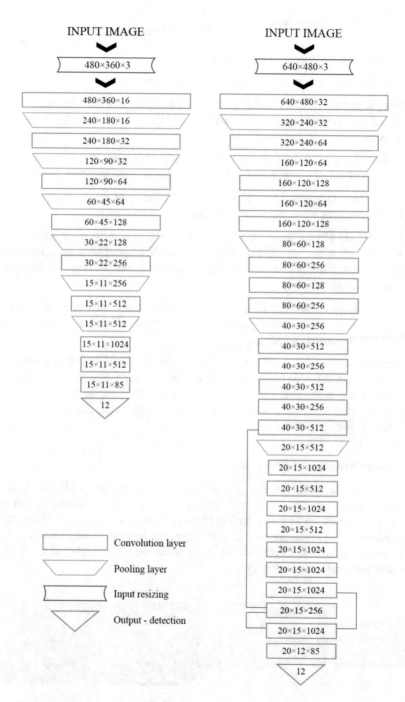

Fig. 1. Network architectures. Small version (left) and regular (right). Numbers in each layer denote the size of given layers output.

5   Data Preparation

5.1   Training Samples Acquisition

The samples for training were gathered as follows:

Data Acquisition. Dedicated vehicles were used to perform test drives by assigned drivers; all test vehicles were equipped with the devices under test - front camera in this case.

Annotation. Logs from test drives were manually annotated in ground truth lab - all objects of interest were marked on all video frames by the lab employees.

Data Selection. Logs with traffic signs present in them were chosen as a basis for the proceedings with automatic traffic sign recognition.

Image Database Preparation. A C++ application was written to save separate video frames with corresponding label file describing positions and sizes of traffic signs; every 30th frame was saved in order to reduce correlation between consecutive images.

Network Training. Training of a convolutional network was performed on this data set but due to significant dissimilarities between different signs, the detection system performed poorly.

Categorization. To address the above issue, the data set was sent again to the lab in order to categorize traffic signs into groups taking into consideration their visual appearance, resulting in 12 object classes. 200 samples of each class were excluded to form a validation set. All further research was based on the samples belonging to these 12 classes.

5.2   Augmentation Methods

Rotation. Images were rotated a random angle in range $[-15, -4] \cup [4, 15]$ degrees. Ranges were chosen in such fashion to provide resemblance to roll angle that may occur in reality. Exclusion of angles near $0°$ is to provide sufficient dis-correlation between original and augmented images. After rotation, black margins that appeared in the images were removed by zooming in the image a minimum possible value (Fig. 2).

Fig. 2. Rotation. Sketch (left) and example on an augmented frame (right)

Horizontal Reflection. Reflection is a simple transformation that doesn't require any explanation. Only vertical reflection along central axis is was used (Fig. 3).

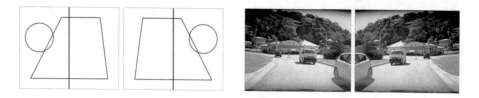

Fig. 3. Horizontal reflection. Sketch (left) and example on an augmented frame (right)

Shear. Shear was, at first, applied to images using Keras ImageDataGenerator class. Unfortunately, this class does not support a transformation of labels, which need to be adjusted to new positions of objects. For label transformation, a manual script had to be written. Viewing the images transformed using ImageDataGenerator with labels on top showed that Keras introduces bias to the image, thus resulting images were not perfectly aligned with corresponding labels. The misalignment was proportional to the shear angle. To address this issue, a custom script was written utilizing OpenCV library. The script performed transformation of image pixels. Outcome images are well aligned with labels (Fig. 4).

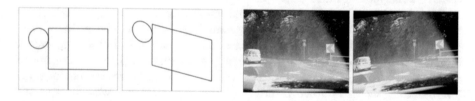

Fig. 4. Shear. Sketch (left) and example on an augmented frame (right)

Hue Shift. is a widely used image augmentation technique which introduces variations in color. In order to perform the hue shift, each image was transformed from RGB to HSV color space, where individual channels are hue, saturation and value (brightness). Values in the first channel were shifted by random value chosen separately for each image of the training set. To avoid an unnatural appearance of the resulting images, hue shift values were chosen from $[-50, 50)]$ range where 50 is equal to approximately 20% of the 8-bit depth of the channel (Fig. 5).

Fig. 5. Hue shift. Two examples of hue shift with original and augmented image. Positive hue shift values (left) and negative (right)

Sample Placement. Sample placement is a newly introduced technique. New images for training are prepared by placing objects of interest, i.e. samples, on background images. The method consists of two steps.

The first step uses already available labeled images to gather statistical information on the position of samples in the training set. Each input image is divided into a matrix of $m \times n$ regions, each holding a counter. Whenever a center of a labeled sample is encountered in a region, the corresponding counter is incremented. Finally, all values are divided by the sum of the used samples, which produces a matrix of probabilities. Each probability is an average chance of encountering an object of interest in a given image region.

The second step consists of extracting objects of interest (i.e. samples) from images and placing them on background images according to probabilities in the matrix obtained in the first step which assures resemblance to reality. In order to exclude any color differences between background and sample image, color transfer is performed on each sample [15].

Five training data sets were prepared employing the technique. Half of the training images in the first set were prepared as described above. In the second set, augmented images were created using equal probabilities for sample placement, which resulted in inserting samples in random places of the image, avoiding overlapping. In the third set, the background was removed from the inserted samples to improve blending of the object images with background and eliminating the possibility that the object detector would learn to look for rectangular shapes instead of traffic signs. The remaining data sets were counterexamples where random parts of images were placed onto background images. Half of the images in the fourth and all images in the fifth set were prepared this way (Fig. 6).

Fig. 6. Three different types of samples inserted into background images. With background (left), with background removed (middle), false samples (right)

6   Training on Multiple Data Sets

A total of 17 different data sets described in Table 2 were prepared and each was used to train a separate network according to the two architectures described in Sect. 4.4, giving 34 networks in total. Training was performed for 80 thousand epochs which is enough to obtain comparable detectors output.

Table 2. Image augmentation data sets

ID	Augmentation technique	Samples per class	Augmented [%]	Description
1	None - all images	Varying	-	All available images. Number of samples in each class: [10996, 3903, 8499, 25478, 9475, 11131, 37844, 3228, 8064, 2195, 34815, 15209]. Totaling 170 837 samples
2	None	1 800	-	12 object classes of the same size
3	Reflection - missing samples	3 600	7	Image augmentation only for three classes with insufficient training images: 964 in class 1, 544 in class 7, 1674 in class 9
4	Rotation - missing	3 600	7	The same as above
5	Reflection	3 600	50	Half of samples reflected
6	Rotation	3 600	50	Half of samples rotated
7	Reflection, rotation	3 600	50	Quarter of samples reflected, quarter rotated
8	Reflection, rotation - equal	5 400	67	One third of samples reflected, one third rotated
9	Shear Keras	3 600	50	Half of samples augmented
10	Shear custom	3 600	50	Half of samples augmented
11	Hue shift	3 600	50	Half of samples augmented
12	Sample placement - random	3 600	50	Half of samples generated using sample placement in random regions
13	Sample placement - statistical	3 600	50	Half of samples generated using sample placement with rectangular samples in statistically evaluated regions
14	Sample placement - statistical, no background	3 600	50	Half of samples generated using sample placement with background-removed samples in statistically evaluated regions
15	Sample placement - false samples	3 600	50	Half of training images generated using sample placement with false samples
16	Sample placement - false samples only	1 800	100	All training images generated using sample placement with false samples
17	Placement, reflection, rotation	7 200	75	Equal groups: original images, sample placement (as in set 14), reflection and rotation

7   Results

7.1   Evaluation Measures

In order to deliver a comprehensive comparison, mean average precision (mAP) measure was computed for each network detections performed on the validation set. mAP was implemented according to the Pascal VOC Challenge results evaluation guidelines (Sect. 4.2 in [16]). This metric is most commonly used when assessing object detectors with multiple classes. In short terms, average precision is computed as an average area below the precision-recall curve for 11 recall levels $(0.0, 0.1, ..., 1.0)$ at the intersection over union (IoU) of 50%. An average for all classes is computed to get mean average precision. Recall and precision are given as:

$$Recall = \frac{tp}{tp + fn} \tag{1}$$

$$Precision = \frac{tp}{tp + fp} \tag{2}$$

where:

tp - total number of detected true positives
fn - total number of detected false negatives
fp - total number of detected false positives

Intersection over union (IoU or Jaccard index) is defined as an area of two sets intersection divided by their union. In object detection, the two sets in the computation are the detected and ground truth region of an object in the image.

$$IoU = \frac{A \cap B}{A \cup B} \tag{3}$$

where:
A, B - ground truth and detected object regions on an image

7.2   Experiment Results

Evaluation of results for all tested neural networks are presented in Tables 3 and 4. In each table, data set ID corresponds to the index in Table 2 and class ID to index in Table 1. The upper table shows results for the deep neural network, the bottom one for the shallow. Three top mAP scores are marked in bold, three worst scores are marked in italics.

Table 3. Deep network architecture (23 convolutional layers, 640 × 480 resolution)

Set ID	Average precision by class ID												mAP
	1	2	3	4	5	6	7	8	9	10	11	12	
1	**55.0**	**52.2**	**79.2**	**76.4**	**79.2**	**88.1**	**72.1**	**58.7**	**80.6**	**72.3**	**77.3**	**78.9**	**72.49**
2	37.8	54.5	68.4	47.2	67.6	78.3	54.8	67.3	70.0	76.2	66.3	67.4	62.97
3	45.7	55.9	69.0	57.5	74.0	79.3	49.1	70.3	80.2	77.3	69.3	76.8	67.03
4	42.2	57.3	74.4	58.7	74.0	79.8	45.0	70.1	79.9	79.0	68.7	74.0	66.91
5	39.3	54.3	72.0	51.6	67.9	80.3	55.0	70.9	77.6	77.6	68.1	75.6	65.84
6	39.9	56.7	68.6	50.5	73.3	79.2	52.9	70.4	72.8	78.6	65.1	76.0	65.34
7	42.9	58.5	69.8	48.1	68.6	79.2	54.3	71.0	72.9	78.9	66.9	72.0	65.25
8	**42.6**	**54.7**	**77.4**	**50.6**	**71.6**	**80.5**	**55.2**	**73.9**	**72.3**	**80.3**	**73.9**	**76.7**	**67.46**
9	*36.5*	*56.6*	*68.0*	*47.7*	*67.2*	*79.1*	*53.1*	*57.1*	*69.9*	*76.9*	*65.5*	*68.1*	*62.14*
10	38.8	54.1	67.8	51.7	68.9	79.3	55.3	69.7	76.1	77.4	64.8	68.6	64.38
11	*40.1*	*53.4*	*65.9*	*45.5*	*67.8*	*79.4*	*54.8*	*70.5*	*67.7*	*77.1*	*65.0*	*66.7*	*62.83*
12	40.1	55.2	71.3	54.9	69.8	80.3	56.3	69.5	74.6	74.5	72.6	74.2	66.09
13	45.6	55.7	78.4	54.1	68.7	80.6	59.3	67.6	78.1	75.9	67.8	68.7	66.70
14	44.9	55.0	76.5	52.8	73.7	80.3	54.5	70.4	79.7	75.3	71.3	74.3	67.38
15	*36.7*	*46.8*	*67.7*	*47.3*	*67.3*	*75.1*	*51.3*	*63.7*	*70.7*	*75.2*	*65.7*	*66.5*	*61.16*
16	0.0	0.0	0.0	0.0	0.0	0.0	0.0	0.0	0.0	0.0	0.0	0.0	0.00
17	**46.0**	**54.7**	**75.2**	**56.7**	**75.9**	**79.8**	**56.8**	**80.1**	**79.0**	**76.7**	**73.7**	**76.3**	**69.24**

Table 4. Shallow network architecture (9 convolutional layers, 480 × 360 resolution)

Set ID	Average precision by class ID												mAP
	1	2	3	4	5	6	7	8	9	10	11	12	
1	**1.0**	**0.6**	**3.0**	**9.1**	**3.0**	**12.5**	**9.1**	**9.1**	**11.6**	**9.8**	**9.1**	**10.1**	**7.32**
2	4.6	1.8	4.6	9.1	9.1	12.1	9.1	4.6	11.0	4.8	1.7	6.7	6.58
3	*3.0*	*1.8*	*2.3*	*2.0*	*3.1*	*14.4*	*1.1*	*1.3*	*11.3*	*9.1*	*5.1*	*10.5*	*5.43*
4	9.1	0.7	1.8	0.7	9.1	12.7	9.1	1.3	9.8	7.4	10.2	10.1	6.83
5	9.1	1.8	9.1	6.1	3.0	10.3	3.0	2.7	10.6	9.4	2.1	10.1	6.45
6	9.1	0.7	3.0	4.6	4.6	12.9	6.1	1.4	10.6	9.1	9.4	7.8	6.59
7	9.1	1.1	0.5	9.1	4.6	14.1	9.1	2.3	11.4	3.6	9.5	7.1	6.79
8	9.1	1.5	1.1	1.4	9.2	15.0	9.1	1.8	10.7	4.9	9.8	9.5	6.92
9	*9.1*	*1.3*	*0.3*	*9.1*	*0.9*	*12.7*	*3.0*	*0.4*	*10.7*	*9.1*	*9.1*	*9.6*	*6.26*
10	4.6	3.0	0.4	4.6	4.7	11.8	9.1	1.8	10.2	9.7	9.7	9.4	6.58
11	9.1	1.1	0.1	9.1	9.1	11.2	6.1	0.6	9.3	9.6	4.5	9.1	6.55
12	**3.6**	**0.1**	**4.6**	**1.3**	**9.1**	**12.2**	**5.5**	**4.6**	**13.0**	**9.3**	**9.7**	**11.8**	**7.05**
13	6.1	0.4	2.6	9.1	3.3	11.3	9.1	1.7	8.2	9.4	4.4	9.8	6.27
14	3.6	2.3	9.1	4.6	1.1	9.9	9.1	2.0	10.9	9.1	9.3	10.2	6.76
15	*4.6*	*0.1*	*1.5*	*9.1*	*9.3*	*11.6*	*9.1*	*1.5*	*4.6*	*4.3*	*4.0*	*10.6*	*5.85*
16	0.0	0.0	0.0	0.0	0.0	0.0	0.0	0.0	0.0	0.0	0.0	0.0	0.00
17	**3.0**	**4.6**	**9.1**	**3.0**	**6.1**	**15.0**	**9.1**	**1.5**	**9.7**	**9.3**	**10.0**	**9.8**	**7.50**

8   Conclusions

When comparing the results from different data sets, it is best to take as a reference point set number 2 with 1800 samples per class and no augmentation.

8.1   Deep and Shallow Network

The deep network is a dependable source of comparison as the mAP score is high as well as the detections quality. The shallow network performs poorly, which might be caused by too short training time with such a simple architecture which makes it unreliable as a source of information.

8.2   Conventional Methods

Not all commonly used augmentation methods proved suitable for the tested data. Reflection and rotation gave a noticeable improvement of mAP by 2.3% (sets 5 and 6 respectively). Shear, when applied correctly (set 10), gave a smaller increase in mAP by only 1.4%. Hue shift, however, resulted in a slight decrease, which proves that it is not suitable for the tested data. It was expected as the original data comes from one type of camera and doesn't manifest many color variations.

8.3   Sample Placement

The newly introduced augmentation technique produced positive results in all its variations Simple sample placement using rectangular samples in random image regions (set 12) already gave result exceeding the ones obtained with reflection or rotation (mAP increased by 3.1%). Introducing the statistical distribution of samples (set 13) further improved the mAP results by another 0.6%. Removal of background in the placed samples gave the best results of all tested augmentation algorithms with mAP gain of 4.4% The surprisingly good results are probably caused by the fact that the method does not just transform already available images but actually introduces new scenes to the image data set. This provides much more diversity than the conventional methods can produce.

8.4   Mixed Techniques

Generating more samples with multiple augmentation techniques improves performance even further (sets 8 and 17). The number of generated samples is more important than diversity of augmentation techniques. Sets 5, 6 and 7 have an equal amount of generated samples, where sets 5 and 6 utilize only 'pure' reflection and rotation respectively and set 7 uses both these techniques. However, mAP for these data groups is almost equal. This outcome is a little counter-intuitive and would probably need more investigation.

8.5   General Conclusions

The most important observation is that not all augmentation techniques are suitable for all kinds of data. While they can be a great aid when building a training data set, they have to be used consciously in order not to decrease the actual performance of the resulting detector. What is also important to notice is that data scientist do not have to limit the techniques to ones that are widely known and well tested. While this approach is understandable, this paper has shown that experimenting with new techniques can prove beneficial. Even algorithms simple in principle can significantly increase the detector's performance. To be successful in this field, the researcher has to know the characteristics of the data that he is working on as well as the augmentation algorithms and their outcomes.

9   Further Research

Augmentation methods in object detection is a vast topic that could not be covered entirely in a limited amount of time. During the research, however, some techniques, e.g. different kinds of style transfer, appeared promising and would be a probable field of further research.

More importantly, presented techniques have to be tested with more objects, not only traffic signs. This includes pedestrian, bike riders, vehicles, traffic signs etc. Techniques that gave positive results will be applied to the extended data set to see if they still yield improvements.

As the tested here shallow network architecture gave unreliable results the authors are tempted to train a different network with higher mAP score to properly validate findings presented in this paper. Topic presented here give the readers focus on the problems that may be faced when building a good training data set for object detection with little amount of initial data. While the presented results and conclusions provide some intuition in the topic, the outcomes would probably be different for any other type of images or network architectures.

Acknowledgment. This work was supported by the Ministry of Science and Higher Education funding for PhD studies under implementation path programme (10/DW/2017/01/1).

References

1. Girshick, R.B., Donahue, J., Darrell, T., Malik, J.: Rich feature hierarchies for accurate object detection and semantic segmentation. In: 2014 IEEE Conference on Computer Vision and Pattern Recognition, pp. 580–587 (2014)
2. Mikołajczyk, A., Grochowski, M.: Data augmentation for improving deep learning in image classification problem. In: 2018 International Interdisciplinary PhD Workshop, Swinoujście, Poland, pp. 117–122 (2018)

3. Girshick, R.: Fast R-CNN. In: 2015 IEEE International Conference on Computer Vision, Santiago, pp. 1440–1448 (2015)
4. Ren, S., He, K., Girshick, R., Sun, J.: Faster R-CNN: towards real-time object detection with region proposal networks. IEEE Trans. Pattern Anal. Mach. Intell. **39**(6), 1137–1149 (2017)
5. Girshick, R.B., Donahue, J., Darrell, T., Malik, J.: Region-based convolutional networks for accurate object detection and segmentation. IEEE Trans. Pattern Anal. Mach. Intell. **38**(1), 142–158 (2016)
6. Uijlings, J.R., van de Sande, K.E., Gevers, T., Smeulders, A.W.: Selective search for object recognition. Int. J. Comput. Vis. **104**(2), 154–171 (2013)
7. Zhao, Z., Zheng, P., Xu, S., Wu, X.: Object detection with deep learning: a review. arXiv:1807.05511 [cs.CV] (2018)
8. Redmon, J.: Darknet: Open source neural networks in c. http://pjreddie.com/darknet/
9. Redmon, J., Divvala, S., Girshick, R., Farhadi, A.: You only look once: unified, real-time object detection. In: 2016 IEEE Conference on Computer Vision and Pattern Recognition, Las Vegas, NV, pp. 779–788 (2016)
10. Redmon, J., Farhadi, A.: YOLO9000: better, faster, stronger. In: 2017 IEEE Conference on Computer Vision and Pattern Recognition (2017)
11. Redmon, J., Farhadi, A.: YOLOv3: an incremental improvement. arXiv:1804.02767 [cs.CV] (2018)
12. Sowmya Shree, B.V., Karthikeyan, A.: Computer vision based advanced driver assistance system algorithms with optimization techniques-a review. In: 2018 Second International Conference on Electronics, Communication and Aerospace Technology, Coimbatore, pp. 821–829 (2018)
13. Skruch, P., Długosz, R., Kogut, K., Markiewicz, P., Sasin, D., Rozewicz, M.: The simulation strategy and its realization in the development process of active safety and advanced driver assistance systems. In: SAE Technical Paper 2015-01-1401 (2015)
14. Sulkowski, T., Bugiel, P., Izydorczyk, J.: In Search of the ultimate autonomous driving simulator. In: 2018 International Conference on Signals and Electronic Systems (2018)
15. Reinhard, E., Ashikhmin, M., Gooch, B., Shirley, P.: Color transfer between images. IEEE Comput. Graph. Appl. **21**, 34–41 (2001)
16. Everingham, M., Van Gool, L., Williams, C.K.I., et al.: Int. J. Comput. Vis. **88**, 303 (2010). https://doi.org/10.1007/s11263-009-0275-4

Machine Failure Prediction Technique Using Recurrent Neural Network Long Short-Term Memory-Particle Swarm Optimization Algorithm

Noor Adilah Rashid[✉], Izzatdin Abdul Aziz,
and Mohd Hilmi B. Hasan

Centre for Research in Data Science, Universiti Teknologi PETRONAS,
Seri Iskandar, Perak, Malaysia
{noor_18001659,izzatdin,mhilmi_hasan}@utp.edu.my

Abstract. This paper proposes a hybrid prediction technique based on Recurrent Neural Network Long-Short-Term Memory (RNN-LSTM) with the integration of Particle Swarm Optimization (PSO) algorithm to estimate the Remaining Useful Life (RUL) of machines. LSTM is an improvement of RNN as RNN faces issues with predicting long-term dependencies. Issues such as vanishing and exploding gradients are the results of backpropagating errors, taking place when the network is learning to store and relate information over extended time intervals. RNN-LSTM is a feasible technique for this research due to its effectiveness in resolving sequential long-term dependencies problems. However, the accuracy can still be enhanced to a satisfactory value considering the optimal network topology has not been discovered yet. Accuracy improvement can be achieved by resorting to hyperparameter tuning. A result of proof of concept validates that by increasing the number of epochs, the accuracy of prediction has improved but increases the execution time. To optimize between the accuracy and execution time, a population-inspired Particle Swarm Optimization (PSO) algorithm is employed. PSO will be utilized to select the optimal RNN-LSTM topology specifically the learning rate instead of using manual search. This optimized hybrid prediction technique is useful to be implemented in predictive maintenance to predict machine failure.

Keywords: Long-Short Term Memory · Particle Swarm Optimization · Machine learning

1 Introduction

1.1 Background of Study

There were 1310 oil and gas related spills in New Mexico in 2016, with machine failures accounted for 52% of being the cause of the spill [1]. A comparative study of the maintenance cost among preventive and predictive for equipment mismanagement, founds that preventive maintenance consumes the highest cost around $54M/year compared to predictive maintenance with $20M/year [2].

© Springer Nature Switzerland AG 2019
R. Silhavy (Ed.): CSOC 2019, AISC 985, pp. 243–252, 2019.
https://doi.org/10.1007/978-3-030-19810-7_24

Predictive maintenance uses machine learning (ML) algorithms to estimate when the machine would likely to fail. ML algorithms such as RNN-LSTM has been proven to make accurate prediction with the help of human domain knowledge to evaluate possible degradation of critical machines [3].

RNN-LSTM has feedback connections to itself to carry inputs from previous and current time-steps. Time-step is the tick of time in the period of data being recorded. To produce an accurate RNN-LSTM, the optimal network topology must be applied. Modification on the RNN-LSTM network topology can improve the accuracy of the algorithm. However, it trades off on another aspect; the execution time. Particle Swarm Optimization (PSO) algorithm is adopted to optimize both accuracy and execution time of the prediction by searching the optimal RNN-LSTM learning rate.

This paper is organized as follows. In Sect. 2, the existing algorithms in predicting machine failure is elaborated in their respective fundamental and theories models. In Sect. 3, a comparative analysis between the algorithms are discussed. In Sect. 4, the proposes hybrid prediction technique is clarified with a comparative study between the optimization algorithms to improve the result. In Sect. 5, the proposed research methodology is described. In Sect. 6, a proof of concept is shown to prove the feasibility of the selected algorithm in this research. This paper concludes with discussion on proposing of new enhancements to the algorithms.

2 Methods

In this section, three machine learning algorithms for machine failure prediction are comprehensively discussed. The methods are Decision Tree, Random Forest and Recurrent Neural Network- Long Short-Term Memory (LSTM). We shall briefly explain the function of the algorithms as well as the mathematical model used in the algorithms.

2.1 Decision Tree (DT)

A decision maker inspired by the structure of a tree as the decision-making process starts from the root where all the attributes will be calculated statistically to find the best attribute, from there, it branches out to 2 nodes and the process repeats until it gets to the final decision at the leaf node. Algorithms for DT such as ID3 and CART employs a top-down greedy search through the space of possible branches with no backtracking. Constructing a DT is all about finding the attribute that returns the highest statistical property, entropy and information gain [4].

$$Entropy = -0.5log20.5 - 0.5log20.5 = 1 \tag{1}$$

Entropy is used to calculate the purity of the attributes as the growing DT partitions the samples into subsets that contain similar instances; also described as homogenous. As shown in Fig. 1, completely pure samples with either zero or one probability have zero entropy while equally distributed samples with 0.5 probability has entropy of 1 [5].

This shows that entropy also measures the level of certainty of splitting and decision making.

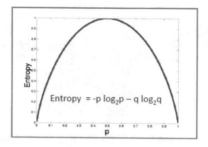

Fig. 1. Graph of entropy VS probability.

$$Gain(T,X) = Entropy(T) - Entropy(T,X) \qquad (2)$$

The desire after splitting is to get a lower entropy value as this means higher information gain, since information gain is the comparison before and after a splitting. A feature with highest information gain will be chosen at the decision node and the proceeding nodes.

2.2 Random Forest

An extension of DT, Random Forest (RF) is an ensemble algorithm that leverages many DTs with some injection of randomness which has been proven to improve the accuracy significantly [6]. The randomness injected is the bootstrap resampling, also known as bagging, to reduce the variance of prediction. A smaller sample is bootstrapped from the whole sample yet in large numbers and same size, with replacement. This approach is implemented in both sample and feature selection. Random feature selection by bagging causes trees to test different attributes and the result which averaged from trees is also different.

2.3 Recurrent Neural Network Long-Short Term Memory

RNN is one of the architectures of neural network (NN) that is suitable for sequential problem due to its special ability to memorize inputs selectively as it has an internal memory [7]. RNN has an internal feedback loop which carries inputs from current state and from the past time-step with backpropagation been done in consideration of time, thus it is called as backpropagation through time (BPTT) [8, 9].

There are two issues in regards of RNN; vanishing and exploding gradient [7]. Gradient is the measure of changes in weight due to the change in error. Exploding gradient happens when the time step is massive, the multiplication of adjustment stages becomes long, resulting multiplication with anything greater than 1 is vast. Vanishing gradient is when multiplication with any weight less than zero, causing the result to become so small; giving no significant change to the model, as if it is reaching plateau. This issue is resolved by Long-Short-Term-Memory (LSTM) [9] (Fig. 2).

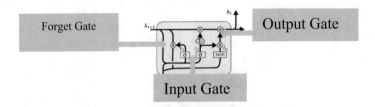

Fig. 2. Long-Short-Term Memory block in RNN.

3   Comparative Analysis on Existing Machine Failure Prediction Algorithms and Previous Research of RNN-LSTM

3.1   Comparative Analysis Between DT, RF and RNN-LSTM

RNN-LSTM is chosen as the algorithm in this research due to the convincing discussion as outlined in Table 1 below.

Table 1. Comparison between RNN-LSTM, DT & RF.

Features		RNN-LSTM	Decision Tree	Random Forest
Application	Prediction	Yes	Yes	Yes
	Regression	Yes	Yes	Yes
	Pattern recognition	Yes	Yes	Yes
	Time-varying pattern	Yes	-	-
Type of Data	Large Dataset	Yes	Yes	Yes
	Complexity	Yes		Yes
	Time-series	Yes	-	-
Interpretability		Less, Black Box nature	Comprehensible	Difficult
Efficiency	Speed	Quite fast	Rapid	Relatively fast
	Real-time prediction	Yes	Yes	Slow & Inefficient
	Accuracy	High	High	High
Memory		Yes	-	-

In term of application, this research aims to predict the RUL of machines. It is based on the time-varying and sequential pattern with two features engineering; time and the measurement of the sensors. It is a regression issue since it predicts numeric value instead of categorical target variables. This is a time-series prediction since it utilizes time-series data. RNN-LSTM fits to perform regression due to its mechanism which utilizes activation and loss functions for calculating the numerical output. It can recognize time-varying pattern due to the ability to capture the relationship of far separated time-series data. It also ensures events in different moments are correlated [11].

RNN-LSTM also handles complex time-series data with long-time lag tasks very well due to its memory structure that consists of forget, input and output gates which enables it to remember or forget the data selectively. This also contributes to RNN-LSTM's dynamic nature that it can observe moving patterns.

RNN-LSTM is fast and efficient in speed due to the ability to process the neurons simultaneously (parallelization) compared to DT and RF that face complexity, also time-consuming when building the trees and forest. Hence, RNN-LSTM is more convenient for real-time prediction. A study also claims that the accuracy of RNN-LSTM is also high [12].

4 Hybrid Prediction Technique

4.1 Hyperparameters Tuning to Improve RNN-LSTM Performance

Hyperparameters are a set of variables that construct the network topology such as network size & initial weights. The structure of a network has a direct effect on the execution time and prediction accuracy. Improving the accuracy can be done by tuning the hyperparameters which uses the optimization algorithms to search for the best combination of hyperparameters that minimizes the generalization error [13].

4.2 Comparative Analysis of Particle Swarm Optimization and Ant Colony Optimization Algorithm

Optimization algorithms can be leveraged with the ML algorithms to improve the accuracy of RNN-LSTM by searching for the optimal hyperparameter setting.

Based on Table 2 [14], PSO has outperformed ACO based in these parameters: (1) application, (2) the ability to reach good solution without local search (3) influence of best solution on population and (4) influence of population size on solution time. PSO is more natural to continuous issue compared to ACO which is applied for discrete optimization problem PSO is feasible for this research as the target variable is continuous-real values [15, 16]. PSO can achieve good solutions without local search due to the dense search space of particles that can find many potential solutions [17]. The best solution has a high influence on the PSO. Plus, the optional local search in PSO is applied only on the best solution and not to the entire population. This would save the searching time as the next iteration of searching would be based on this good solution finding. The population size influences PSO solution time linearly in contrast to ACO; exponentially due to the absence of local search in PSO.

A study proved that PSO exhibits computational efficiency over ACO in terms of the training time duration and an outstanding accuracy compared to ACO despite different functions and number of variables used [16]. PSO is a high accuracy and less time-consuming optimization algorithm, hence, it is feasible for this research where optimization between accuracy and execution time is desired.

Table 2. Comparison between ACO & PSO optimization algorithms.

Parameters	ACO	PSO
Application	Discrete	Continuous
Ability to reach good solution without local search	Medium	High
Influence of best solution on population	Medium	High
Influence of population size on solution time	Exponential	Linear
Density of search space	High	High

5 Proposed Research Methodology

The methodology of the proposed technique is identified to give an insight of how this research is going to be conducted as in Fig. 3 below.

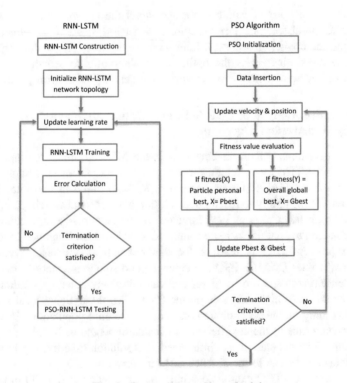

Fig. 3. Proposed research methodology

(1) During RNN-LSTM construction, RNN-LSTM is constructed, and the learning rate of RNN-LSTM will be initialized from the searching of PSO. (2) In PSO Initialization, inertia weight λ, learning factors μ_1 and μ_2, the number of particles M and the number of iterations N are set. Fitness value is set as the technique accuracy. The locations and velocities of particles are initialized randomly. (3) Training data is split and is inserted into PSO during data insertion. (4) During particles' speed and locations update, at the 1st iteration of PSO, speed and position of particles are set as in PSO initialization. Speed and position of particles are updated based on the fitness value on the subsequent iterations. (5) Fitness value based on the speed and location of particles updated in Step (4) is calculated by using a cost function during fitness value evaluation. (6) Get the best fitness value of each particle, Pbest and the global optimal solution, Gbest of the searching based on the weights used in Step (7). If particles' fitness > global best fitness, update Gbest. If particle fitness > particle best fitness, update Pbest. (7) Termination criterion for PSO are the maximum number of iterations and Gbest value. If the maximum number of iterations of PSO algorithm has been reached and Gbest value is no longer change, continue to Step (8). Otherwise, loop back to Step (4). (8) Gbest return- Get the optimal learning rate of RNN-LSTM from the Gbest of PSO algorithm for updating Step (9) for PSO-RNN-LSTM training. (9) Learning rate update- The optimal learning rate is updated from the result of PSO searching. (10) PSO-RNN-LSTM Training- RNN-LSTM is trained by using the selected the learning rate in (9). Error is calculated based on the rate used and back-propagated for error modification.

6   Proof of Concept (POC)

This chapter will provide the preliminary result as the proof of feasibility of RNN-LSTM to be used in this research. Section 6.1 explains the structure of the LSTM used. Section 6.2 describes the objectives of the proof of concept. Section 6.3 explains the procedure for POC, while Sect. 6.4 presents the result obtained from the experiment and hence, determines whether the objectives are met.

6.1   Prediction Technique

RNN-LSTM is made by using Python and Keras. It comprises of 12 connections including to themselves. The input and output layers contain only one node meanwhile the hidden layer consists of 4 LSTM nodes. Loss function used is the mean squared error. The number of epochs used will be determined from the experiment of proof of concept since different epoch number produces different prediction result. Result and explanation of the experiment is in Sect. 6.4.

6.2   Results and Discussion

The proof of concept aims: (1) To prove that RNN-LSTM can be used to make time-series prediction in this research by achieving a feasible result in terms of accuracy and time efficiency, (2) To prove that the accuracy of RNN-LSTM can be improved by tuning the hyperparameters. (3) To prove that the proposed methodology in Sect. 4.2 can be executed in this research to optimize between accuracy and time efficiency.

Results from the experiments on different epoch numbers and LSTM nodes are visualized and tabulated in Fig. 4 and Table 3 respectively.

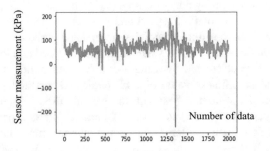

Fig. 4. Graph of actual vs training and testing prediction result.

Figure 4 plots the prediction result of sensors measurement (kPa) versus the number of data. The actual sensors reading are in blue, training prediction in orange and testing prediction in green for 4 LTSM nodes at 500 epochs. This graph shows that the prediction results for both training and testing follows the sensors reading pattern closely except for the lowest unexpected outlier reading. This proves that RNN-LSTM can recognize and map the non-linear relationship between input and output very well. It is still needed to be improved due to the error in prediction of the outlier. Deviations between the prediction and actual reading, also known as errors are tabulated in Table 4 below.

Table 3. Prediction performance of different LSTM nodes, number of epochs.

LSTM nodes	Number of epochs	RMSE		R2		Time (min)
		Training	Testing	Training	Testing	
2	100	17.10	13.15	0.73	0.54	
	500	16.49	12.99	0.75	0.55	
	1000	16.55	13.17	0.75	0.54	
4	100	17.22	13.42	0.73	0.52	3.02
	500	16.37	13.03	0.75	0.55	17.22
	1000	16.28	12.93	0.76	0.55	30.58
6	100	16.87	12.95	0.74	0.55	
	500	16.39	13.07	0.75	0.55	
	1000	16.71	13.34	0.74	0.53	

By referring to Table 3, at 1000 epoch number, the best prediction is observed at 4 number of LSTM nodes with 0.76 training R^2 value, as compared to the other 2 and 6 LSTM nodes with 0.75 and 0.72 training R^2 value respectively. This can be recognized as an optimal number of epochs since it produces the least error among other two.

At constant 4 LSTM nodes, it seems that RMSE decreases and R^2 score increases when the number of epochs rises. R^2 is a coefficient of determination which shows the percentage of the variance in the dependent variable that the independent variables can explain. It measures the correlation between the actual reading and the prediction result on a convenient 0–1 scale. Higher R^2 means the technique fits the data better. RNN-LSTM has reached the highest R2 score of 0.76 which is highly correlated prediction.

RNN-LSTM is proven feasible to be used as a prediction technique in this research. Hence, the first objective of this proof of concept is met. The second objective for this proof of concept also has been fulfilled when the experiment results prove that, number of LSTM hidden nodes and epoch numbers are the hyperparameters on RNN-LSTM which when are tuned would either improve or worsen the prediction. In conclusion, RNN-LSTM can be made by using 4 number of LSTM nodes and 1000 epoch number as the optimal network topology for this proof of concept.

The trend of training time in Table 4 is observed to be linearly increases as the number of epochs increases. This shows that by tuning the number of epochs, it would improve the accuracy of RNN-LSTM and increases the execution time. Hence, the third objectives of this proof of concept is met since the proposed methodology in Sect. 5 which suggests implementing an optimization algorithm, PSO, to improve the accuracy and time efficiency of prediction is highly recommended in this research.

7   Conclusion

The novelty of this research is to improve the RNN-LSTM on predicting machine failure by integrating with an optimization algorithm. To increase the accuracy, hyperparameter specifically the learning rate tuning has been done. Tuning has also increased the execution time. To optimize between accuracy and execution time, optimization algorithm, PSO will be hybridized with RNN-LSTM to find for the optimal RNN-LSTM topology.

A proof of concept was conducted to prove the feasibility of RNN-LSTM. The results proved: (1) RNN-LSTM is feasible to be used as a prediction technique in this research, (2) improvement on RNN-LSTM topology contributes to accurate prediction and increases the execution time and (3) the proposed research methodology is highly recommended to optimize both accuracy and execution time of the prediction technique. The future work will consist of developing RNN-LSTM algorithm with PSO algorithm. Finally, all the research objectives will be met by continuing the research along the proposed path.

References

1. National Instruments: The Cost of Oil and Gas Equipment Mismanagement (2015). http://www.ni.com/white-paper/52604/en/. Accessed 1 Oct 2018
2. Nadai, N., Melani, A.H., Souza, G.F., Nabeta, S.I.: Equipment failure prediction based on neural network analysis incorporating maintainers inspection findings. In: 2017 Annual Reliability and Maintainability Symposium (RAMS), p. 7 (2017)
3. Kusiak, A., Verma, A.: Prediction of status patterns of wind turbines: a data-mining approach. J. Sol. Energy Eng. **133**(1), 011008 (2011). https://doi.org/10.1115/1.4003188
4. Sayad, D.S.: An introduction to data science (2010). https://www.saedsayad.com/decision_tree.html
5. Rokach, L., Maimon, O.: Data Mining With Decision Trees: Theory and Applications. World Scientific Publishing Co. Pte. Ltd., Singapore (2015)
6. Cutler, A., Cutler, D.R., Stevens, J.R.: Random forests. In: Ensemble Machine Learning, pp. 157–175. Springer (2012)
7. Medsker, L., Jain, L.C.: Recurrent Neural Networks: Design and Applications. CRC Press, Boca Raton (1999)
8. Skymind: A Beginner's Guide to LSTMs and Recurrent Neural Networks. https://skymind.ai/wiki/lstm. Accessed 5 Oct 2018
9. Hochreiter, S., Schmidhuber, J.J.N.: Long short-term memory. **9**(8) 1735–1780 (1997)
10. Kohli, S., Miglani, S., Rapariya, R.: Basics of Artificial Neural Network (2014)
11. Brownlee, J.: When to use MLP, CNN, and RNN neural networks (2018). https://machinelearningmastery.com/when-to-use-mlp-cnn-and-rnn-neural-networks/. Accessed 5 Oct 2018
12. Bhandare, A., et al.: Applications of convolutional neural networks. (IJCSIT) Int. J. Comput. Sci. Inf. Technol. **7**(5), 2206–2215 (2016)
13. Kennedy, J., Eberhart, R.: Particle swarm optimization. In: Proceedings of IEEE International Conference on Neural Networks IV (1995)
14. Kachitvichyanukul, V.: Comparison of three evolutionary algorithms: GA, PSO, and DE. Ind. Eng. Manage. Syst. **11**(3), 215–223 (2012). http://dx.doi.org/10.7232/iems.2012.11.3.215
15. IRIDIA – Technical Report Series ISSN 1781-3794. Published by: IRIDIA, Institut de Recherches Interdisciplinaireset de Développements en Intelligence Artificielle Université Libre de Bruxelles Av F. D. Roosevelt 50, CP 194/61050 Bruxelles, Belgium. Technical report number TR/IRIDIA/2006-010
16. Raja, V.S., Rajagopalan, S.: A comparative analysis of optimization techniques for artificial neural network in bio medical applications (2014)
17. Blum, C.: Ant colony optimization: introduction and recent trends. Elsevier Phys. Life Rev. **2**(4), 353–373 (2005). https://doi.org/10.1016/j.plrev.2005.10.001

Assessing the Small Satellites Resilience in Conditions of Anomalous Flight Situation

Alexander N. Pavlov[1,2(✉)], Dmitry A. Pavlov[2], Evgeny V. Kopkin[2], and Alexander Yu. Kulakov[1]

[1] Saint Petersburg Institute for Informatics and Automation of the Russian Academy of Sciences, Saint Petersburg, Russia
Pavlov62@list.ru
[2] Mozhaisky Military AeroSpace Academy, St. Petersburg, Russia

Abstract. The analysis of the modern small satellite on-board systems resilience assessment methods while managing their configuration and reconfiguration under the conditions of predictable emergency flight situations has shown that in case of conditions of unpredictable emergency flight situations these methods are not acceptable. It should be noted that they are not acceptable in case of the design and creation of the on-board system completely differing in composition and structure from known on-board systems. This situation requires the development of conceptually new methodological foundations for assessing the resilience of the small satellite on-board systems, conducting an analysis of such an important feature as the structural resilience of the small satellite on-board systems configuration. The article presents possible ways of assessing the structural resilience of the small satellites on-board system configurations, reflecting the implementation of its own functions by the elements and subsystems of the small satellites on-board systems, as well as their participation in the technological operations of the small satellites control under various scenarios of the implementation of the flight program.

Keywords: Small satellites · Predictable and unpredictable disruptions · Functional structural resilience · Technological structural resilience

1 Introduction

The expansion of the range of functional tasks which are being solved by modern small satellite, and an increase in the duration of their active existence is possible by ensuring the effective and high-quality functioning of their on-board systems both in nominal operating conditions and in the event of predictable and unpredictable emergency situations. It is possible to ensure the fulfilment of the small satellite target functions in the event of anomalous flight situations by ensuring the resilience of the small satellite on-board systems. Moreover, the resilience is understood as the property of the system to preserve and restore its characteristics (the vector indicator of the quality of the small satellite functioning) under disastrous environmental impacts. For example, the following targets for the Earth remote sensing small satellites include [1]: linear terrain resolution; efficiency of obtaining targeted information; performance in nominal

© Springer Nature Switzerland AG 2019
R. Silhavy (Ed.): CSOC 2019, AISC 985, pp. 253–265, 2019.
https://doi.org/10.1007/978-3-030-19810-7_25

operating conditions; the small satellite efficiency in the specified area of sensing; active existence duration (functioning). In case of predictable disruptions or "normal" flight conditions, the following approach, as a rule, is used to assess the resilience of the small satellite on-board systems [1–3]. An imitation of the small satellite orbital motion and its operation for the intended purpose is performed. An imitation of failures of elements of the target equipment and on-board support systems is also organized, depending on the simulated level of reliability. For each time point of imitation, the operability of the target equipment and on-board support systems is checked. In case of a failure, the random time of forced breaks in the operation of a particular on-board system and the values of the target indicators are estimated. The calculation is terminated if the small satellite fails, when it is impossible to continue functioning for the intended purpose (the occurrence of critical failures), or when a time equal to the period of active existence is reached. Such calculations are carried out for different levels of reliability of predictable emergency situations. At each level, a predetermined number of statistical tests or an amount that provides the specified simulation accuracy is produced. The calculated data are displayed on the Kiviat diagram. To determine the indicator of the small satellite on-board systems resilience, the area of the figure in the diagram is compared with the areas of the figures reflecting the initial and maximum allowable values of the target indicators. If at least one of the target indicators is less than the acceptable value, then this corresponds to the loss of the small satellite resilience, which requires a decision on the nature of the further functioning of the small satellite. However, the problem with this approach is that at present such dependencies are obtained only as a result of the operation of already created satellite. When designing the same small satellite, the results of statistical processing of previous or operating small satellite are used. It is normal if the new small satellite on-board system is similar in structure and composition with the previous system. But, if the developed small satellite differs significantly from the previously created ones, then this approach is not always acceptable. In addition to the predictable disruptions, there are unpredictable, such that no one can foresee in advance, and therefore it is impossible to prepare in advance for them. And not least in real flight these unpredictable emergency situations occur, if not more often, then, at least, in frequency, they appear commensurate with the predictable ones. Under these conditions, the models and methods used in the theory of reliability become inapplicable to ensure the resilience of the small satellite on-board systems, which requires the development of a conceptually new approach to ensuring the resilience of the small satellite on-board systems.

2   Analysis of the Known Approach to the Assessment of the Structural Resilience of the Small Satellite On-Board Systems in the Context of Unpredictable Disruptions

In the framework of studies on the development of methodological and methodological foundations for ensuring the resilience of the small satellite on-board systems, an analysis of such an important feature of the small satellite as the structural resilience of their configuration is required. In a broad sense, the structural resilience of a small satellite is understood to be such an ability of the object in question that allows it to maintain, within certain limits, the quality of its target functioning (or restore such ability) by changing (shaping) the corresponding structures (configurations).

The change in the structural states of the on-board systems small satellite is associated both with the on-board systems component failures, functional elements, and with the execution of the flight program. An element is failed (inoperable) if it is not able to perform all the technological operations assigned to it. A functional element will be considered partially able to work if it can perform at least one of the assigned technological operations. It is obvious that the values of particular indicators of the quality of the small satellite on-board systems functioning in each condition depend on: many failed, healthy, partially workable functional elements; distribution of technological operations; reallocation of these operations between workable or partially workable functional elements.

An important and integral condition for studying the capabilities of the small satellite on-board systems [4–8] is to analyze and evaluate the architecture of its structural states, reflecting both the functional and technological features of the management of the small satellite.

Structural models of the most complex technical systems functioning can be correctly described [9, 10] by block diagrams, fault trees and events, connectivity graphs, multi-terminal networks, etc. However, these structural models can describe the functioning of only monotonic systems. In monotonous models, it is impossible to take into account the logically complex and contradictory relationships and relationships between functional elements, for example, which in some structural states of the system increase, and in others, decrease the indicator of the effectiveness of its functioning. Also, monotonous models do not represent systems in which elements simultaneously operate, some of which provide an increase, for example, reliability or resilience, and another part causes failures or accidents, i.e. has the opposite, detrimental effect on the security of the system as a whole.

When studying the small satellite on-board systems resilience, the structure of which is described using graph models (monotone system of type 1 [8]), the small satellite on-board system is considered «destroyed», if in case of the graph elements deleting, it satisfies one or more of the following conditions [8]: the graph consists of at least two connected components; there are no directed paths for certain sets of vertices; the number of vertices in the largest component of the graph is less than some predetermined number; the shortest path exceeds a given value. Accordingly, the small satellite on-board systems are considered to be tenacious if these conditions are not met.

To analyze the structural resilience properties of the small satellite on-board systems under these conditions, as well as to synthesize a system with the required structural resilience property, it is necessary to introduce a quantitative assessment that adequately reflects the property in question.

In the study of the structural resilience of the small satellite on-board systems, the concept proposed in [8] introduces generalized failure of the i multiplicity, which considers the structural conditions of the small satellite on-board systems resulting from the sequential failure of various combinations (C_n^i) of the entire set of functional elements of structures for i different functional elements ($i \leq n$, where n is the number of functional elements of the considered structure of small satellite on-board systems). Among the set of structural states for a given generalized failure is determined by the set of working states, the power of which we denote R_i, or the set of unworkable states, the power of which we denote $N_i (N_i + R_i = C_n^i)$.

For comparison of various structures, the relative function of the structural resilience of the small satellite on-board systems $\Psi(i/n)\left(\Psi(i) = \frac{R_i}{C_n^i} = 1 - \frac{N_i}{C_n^i}\right)$ is determined, its linear interpolation is performed by a piecewise linear function $\tilde{\Psi}(x), x \in [0, 1]$ and the integral indicator of the structural resilience of the small satellite on-board systems is introduced in the form of the following functional $F_g = \int\limits_0^1 \tilde{\Psi}(x)dx$.

We assume that the small satellite on-board systems are in an inoperable structural state if, in a generalized failure, all elements that are at least at least one of the minimum cuts of the small satellite on-board systems structure are removed.

In the most general case, the structure of the small satellite on-board systems is characterized k by minimum cuts, each of which consists of $m_j (j = 1, \ldots, k)$ elements. Moreover, the minimum cuts have common elements.

In this situation, the number of inoperable structural states with a generalized failure of the i multiplicity will take the following form [8]:

$$
\begin{aligned}
N_i &= \sum_{j=1}^{k} \delta(i - m_j)C_{n-m_j}^{i-m_j} - \sum_{j_1=1}^{k}\sum_{j_2 > j_1}^{k} \delta(i - m_{j_1} - m_{j_2} + m_{j_1 j_2})C_{n-m_{j_1}-m_{j_2}+m_{j_1 j_2}}^{i-m_{j_1}-m_{j_2}+m_{j_1 j_2}} + \\
&+ \sum_{j_1=1}^{k}\sum_{j_2 > j_1}^{k}\sum_{j_3 > j_2}^{k} \delta(i - m_{j_1} - m_{j_2} - m_{j_3} + m_{j_1 j_2 j_3})C_{n-m_{j_1}-m_{j_2}-m_{j_3}+m_{j_1 j_2 j_3}}^{i-m_{j_1}-m_{j_2}-m_{j_3}+m_{j_1 j_2 j_3}} - \ldots \\
&\ldots (-1)^{k-1}\delta(i - m_{j_1} - m_{j_2} - \ldots - m_{j_k} + m_{j_1 j_2 \ldots j_k})C_{n-m_{j_1}-m_{j_2}-\ldots-m_{j_k}+m_{j_1 j_2 \ldots j_k}}^{i-m_{j_1}-m_{j_2}-\ldots-m_{j_k}+m_{j_1 j_2 \ldots j_k}}
\end{aligned}
\tag{1}
$$

where $\delta(x) = \begin{cases} 1, & x \geq 0 \\ 0, & x < 0 \end{cases}$ is a discrete form of the Heaviside step function.

In formula (1), the values $m_{j_1 j_2 \ldots j_k}$ represent the total number of common elements in the minimum cuts with numbers $j_1, j_2, \ldots, j_k$. Using formulas (1), it is possible to calculate the relative function of the structural resilience of small satellite on-board systems having a monotonic structure, and accordingly determine the integral indicator of the structural resilience of the system $F_g = \int\limits_0^1 \tilde{\Psi}(x)dx$.

To calculate the structural resilience, a set of minimum cuts is needed, as well as the definition of common functional elements in these sections. In general, finding the minimum failure rates is NP difficult. In this case, the calculation of the structural resilience index using the generalized formula (1) is a super-complicated combinatorial problem. At the same time, it should be noted that not all monotonic structures can be described using graph models.

3   The Method of Assessing the Structural and Functional Resilience of Small Satellite On-Board Systems in Conditions of Unpredictable Disruptions

To overcome these features of assessing the structural resilience of the small satellite on-board systems, we propose the following approach, which is based on the concept of the genome structure [8]. As a rule, the structural analysis of the small satellite on-board systems functioning begins with the construction of a functional integrity scheme of an object [10]. The functional integrity scheme is a logically universal graphical tool for the structural representation of the studied properties of objects. The functional integrity schemes allow to correctly represent both all traditional types of structural schemes (flowcharts, failure trees, event trees, graphs of connectedness with cycles) and a fundamentally new class of non-monotonic (non-coherent) structural models of various properties of the systems under study. The development of the small satellite on-board systems functional integrity scheme means, first of all, a graphical representation of the logical conditions for the implementation of its own functions by elements and subsystems of the small satellite. The second important aspect of building and further using the functional integrity scheme is an indication of the specific purpose of the simulation - the logical conditions for implementing the investigated system property, for example, reliability or failure of the small satellite on-board systems, safety or the occurrence of an accident, the implementation of certain modes of functioning of the small satellite on-board systems, etc.

It is known that the structure genome $\vec{\chi} = (\chi_0, \chi_1, \chi_2, \ldots, \chi_n)$ [8], which is a concentrated representation of the structural state of an object, contains and allows to determine the following information in the process of structural research of complex objects: first, information about the topological properties of the structure of a monotone system; secondly, information on the belonging of the object under study to the class of monotone or non-monotonic systems; thirdly, to assess the indicators of the structural and functional resilience of the system.

For the formal description and analysis of the process of degradation (restoration) of small satellite on-board systems as factors with which you can change the structure of small satellite on-board systems, we will consider the operation of removing (restoring) critical elements $\{ P_{j_1}, P_{j_2}, \ldots, P_{j_N} \} = \tilde{P}$ from the functional integrity scheme. In general, all functional elements of small satellite on-board systems can be considered as critical elements. In the process of removing (restoring) elements, the structure of the small satellite on-board systems can be in one of its intermediate states S_α.

According to the concept of the genome structure, the structural states S_α (initial, final, intermediate) are characterized by their genomes $\vec{\chi}_\alpha$ (by this material $\vec{\chi}_\alpha$ we mean the dual analogue of the genome), while the indicators of the structural and functional resilience of the small satellite on-board systems, consisting of homogeneous, non-uniform functional elements, according to the reliability of their functions can be calculated by the following formulas [8]:

$$F_{hom}(\vec{\chi}_\alpha) = \vec{\chi}_\alpha \cdot (1, \frac{1}{2}, \frac{1}{3}, \ldots, \frac{1}{n+1})^T,$$

$$F_{het}(\vec{\chi}_\alpha) = \vec{\chi}_\alpha \cdot (1, \frac{1}{2}, \frac{1}{2^2}, \ldots, \frac{1}{2^n})^T, \tag{2}$$

$$F_{possib}(\vec{\chi}_\alpha) = \sup_{\mu \in [0,1]} \min \{\vec{\chi}_\alpha \cdot (1, \mu, \mu^2, \ldots, \mu^n)^T, \ g(\mu)\}.$$

We assume that the structural state S_α characterized by the genome $\vec{\chi}_\alpha$ is directly related to the structural state S described by the genome $\vec{\chi}$, if there is a functional elements ($\exists \ P_j \in \tilde{P}$), the failure (restoration) of which ($P_j = 0$ or $P_j = 1$) takes the system from state S to state S_α (from state S_α to state S).

Let us designate this variation of the structural state of the small satellite on-board systems as follows: $\vec{\chi} \overset{P_j}{\leftrightarrow} \vec{\chi}_\alpha$. The set of all structural states directly associated with the state $\vec{\chi}$ is denoted by $X(\vec{\chi})$.

One of the possible trajectories of the reconfiguration of the structure of the small satellite on-board systems in the process of occurrence of failures (recovery) can be described by the following chain of transitions

$$\vec{\chi}_{\alpha_0} \overset{P_{j_1}}{\leftrightarrow} \vec{\chi}_{\alpha_1} \overset{P_{j_2}}{\leftrightarrow} \vec{\chi}_{\alpha_2} \overset{P_{j_3}}{\leftrightarrow} \ldots \overset{P_{jN-1}}{\leftrightarrow} \vec{\chi}_{\alpha_{N-1}} \overset{P_{jN}}{\leftrightarrow} \vec{\chi}_{\alpha_N},$$

where $\vec{\chi}_{\alpha_0} = \vec{\chi}_0$, $\vec{\chi}_{\alpha_N} = \vec{\chi}_f$, the set $\{P_{j_1}, P_{j_2}, \ldots, P_{jN}\} = \tilde{P}$, i.e. the set of failed (restored) functional elements small satellite on-board systems in the transition chain is a permutation of the elements of the set $\tilde{P}$.

Structural changes occurring in the intermediate state $\vec{\chi}_\alpha$ on the reconfiguration trajectory will be evaluated by one of the indicators of the structural and functional resilience of the small satellite on-board systems (2) included in the considered set: $F_{failure}(\vec{\chi}_\alpha) \in \{F_{hot}(\vec{\chi}_\alpha), F_{het}(\vec{\chi}_\alpha), F_{possib}(\vec{\chi}_\alpha)\}$. In addition, in each intermediate structural state $\vec{\chi}_\alpha$, the small satellite on-board systems is characterized by a certain set of structural and topological constraints $\Psi_l(\vec{\chi}_\alpha) \leq 0$, $l = 1, 2, \ldots, L$, formally defined and quantified by using [8] corresponding indicators of structural resilience, flexibility, reachability, structural complexity, etc. In other words, these restrictions define the range of allowable variations, which will be denoted in the following Ξ.

Then the task of building an optimistic (pessimistic) scenario for reconfiguring the small satellite on-board systems can be represented as the following optimization problems (3).

$$\sum_{j=0}^{N} F_{failure}(\vec{\chi}_{\alpha_j}) \rightarrow \quad \max(\min) \tag{3}$$
$$\vec{\chi}_{\alpha_j} \in X(\vec{\chi}_{\alpha_{j-1}})$$
$$\vec{\chi}_{\alpha_0} = \vec{\chi}_0, \ \vec{\chi}_{\alpha_N} = \vec{\chi}_f,$$
$$\Psi_l(\vec{\chi}_{\alpha_j}) \leq 0, \ l = 1, 2, \ldots, L$$
$$\{P_{j_1}, P_{j_2}, \ldots, P_{jN}\} = \tilde{P}$$

In [8], a combined method of randomly directed search for solutions to the problem was substantiated and an algorithm was developed that implements the above method. The combined method and the corresponding algorithm allows you to search for both optimistic and pessimistic trajectories, as well as intermediate trajectories chosen randomly.

Then, as a generalized indicator of the structural and functional resilience of the small satellite on-board systems in the process of its structural reconfiguration according to the scenario $\mu_\varsigma^{(k)}$, a relationship can be proposed $J^k = S_0^k/S^k$. Here $S_0^k = \sum_{j=0}^{N-1} \frac{F_{failure}(\vec{\chi}_{\alpha_j}^{(k)}) + F_{failure}(\vec{\chi}_{\alpha_{j+1}}^{(k)})}{2}$ it is equal to the total structural and functional resilience of the functioning of the small satellite on-board systems in the process of reconfiguration within the framework of the scenario $\mu_\varsigma^{(k)}$, and $S^k = \max_{j=0,1,...,N} \{F_{failure}(\vec{\chi}_{\alpha_j}^{(k)})\} \cdot N$ is proportional to the total index of the structural and functional resilience of the functioning of the small satellite on-board systems along the trajectory in case of preservation of the possible maximum resilience of functioning during the development of the considered scenario.

It should be noted that the maximum value of the generalized index of structural and functional resilience $J^{max} = \max_k\{J^k\}$ will be achieved in the optimistic scenario of on-board systems reconfiguration, and the minimum value $J^{min} = \min_k\{J^k\}$ in the pessimistic one. We will conduct M simulation experiments. On each k experiment, a sequence $\mu_\varsigma^{(k)} = \left[\vec{\chi}_{\alpha_0}, \vec{\chi}_{\alpha_1}^{(k)}, \vec{\chi}_{\alpha_2}^{(k)}, ..., \vec{\chi}_{\alpha_{N-1}}^{(k)}, \vec{\chi}_{\alpha_N}\right]$ is constructed (where $\vec{\chi}_{\alpha_0} = \vec{\chi}_0, \vec{\chi}_{\alpha_N} = \vec{\chi}_f$) corresponding to the path of the small satellite on-board systems reconfiguration. For the constructed trajectory, the value of the generalized index of structural and functional resilience is calculated $J^k = S_0^k/S^k$. Next, we find the average value of the structural resilience of all tests $J^0 = \frac{1}{M}\sum_{k=1}^{M} J^k$. Then it can be argued that the real values of the generalized index of the structural and functional resilience of the small satellite on-board systems J_{SG} lie in the interval $\left[J^{min}, J^{max}\right]$ and the most expected value is J^0. In this case, the predicted values of the indicator J_{SG} can be set with a fuzzy triangular number (a, α, β), where $a = J^0$, $\alpha = J^0 - J^{min}$, $\beta = J^{max} - J^0$.

In addition, the calculation of structural and functional resilience index values $F_{failure}(\vec{\chi}_\alpha) \in \{F_{hot}(\vec{\chi}_\alpha), F_{het}(\vec{\chi}_\alpha), F_{possib}(\vec{\chi}_\alpha)\}$ can be carried out on the assumption that the structure of the small satellite on-board systems consists only of elements that are uniform in reliability of their functions, only elements that are not uniform in reliability in their functions, and finally there are potential failures to perform their functions. For each of these three cases, by calculating the values of the indicator J_{SG}, we obtain, respectively, three fuzzy triangular results: (a^o, α^o, β^o), (a^n, α^n, β^n), (a^b, α^b, β^b). Then, as the value of the generalized indicator of the structural and functional resilience of the small satellite on-board systems J_{SG}, we will assume the average value of the results obtained $J_{SG} = \frac{(a^o, \alpha^o, \beta^o) + (a^n, \alpha^n, \beta^n) + (a^b, \alpha^b, \beta^b)}{3}$.

Thus, the task of calculating the value of the generalized indicator of the structural and functional resilience of the small satellite on-board systems has been reduced to the analysis of optimistic, pessimistic or random (arbitrary) trajectories of the structural and functional reconfiguration of the object caused by failures (restoration) of the functional elements of the small satellite on-board systems.

It should be noted that the failure (restoration) of an element leads to the failure (restoration) of the other functional elements of the small satellite associated with it logically associated with it. Therefore, in addition to the introduced generalized indicator of the structural and functional resilience of the small satellite on-board systems J_{SG}, it is possible to introduce an absolute indicator of the structural and functional resilience of the small satellite on-board systems. Each trajectory of the reconfiguration of the on-board systems structure of the small satellite is characterized by a number of degradation levels J_D, the last of which corresponds to the on-board systems transfer of the on-board systems into an inoperable state. So for a pessimistic trajectory the number of levels is minimal and equal $J_D^{\min}$, for an optimistic trajectory it is maximal - $J_D^{\max}$. The values of the absolute index of the structural and functional resilience of the small satellite on-board systems J_{AG} will lie in the interval $[J_D^{\min}, J_D^{\max}]$, and you can also calculate the most expected value equal J_D^0. In this case, the values of the index J_{AG} are similar, as well as J_{SG}, can be set with a fuzzy triangular number (a_A, α_A, β_A), where $a_A = J_D^0$, $\alpha_A = J_D^0 - J_D^{\min}$, $\beta_A = J_D^{\max} - J_D^0$.

4   The Method of Assessing the Structural and Technological Resilience of Small Satellite On-Board Systems in Conditions of Unpredictable Disruptions

Different functional elements of small satellite on-board systems are far from identical roles, their failures can lead to consequences of varying degrees of influence on the state of the system. This is due to both technical differences (throughput, performance, etc.), as well as differences in facility management technology.

In order to determine the role of specific on-board systems components in ensuring the operability of the small satellite on-board systems, in general, an analysis of the structural and technological resilience indicators should be carried out with the participation of the on-board systems functional elements in various scenarios of the small satellite flight program implementation.

So, for example, the typical small satellite AIST-2d control technology presented in [11], with its normal operation, includes, as a rule, the following technological modes (operations) of small satellite orientation (Table 1): the mode "Bringing the small satellite into an OCS-oriented position" - "rough orientation", the mode "Bringing the small satellite into an OCS-oriented position" - "accurate orientation", the mode "Bringing the small satellite into the sun-oriented position" (OCS - orbital coordinate system). To perform these modes, various sensitive elements and executive bodies are used (Table 1), which include individual angular velocity meters (IAVM), flywheel control motors (FCM), Earth orientation instruments (EOI), optical star sensors (OSS), magnetometers (Mag), electromagnets (El), satellite navigation system(SNS).

Table 1. Small satellites orientation modes

Element type	IAVM				FCM				SNS	OSS		EOI		Mag		El		
Element No	1	2	3	4	5	6	7	8	9	10	11	12	13	14	15	16	17	18
Rough orientation mode	1 out of 4				2 out of 4				1	1 out of 2		1 out of 2		1 out of 2		2 out of 3		
Accurate orientation mode	3 out of 4				3 out of 4				1	1 out of 2		1 out of 2		1 out of 2		1	1	1
Sun oriented mode	3 out of 4				2 out of 4				1	1 out of 2		1 out of 2		1 out of 2		2 out of 3		

At the same time, the values of the frequency of application of the technological modes of orientation of the small satellite when performing typical operations on receiving (transmitting) various information (special, telemetric, ballistic, command-program) circulating in the control circuits of the small satellites ICS during the typical small satellite control cycle are known: the "rough" orientation mode is used in 40%, the "accurate" orientation mode in 52%, the "Sun-oriented" mode in 8%.

To analyze the indicator of the structural and technological resilience of the small satellite on-board systems, it is proposed to use the approach proposed in [8], which implies the description of the participation of the functional elements of the small satellite on-board systems in the technological operations in the form of hypergraphs and the calculation of their derivatives.

A possible geometric interpretation of the considered hypergraph is shown in Fig. 1.

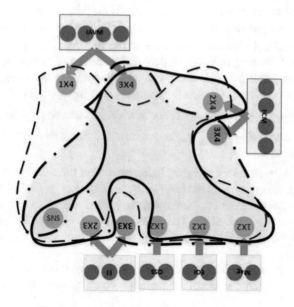

Fig. 1. Hypergraph of technological operations

Using this approach allows analyzing the structural and technological properties of the small satellite on-board systems and calculating various indicators - the intensity of the application of elements in various technological processes, the uneven participation of elements in various technological management cycles, the structural and techno-logical resilience of the small satellite on-board systems, etc., which allow to evaluate structurally - technological resilience of both individual functional elements on-board systems and on-board systems as a whole in the development of new technologies for managing the configuration and reconfiguration of the small satellite on-board systems.

For a given technological structure of the small satellite control, the functional elements of the small satellite on-board systems are known that are involved in various technological control operations. Each technological operation is the edge of the hypergraph, and the functional elements of the small satellite on-board systems rep-resent its vertices. The specified hypergraph can be formally represented by an inci-dence matrix Q.

In order to analyze the intensity of the use of the elements of the resulting hypergraph, when performing technological operations of the small satellite control, it is necessary to calculate the frequency matrix of relations F and the derivative of the hypergraph. The matrix F can be calculated from the incidence matrix Q using the formula $F = Q^T \times Q$, where Q^T is the transposed incidence matrix.

The derivative of the hypergraph of the technological operations of the small satellite is a graph, each pair of vertices is weighted by the ratio of the frequency $(f_{ii} - f_{ij}) + (f_{jj} - f_{ij})$ of their inconsistent participation in the technological operations of the management of the small satellite to the frequency f_{ij} of joint participation $(G = \left\| g_{ij} \right\| = \left\| \frac{f_{ii} - 2f_{ij} + f_{jj}}{f_{ij}} \right\|)$.

In other words, the derivative of the hypergraph will characterize the degree of intensity and uniformity of the participation of elements in the performance of tech-nological operations of small satellite control. To do this, using the derivative of the hypergraph, by normalizing the values of its adjacency matrix $G = \left\| g_{ij} \right\|$, a fuzzy relation of the technological independence of the functional elements is constructed according to the following rule. Let us denote the number of technological operations in which the functional elements are involved through n. In this case, the maximum possible value of the matrix element $G = \left\| g_{ij} \right\|$, other than ∞, is $\frac{f_{ii} - 2f_{ij} + f_{jj}}{f_{ij}} = \frac{n-2+1}{1} = n - 1$. Then the fuzzy relation of technological independence can be defined by a matrix $Z = \left\| z_{ij} \right\|_{m \times m}$, the values of the elements of which are calculated as follows $z_{ij} = \frac{g_{ij}}{n-1}$.

To determine the final technological independence of the elements, we calculate the transitive closure matrix $Z^* = Z \circ Z^2 \circ \ldots \circ Z^{m-1}$ of a fuzzy graph. For this, we use the max min composition operation. Moreover, the greater the value of the coefficient of technological independence of the functional elements, the less its influence on the conduct of technological operations, i.e. all the more tenacious, in terms of techno-logical operations.

Therefore, it is proposed to calculate the indicator x_i of the structural and techno-logical resilience of the functional elements on-board systems small satellite by the formula (4).

$$h(x_i) = \frac{\sum_{j=1}^{m} z_{ij}^*}{m} \qquad (4)$$

To assess the structural and technological resilience of the small satellite on-board systems, two indicators are proposed. One of them characterizes the average technological independence of all functional elements on-board systems small satellite and can be calculated as follows (5).

$$h_1 = \frac{\sum_{i=1}^{m} \sum_{j=1}^{m} z_{ij}^*}{m^2} \qquad (5)$$

To calculate the second indicator of the structural and technological resilience of the small satellite on-board systems, you can use the structural and topological indicator (6) of decentralization (flexibility) of a fuzzy graph of technological independence [8]

$$h_2 = \max\{0, 1 - \frac{1}{2(m-1)(m-2)} \sum_{j=1}^{m} \left[\max_{j=1,\dots,m} a(j) - z^*(j) \right]\}, \qquad (6)$$

where

$$z(j) = \sum_{i=1}^{m} z_{ji} + \sum_{i=1}^{m} z_{ij}, a(j) = \sum_{i=1}^{m} a_{ji} + \sum_{i=1}^{m} a_{ij}, \quad a_{ij} = \begin{cases} 1, z_{ij}^* > 0 \\ 0, z_{ij}^* = 0 \end{cases}, \quad (i,j = 1,2,\dots,m).$$

5 Conclusions

A distinctive feature and novelty of the proposed approach is that on a single methodological basis (the original concept of the genome of the structural construction of structurally complex objects) it is possible to carry out a study of structural and functional properties and carry out the operational calculation of interval, optimistic and pessimistic estimates of structural resilience indicators as monotonous, non-monotonic and homogeneous, inhomogeneous structures on-board systems small satellite on-board systems.

A new indicator of the structural and technological resilience of both the functional elements and the on-board systems total small satellite during the technological operations of the small satellite control has been introduced. For the evaluation and analysis of the structural and technological resilience of the small satellite on-board systems, an original approach has been proposed, which is based on the concept of the hypergraph differentiation of the small satellite technological control modes.

The proposed indicators of the structural-functional and structural-technological resilience of the small satellite on-board systems in the case of calculated, and especially unpredictable emergency flight situations on board the small satellite will allow to analyze and evaluate the resilience of a particular configuration of the small satellite on-board systems.

Acknowledgements. The research described in this paper is partially supported by the Russian Foundation for Basic Research (grants 16-29-09482-ofi-i, 17-08-00797, 17-06-00108, 17-01-00139, 17-20-01214, 17-29-07073-ofi-i, 18-07-01272, 18-08-01505, 19-08-00989) state order of the Ministry of Education and Science of the Russian Federation № 2.3135.2017/4.6, state research 0073–2019–0004, International project ERASMUS+, Capacity building in higher education, № 73751-EPP-1-2016-1-DE-EPPKA2-CBHE-JP, Innovative teaching and learning strategies in open modelling and simulation environment for student-centered engineering education.

References

1. Akhmetov, R.N., Makarov, V.P., Sollogub, A.V.: Baipasnost kak atribut zhivuchesti avtomaticheskikh kosmicheskikh apparatov v anomalnykh poletnykh situatsiiakh [Bypass as an attribute of unmanned satellite operability in anomalous flight situations]. Vestnik Samarskogo aerokosmicheskogo universiteta, no. 4(44), pp. 9–21 (2015). (In Russian)
2. Small Satellites: A Revolution in Space Science/Final Report, 83 p. Keck Institute for Space Studies, California Institute of Technology, Pasadena, CA, July 2014
3. Sandau, R., Röser, H.-P., Valenzuela, A.: Small Satellites for Earth Observation: Selected Contributions, 406 p. Springer, Heidelberg (2008)
4. Small Spacecraft Technology State of the Art/Mission Design Division Staff, 168 p. Ames Research Center, Moffett Field, California, December 2015
5. Huges, P.C.: Spacecraft Attitude Dynamics, 585 p. Dover Publications, New York (2004)
6. Manuilov, Ju.S., Novikov, E.A., Pavlov, A.N., Kudriashov, A.N., Petroshenko, A.V.: Sistemnyi analiz i organizatsiia avtomatizirovannogo upravleniia kosmicheskimi apparatami [System analysis and organization of automated satellite control], 266 p. VKA imeni A.F. Mozhajskogo, SPb (2010). (In Russian)
7. Pavlov, A.N.: Nechetko-vozmozhnostnyi podkhod k analizu i otsenivaniiu bezopasnosti slozhnykh organizatsionno tekhnicheskikh sistem [Fuzzy – possibility approach to the analysis and assessment of the security of complex organizational and technical systems]. Materiali XI Sankt–Peterburgskoi mejdunarodnoi konferencii RI–2008, pp. 48–49. Sankt-Peterburgskoe obshchestvo informatiki vychislitelnoi tekhniki i sistem upravleniia, Spb (2008). (In Russian)
8. Aleshin, E.N., Zinovev, S.V., Kopkin, E.V., Osipenko, S.A., Pavlov, A.N., Sokolov, B.V.: Sistemnyi analiz organizatsionno-tekhnicheskikh sistem kosmicheskogo naznacheniia [System analysis of organizational and technical systems for space applications], 357 p. VKA imeni A.F. Mozhajskogo, Spb (2018). (In Russian)
9. Pavlov, A.N., Osipenko, S.A.: Metodika otsenivaniia tekhnicheskogo sostoianiia sredstv nazemnogo avtomatizirovannogo kompleksa upravleniia kosmicheskimi apparatami [Methods of assessing the technical condition of ground-based automated control systems for satellites]. In: Trudy Voenno-kosmicheskoj akademii im. A.F. Mozhajskogo, vol. 27, pp. 159–169 (2008). (In Russian)

10. Polenin, V.I., Riabinin, I.A., Svirin, S.K., Gladkova, I.A.: Primenenie obshchego logiko veroiatnostnogo metoda dlia analiza tekhnicheskikh voennykh organizatsionno funktsion- alnykh sistem i vooruzhennogo protivoborstva [The use of a common logical – probabilistic method for the analysis of technical, military organizational and functional systems and armed confrontation], 416 p. The Russian Academy of Natural Sciences, Spb (2011). (In Russian)
11. Opytno-tekhnologicheskiy malyy kosmicheskiy apparat «AIST-2D» [Experimental techno- logical small satellite "AIST-2D"], 324 p. Samarskiy nauchnyy tsentr Rossiyskoy akademii nauk, Samara (2017). (In Russian)

Audio Gadget Recommendation by Fuzzy Logic

Md. Mokarram Chowdhury, Farhan Tanvir, Md. Shakilur Rahman,
Md. Motiur Rahman, Md. Al-Sahariar, and Rashedur M. Rahman[✉]

Department of Electrical and Computer Engineering, North South University,
Plot-15, Block-B, Bashundhara Residential Area, Dhaka, Bangladesh
{mokarram.chowdhury, tanvir.farhan, motiur.rahman,
rashedur.rahman}@northsouth.edu,
shakilur.nsu@gmail.com, sahariarnsu@gmail.com

Abstract. The primary objective of this research is to search and sort audio gadget list according to user preference with fuzzy logic. The system has a database that contains the price of the product and necessary information along with reviews. This information is considered as the parameters of the fuzzy system. In this system, the user can specify approximate budget, genres, sound quality, and review of the audio gadget that are used to purchase proper audio devices. Each item's metadata is converted into a fuzzy parameter and passed to fuzzy recommendation system. After calculation, the fuzzy recommendation system assign's a value to each item. Those values are sorted in descending order, and the highest value is counted as the most preferable item for the user. The system also allows users to give feedback for different attributes such as genre, sound quality, user satisfaction as the review. Therefore, the system becomes user-adaptive when a user uses the system for a while.

Keywords: Audio gadget · Recommendation · Fuzzy system · ANFIS

1 Introduction

Nowadays, audio gadget is a highly used product by people. Many types of audio gadgets are introduced with various qualities. Wired, wireless with multiple kinds of designs are built for user satisfaction. Digital communication, listening to music, etc. are the usage of these products. Different types of audio gadgets are manufactured based on consumer necessity. Audio gadgets are divided into the categories depending on size and form including many types of features. These are also categorized based on user preference. Based on users' need, the production volume of these products is increasing day by day. When any user needs to choose his/her desired audio gadgets, it is complicated to find out the desirable product from the huge varieties. In this research, we try to help the user to choose the popular audio device. We use fuzzy logic to provide an easy solution for a satisfactory outcome. The desired preferences of audio gadgets cannot be determined using a linear scale. Highly desirable audio gadget based on user preference will be met as output using fuzzy recommendation system.

© Springer Nature Switzerland AG 2019
R. Silhavy (Ed.): CSOC 2019, AISC 985, pp. 266–276, 2019.
https://doi.org/10.1007/978-3-030-19810-7_26

2   Related Works

Our system works over different types user inputs, and we take a review from online user rating which is adaptive. We gain knowledge over it from some related research papers. We will discuss some of those works here. The authors in [1] offered a recommendation system based on fuzzy Bayesian networks with utility theory. They made the system based on different contexts of music. In the next paper [2], an intelligent fuzzy-based recommendation system has been implemented for consumer electronic products. Zhao, Wang, and Zhong [3] proposed a fuzzy recommendation system which was mainly based on customer online review. According to the paper, the system takes user review as one of its inputs which is mostly in natural language. Jang [4] introduced the architecture and learning procedure underlying ANFIS for their work. There was another research on Rule-base structure identification in an adaptive-network-based fuzzy inference system by Sun et al. [5]. The authors summarized the architecture presented based on fuzzy inference rule and Kalman filtering algorithm to identify the system parameters. Jang, Sun, and Mizutani [6] focused on theoretical aspects and also empirical observations and verifications of various applications in practice. Wang [7] described an approach toward Stable adaptive fuzzy control of nonlinear systems where the research showed that there is no need for a mathematical model for the adaptive system. The authors used directly if-then rules and sent it to the fuzzy controller. Finally, the approaches studied in fuzzy logic methods in recommender systems [8] differ from collaborative filtering, they were based solely on the preferences of the single individual for whom they were providing the recommendation and make no use of the choices of other collaborators.

3   Theory

In this research, our target is to build an adaptive fuzzy system that will be able to recommend audio gadgets based on various input parameters taken as a search query from the user. Expert reviews are collected from multiple sources, e.g. blogs, forums, facebook groups, etc. The system will also take surveys from the users who will use the system and store those reviews as a fuzzy value in the database.

3.1   Inputs

The input variables are discussed in the following sections.

Budget: Our first variable of this system is budget which is an integer value. Users will specify their preferable budget. We have divided this variable in 5 levels which are more preferable, preferably, less preferable, cheap and expensive. We have considered more preferable as $|x \pm 10|$, preferable as $|x \pm 20|$, less preferable as $|x \pm 50|$, cheap as less than $|x - 50|$ and expensive as greater than $(x + 50)$ where x is an integer that represents user's budget.

Suppose, a user has given $100 as budget. Then more preferable will be $90 to $110 (100 ± 10), preferably will be $80 to $120, less preferable will be $50 to $150, cheap will be $50 and expensive will be $150.

When the user will give his/her budget preference, he/she is very much able to afford that amount. In our less preferred level, we are considering a range of 100 dollars keeping his/her preferred amount in the middle. So we can think that they can afford this amount of money. The more preferred amount will get more score than any other preference level. So we are providing a score of 3 to a more preferred amount. In the same way, we are giving 2 and 1 to favorite and less preferred amount respectively. As cheap and expensive is far from his budget, we give them score 0. We have used the triangular curve for the budget in the fuzzy system which is shown in Fig. 1.

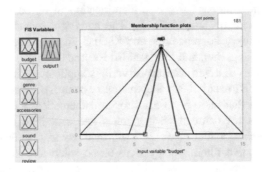

Fig. 1. Input membership function (Budget)

Genre: This variable defines the type of music that the user would prefer to listen in their audio gadget. There are different types of music genres like rock, pop, jazz, metal, etc. We have defined three levels of membership for this variable, more preferred, preferred and less preferred. This scale determines the level of preference for specific genre of the user. The user will be able to choose a single genre like rock or multiple genres like rock and pop as their preferred genre as input. The system will assign scores to the audio gears from the fuzzy database for each genre. Then it will take the average of those scores and assign it as a final score to that item from the database.

Suppose that a user wants to purchase an audio gear that is best suited for listening to rock and pop at the same time. The user will give the input as "Rock and Pop". Upon searching, the system fetches all the items that are marked as suited for rock and pop music. The fuzzy system will then assign scores for rock and pop to each piece. As the fuzzy system cannot take multiple parameters for a single variable, it will first pass rock and then pop to the fuzzy system. Then it will make the average of those scores and assign the final score to the fetched items of the database.

Item1 genre1 $=$ fuzzysystem (genre1)
Item1 genre2 $=$ fuzzysystem (genre2)

$\vdots$
$\vdots$
$\vdots$

Item1 genreN $=$ fuzzysystem (genreN)
Item1 genre score $=$ (Item1 genre 1 $+$ Item1 genre 2 $+$ ... Item1 genre N)/N

Sound Quality: Next variable in our system is sound quality. This is the most crucial variable in our system. The primary objective of any audio gadget is to listen to music with the best quality output. This variable determines a preferred sound quality to the user. This is a compound variable containing these sub variables:

(a) Frequency response: Frequency response is further divided into three sub-categories-High, mid and low frequency. It defines how a specific gadget would answer to the high, low or middle frequencies of music. In music listening, the response of these frequencies establishes the quality of sound output. In our system user will be able to enter scores for these subcategories in a scale of 10. Users can give the desired rating to three frequencies individually.

(b) Response Type: Another attribute is the response type of the audio gadget. There are two types of responses of any audio device- (i) Flat response, (ii) Colored response

A flat response produces a sound frequency that generates sound that was recorded or source music. Whereas colored response boosts/tunes the frequency, as a result, produced sound is slightly different than the source. The user can choose one response type from 3 categories for any response type which are low, medium and high. For this attribute, the score will be out of 10. Low will represent 3; medium will serve 6 and high will represent 10.

(c) Soundstage & Imaging: In this subcategory, we will take scores for both soundstage and imaging. Users will be given three options for both soundstage and imaging. For soundstage, there will be three options normal, wide and very wide. For imaging, there will also be three options normal, good and great. The numeric value 3, 7 and 10 are representing the score for both attributes accordingly.

Now we will add the scores from frequency response, response type, soundstage and imaging that will produce a score in the scale of 60. Then we will compare that score from the items in our database. We will take the absolute difference between them. Let, x be the absolute difference between the scores. Then the system will subtract that value from a constant value 15, $|15 - x|$. We have divided this difference into three membership levels- High quality, medium quality, and low quality. If the difference is between 0–6 we will consider it as low quality, for 4–11 it will be medium quality, and for 9–15 it will be regarded as high quality.

Suppose, a specific user gives input in the sound quality variable. The user provides high-7, medium-8, and low-5 in frequency subcategory; he selects high for flat response in response type, wide for soundstage and great under imaging attribute that represents 7 and ten as scores. So the final score will be 7 + 8 + 5 + 10 + 7 + 10 = 47, out of a total score of 60. Some sample with scores is given in Table 1.

Table 1. Sound quality

Name	Sound quality
Beats In-Ear by Dr. Dre	45
KZ ate	37
Sennheiser IE80	55
Shure 835	58

We have used the trapezoidal curve for input membership function. Here the curve of sound quality is shown in Fig. 2.

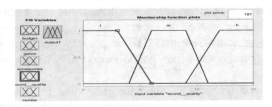

Fig. 2. Input membership function (Sound Quality)

Use Cases: Use cases will be defined by the user that will indicate the scenario in which the user will use the audio gadget. The user will be given some categorized options of scenarios that cover the most extensive range of audio gadgets. In the system, there will be options- use while traveling, work out use, children use, etc. The user will be able to choose from these use cases to match their requirements. Both single and multiple options can be chosen, and the system will output scores based on the options selected. Here the score will be out of 10. For multiple options selected the score will take the average of the scores from the options and give it as a score.

Accessories: Next variable in our system is accessories. The user can specify what type of accessories they wish to get with their audio devices. Audio gadgets offer several types of accessories depending on the gadget. Suppose the user is searching for In-ear monitor which may or may not provide extra earbobs, cables, and some other accessories. There are three membership levels high, medium and low. Accessories variable is stored in our database as a value out of 10. If the value is between 0–4 then its low, for 3–7 medium and 6–10 is high.

Review: Users can search according to user reviews in this system. Users have access to put three levels of satisfaction as a review. Review is defined by 1 to 5 scale. While the user will give input by review scale as an attribute, the system will find all the audio gadgets that are one more or less of the user-defined review. We have assigned 3 levels of fuzzy membership as more preferable, preferable, and less preferable for the user review variable.

3.2   Output Variable

In this system, the output variable will be defined by the preference level. The preference level will be fetched from the summation of all user inputs level. There are 6 user level inputs, and they have three degrees in membership function. So, if all the information has the highest value which is 3, the summation of all input value will be 18, and if all input variables have the lowest score, the summation of the variable will be 0. There will be 6 preference levels. The preference levels are shown in the following Table 2.

Table 2. Output variable

Rule number	Preference level	Range
1	Level 1	0–3
2	Level 2	3–6
3	Level 3	6–9
4	Level 4	9–12
5	Level 5	12–15
6	Level 6	15–18

Let us consider an example, if Budget score = 2, Genre score = 2, Sound quality score = 3, Use case score = 2, Accessories score = 2, Review score = 3. Then the summation = 2 + 2 + 3 + 2 + 2 + 3 = 14. The summation is 14. So, it is in level 5. That means rule 5 will be triggered. So the 5th level membership will be used to generate the output.

3.3   Output Curve

Triangular function is used for the output curve. This engine will generate a single output. We have considered five output level. So, we have assigned six output membership function f1, f2, f3, f4, f5, f6. The curves of the output membership function have been demonstrated in Fig. 3. We have used a triangular curve to represent the preference level of the membership function of output. It will be simple in computing. The figure is shown in Fig. 3.

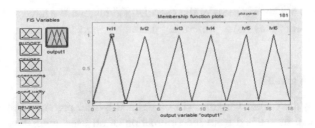

Fig. 3. Output membership function

3.4   Feedback

When any user gives any input, and the system assigns the level of membership value, it is necessary that the value is reliable. That is why the system will take the feedback values and use these feedbacks for assigning the value of the membership function for user-level inputs. The system will also receive feedback upon using our recommending system to make data more precise. The system will take data for sound quality, review, and genre as they are the most complex and vary from different person to person. As a result, the stored data may not deliver the highest satisfaction level to maximum users. This feedback will enable us to caliber the data and help the users to get their desired audio gadget. We will take the feedback of the users and will compare those with the stored data and adjust accordingly.

3.5   Adaptive System

User Weight Based Adaptation: In our fuzzy inference system, the user gives sound quality, genre, and review as user input. To get the scores for input parameters, an average of scores has been taken in the input variable area. The scores can be more precise by using the neural network's adaption techniques. For this adaption technique, a *weight$_i$* has been assigned to each user. W_i will be calculated from a function of *parameter$_i$* where i is the number of reviews that users have given in the system. To calculate the score for given parameters of the individual user, W_i and U_i will be multiplied where U_i is user given score. The final score will be calculated by taking the average of all scores for the given parameters. The following formula will calculate the adaptive score:

$$w_i = \frac{2\arctan(r_i)}{\pi}, r_i > 0 \tag{1}$$

$$\text{Total Score} = \frac{\sum_{i=1}^{n} WiUi}{n} \tag{2}$$

In Fig. 4, we have shown the neural network logic graph which is has been used for the system adaptive.

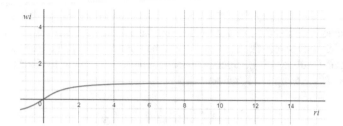

Fig. 4. Adaptive system based on user's weight

For example, 5 users gave the feedbacks on review. The users are U_1, U_2, U_3, U_4, U_5. To get the weight of the user, the number of each user's review will be necessary that means the r_i value of each user. Suppose, the value of U_1, U_2, U_3, U_4, U_5. are respectively 4, 3, 3, 2 and 1. The number of reviews of the users are 1, 3, 5, 9 and 20 respectively. So the weight per user will be- For u_1, $w_1 = \frac{2\arctan(1)}{\pi} = 0.5$;

For u_2, $w_2 = \frac{2\arctan(3)}{\pi} = 0.795$; For u_3, $w_3 = \frac{2\arctan(5)}{\pi} = 0.874$;
For u_4, $w_4 = \frac{2\arctan(9)}{\pi} = 0.93$; For u_5, $w_5 = \frac{2\arctan(20)}{\pi} = 0.968$;

Here, the weight of each user has been calculated. Suppose, for Sennheiser HD201, the review of U_1, U_2, U_3, U_4, U_5. are 3, 2, 3, 2 and 1. So, the total score of the evaluation will be

$$\text{Total Score} = \frac{0.5 * 3 + 0.795 * 2 + 0.874 * 3 + 0.93 * 2 + 0.968 * 1}{5} = 1.708$$

After calculating this result, the system will pass this value into the fuzzy inference system, and the system will consider 1.708 as the level of review of Sennheiser HD201. Same as, if these users give feedback on the genre for AKG Y50 the weight will be regarded as identical. Suppose, they put the feedback on genre for this gadget respectively 2, 1, 2, 1 and 3. Then the total score will be

$$\text{Total Score} = \frac{0.5 * 2 + 0.795 * 1 + 0.874 * 2 + 0.93 * 2 + 0.968 * 3}{5} = 1.6614$$

Here, the system will consider 1.6614 as the level of the genre for this particular audio gadget.

User Specific Adaption: The parameters: sound quality, genre, and review vary for different users. For example, sound quality varies from user to user. Some user can be satisfied with one audio gadget's sound quality; some user may be not. To solve this issue, the user-specific adaptation has been considered in this system. To overcome this problem, we have defined a user-specific preference scale for these attributes to match that user's personal preferences. This preference will be updated after giving a new user review. This system keeps track of the user and takes that user's at most five feedback. The system will consider only those variables which are based on user preference. The average user feedback will be subtracted from each feedback of that user and make a mean value of those differences. This value is a constant value. For the equation, this constant value is denoted by α. That value will be subtracted from the new feedback of that particular user feedback. That result will be considered as adaptive user feedback.

Averaged feedback: f_{avg}, Specific user feedback: f_{spe}; New feedback of that particular user: f_{new}, Adaptive feedback: f_{adv}

$$\alpha = \sum_{i=1}^{n} \frac{f_{spe} - f_{avg}}{n}; n \leq 5 \tag{3}$$

$$f_{adv} = f_{new} - \alpha \tag{4}$$

For example, a user has given feedback on some audio gadgets. Suppose, he or she gave 3 or high quality on sound quality for Noble Audio Katana, 2 or preferred on the genre for Jaybird X3 and 3 or more preferable on review for Sony MDR-HW700. That means, the feedback from that user is

Sound Quality = 3, for Noble Audio Katana; Genre = 2, for Jaybird X3; Review = 3, for Sony MDR-HW700.

For instance, all other users' averaged feedbacks on those audio gadgets are

Sound Quality = 2.3, for Noble Audio Katana; Genre = 1.5, for Jaybird X3; Review = 2.1, for Sony MDR-HW700.

From this information, the constant value α will be

$\alpha = \frac{3-2.3}{3} + \frac{2-1.5}{3} + \frac{3-2.1}{3} = 0.7$

For that user, if he or she gives 2 as feedback, the user-specific adaptive system will give the output

$$f_{adv} = 2 - 0.7 = 1.3$$

This output will be saved in this particular user account. After that, when the user will search for any audio gadget which score level is 2, the system will give the output which audio gadget has the preference level of 1.3.

4 Results and Discussion

To populate the database we have collected audio gadget data from various blogs, websites, expert reviews from facebook groups, etc. We have received 150 items and their data to test the database. For example, suppose a user is looking for an audio gadget and his budget is $130, he likes to listen to 'Rock' songs mostly. So the user gives $130 in the budget variable and rock in the genre variable. In the sound quality variable, the user gives 51 in total and in the accessories variable user gives 'high quality' and the use while travelling valued 7. Finally, the user wants to get the results that have a user review of 4. Now in our database, we have an item 'Yamaha EPH-100' which has a price of 133 dollars, the sound quality score of 45, genre score of 8, accessories score of 12 and review of 4.6. The system takes the inputs and passes them to the fuzzy controller. Fuzzy controller Then calculates the score using Mamdani min and outputs the preference level of that item. In this process, our controller goes through the stored records from the database and sorts them in descending order. Let us consider Table 3 as the inputs given by a user.

Table 3. Sample input

Input variable	Value
Budget	130
Genre	Rock
Sound quality	9 + 8 + 9 + 9 + 7 + 9 = 51
Accessories	8
Use case	7
Review	4

The system will then pass these parameters into the fuzzy inference system. For the genre system will take inputs two times as it can only make a single input. The fuzzy system will then calculate the score for the preference level outputs in descending order. Table 4 gives the sample output

Table 4. Sample output

Name	Budget	Sound quality	Accessory	Review	Genre	Use case	Score
Yamaha EPH-100	133	45	6	4.6	8	4	12.1
Etymotic HF3	146	57	7	4.2	8.2	7	11.8
Sennheiser IE80	153	55	8	4.5	7.5	7	11.6

Now, this output is the result without applying the adaptiveness based on the user weight and user-specific adaptiveness. User wight based adaptation scales the values stored in the database for better precision of the score throughout the system. User-specific adaptation adjusts scores according to the user's preferences. Table 5 shows the difference between the output without adaptiveness and with adaptiveness

Table 5. Differenc between without user-specific adaptation and user weight based adaption and with user-specific adaptation and user weight based adaption

Result without user-specific adaptation and user weight based adaptation	Final score	Result with user-specific adaptation and user weight based adjustment	Final score
Yamaha EPH-100	12.1	Yamaha EPH-100	11.8
Etymotic HF3	11.8	Etymotic HF3	11.4
Sennheiser IE60	11.8	Sennheiser IE60	11.7

Suppose, after using our service a user felt that the sound quality of Sennheiser IE60 is better than that of Yamaha EPH-100's sound quality. So in the feedback part user gave IE60's sound quality more score. And suppose that these particular users use this system frequently. As a result, the system calculates the weight and adjusts the values in the database according to that user's feedback. So after some time when a user uses our system multiple times the output for the same input will be as shown in Table 6.

Table 6. Result with user-specific adaptation and user weight based adaption

Name	Budget	Sound quality	Accessory	Review	Genre	Score
Sennheiser IE80	153	48	12	4.7	8	11.7
Yamaha EPH-100	133	43	14	4.2	8.2	11.6
Etymotic HF3	146	44	11	4.4	7.5	11.4

5 Conclusion and Future Work

We have developed this system to recommend audio gadgets. Currently the system is user adaptive so it can consider each user's preferences individually and update the system's database by the user given reviews on different used items. As audio is a huge platform, we could not cover all the attributes is our research. In future, we want to improve our system and also planning to develop a mobile app for this system for easy accessibility and convenience of users.

References

1. Park, H.-S., Yoo, J.-O., Cho, S.-B.: A context-aware music recommendation system using fuzzy Bayesian networks with utility theory. In: Wang, L., Jiao, L., Shi, G., Li, X., Liu, J. (eds.) FSKD 2006. LNCS, vol. 4223, pp. 970–979. Springer, Heidelberg (2006)
2. Cao, Y., Li, Y.: An intelligent fuzzy-based recommendation system for consumer electronic products, 4 May 2006. https://www.sciencedirect.com/science/article/pii/S0957417406001369
3. Zhao, N., Wang, Q.H., Zhong, J.F.: Research on fuzzy intelligent recommendation system based on consumer online reviews. In: Wang, M. (ed.) KSEM 2013. LNCS, vol. 8041, pp. 173–183. Springer, Heidelberg (2013). https://doi.org/10.1007/978-3-642-39787-5_14
4. Jang, J.: ANFIS: adaptive-network-based fuzzy inference system. IEEE Trans. Syst. Man Cybern. 23(3), 665–685 (1993). https://doi.org/10.1109/21.256541
5. Sun, C.: Rule-base structure identification in an adaptive-network-based fuzzy inference system. IEEE Trans. Fuzzy Syst. 2(1), 64–73 (1994). https://doi.org/10.1109/91.273127
6. Jang, J., Sun, C., Mizutani, E.: Neuro-fuzzy and soft computing: a computational approach to learning and machine intelligence [books in brief]. IEEE Trans. Neural Netw. 8(5), 1219 (1997). https://doi.org/10.1109/tnn.1997.623228
7. Wang, L.: Neural networks and fuzzy systems. IEEE Trans. Fuzzy Syst. 1(2), 146–155 (1993). https://doi.org/10.1109/91.227383
8. Yager, R.: Fuzzy logic methods in recommender systems, 4 June 2002. https://www.sciencedirect.com/science/article/pii/S0165011402002233

Performance Analysis of Different Recurrent Neural Network Architectures and Classical Statistical Model for Financial Forecasting: A Case Study on Dhaka Stock Exchange

Akash Bhowmick, Asifur Rahman, and Rashedur M. Rahman[✉]

Department of Electrical and Computer Engineering, North South University,
Plot-15, Block-B, Bashundhara Residential Area, Dhaka, Bangladesh
akashnsu2016@gmail.com, arnabrahman@hotmail.com,
rashedur.rahman@northsouth.edu

Abstract. In recent years the advancement in neural network architecture and introduction of recurrent neural network has attracted a lot of interest to work with sequence data. LSTM is derived from the basic architecture of Recurrent Neural Network. It has memory units which extends the power of Recurrent Neural Network. In this paper, we analyze the performance of different advanced neural network architectures and classical time series forecasting method, e.g., ARIMA on selective stock prices from Dhaka Stock Exchange (DSE). Our experimental results show that the neural network models perform better than the ARIMA model in reducing RMSE (Root Mean Square Error).

Keywords: Recurrent neural network · Long short term memory ·
Gated recurrent unit · Time series forecasting · ARIMA model ·
Stock price forecasting

1 Introduction

Predicting the stock price movement is a challenging task due to high volatility, noise and dynamic nature [1, 2]. But in recent years due to advancements in computational power, complex algorithms and emergence of big data, the traditional forecasting method had shifted its dynamics. The investment strategies are now being assisted by the decision-making processes harnessing the power of deep learning. A lot of work has been done using machine learning algorithms and deep learning techniques. These methods have shifted the dynamics of how conventional technical analysis used to work. Deep learning techniques have addressed both the fundamental and technical analysis. Advances in natural language processing have assisted in the analysis of financial statements. In Bangladesh the stock market is considered as very volatile and sparse. The semi strong form of market efficiency makes it very challenging for the traders to predict short and long term movement.

Most of the traders in investment or portfolio management companies tend to use the classical time series forecasting method known as Auto Regressive Integrated Moving Average (ARIMA) model [3]. In this paper, predictions are done using four

© Springer Nature Switzerland AG 2019
R. Silhavy (Ed.): CSOC 2019, AISC 985, pp. 277–286, 2019.
https://doi.org/10.1007/978-3-030-19810-7_27

different neural network architectures that is vanilla recurrent neural network, long short term network, long short term network with peephole, and gated recurrent unit. We also use ARIMA model. The experiment has been done on some popular companies listed in Dhaka stock exchange and the predicted result is compared with the actual value. The rest of the paper is organized as follows. In Sect. 2, related works in recent years on different capital markets has been discussed. Section 3 deals with the methodologies we have adapted to investigate the experiment. In Sect. 4, we present the result of the conducted experiments and analyze the performance of different neural architectures with ARIMA model. In the last section we discuss future direction where the current research can serve as a framework for deep portfolio optimization.

2 Related Works

Deep learning with long short term memory networks for financial market predictions was discussed in Fischer, Thomas, Krauss, and Cristopher [4]. In this paper the authors conducted the experiment in different parts, in the first part they gave a subtle introduction of LSTM. In the second part they tried to unveil the black box of LSTM network and justified why it was a good architecture which could be efficient for forecasting methods. Lastly they developed a rule based trading strategy that gave the decision of winning or losing the stock. A LSTM-based method for stock returns prediction for China stock market was discussed in [5]. In this paper the authors conducted the forecasting techniques on Chinese stock market data. They trained the model with stochastic gradient descent and RMSprop, they chose theano and keras as deep learning platform. They applied several methods with different parameters and their experiment revealed that normalization was very helpful for improving accuracy. Guidelines for financial forecasting with neural networks could be found in [6]. From this paper we adapt the method to conduct the experiment. Here the authors described 7 steps that should be considered for neural network forecasting. Their steps included both preprocessing and post processing methods needed for NN forecasting. They encouraged the genetic algorithm based training process. In the paper [7] the authors conducted experiment with different machine learning approaches and formulated output in three sections. First, the next period direction, second, the next period price change and lastly next period actual price. In this paper the authors concluded that their result could be enhanced using deep learning techniques which we have also tried in our paper. The paper [8] addressed time series forecasting of the Forex market and then outlined and analyzed deep learning techniques. Lastly they "deployed and analyzed the deep learning model using H2O library as an agent of A-trader system." A Recurrent Neural Network Approach in Predicting Daily Stock Prices for the Sri Lankan Stock Market was discussed in [9]. In this paper the authors ran experiment on Sri Lankan stock market using recurrent neural network and analyzed the performance of different recurrent neural networks on the dataset.

3   Methodology

We adopted two different methodologies, one for the deep learning approach and the other for ARIMA model. We discuss the deep learning approach first followed by the ARIMA model.

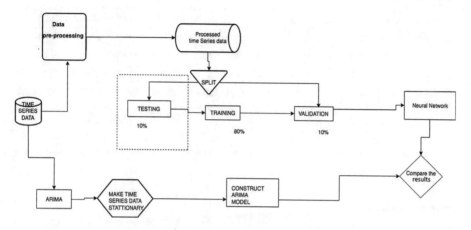

Fig. 1. Complete workflow of the experiment.

Figure 1 depicts the flowchart of our system. To conduct deep learning experiments, we divided the experiment into five steps.

Step 1. We imported the libraries and settings required for running experiment. We used Google's tensor flow and keras for quick prototyping of the neural network.

Step 2. Data collected from Dhaka stock exchange was loaded into the pandas [12] dataframe for analyzing and visualizing. We dropped the NAN valued rows and turned the trading date into index of each row. Five companies: Beximco Pharmaceutical, Grameenphone, Eastern Bank Limited, British American Tobacco and Apex Footwear, are chosen in this research. The earliest data available to us was from the year of 2000 for Beximco pharmaceutical. We kept all the data points for the sequential modeling and prediction as every observation window plays a vital role. Figure 2 depicts a sample dataset. Figure 3 depicts the closing price variation of a stock, e.g. Beximco for Dhaka Stock Exchange.

Step 3. We have normalized the data using the scikit learn's minmax library.

$$\text{Min\_max\_scalar} = \text{sklearn.preprocessing.MinMaxScalar}()$$

MinMax normalization uses linear transformation of x to map

$$Y = (x - \min)(\max - \min) \tag{1}$$

Here, x = observed value,
Min = minimum value,
Max = max value

TRADEDATE	COMID	LOWPRC	HIPRC	AVGPRC	CLSPRC	TRDNO	TRDVOL	TURNOVER	COMPANYNAME
2000-07-15	453	78.5	82.0	79.63	78.7	2537	769775	61278365.0	BEXIMCO PHARMACEUTICALS LIMITED
2000-07-16	453	69.6	79.0	73.25	71.4	4590	1627125	118776810.0	BEXIMCO PHARMACEUTICALS LIMITED
2000-07-17	453	70.0	75.7	74.01	74.3	4489	1422375	105223510.0	BEXIMCO PHARMACEUTICALS LIMITED
2000-07-18	453	72.0	77.8	74.32	73.2	3985	1361275	101138340.0	BEXIMCO PHARMACEUTICALS LIMITED
2000-07-19	453	68.7	72.6	70.52	69.9	4899	1671075	117673305.0	BEXIMCO PHARMACEUTICALS LIMITED

Fig. 2. A sample of the raw dataset.

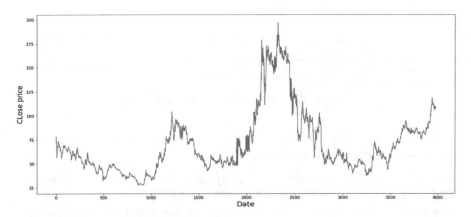

Fig. 3. Visualization of closing price against number of days for Beximco Pharmaceuticals.

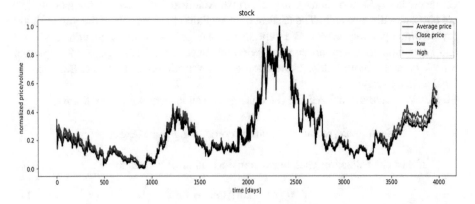

Fig. 4. Visualization of data after normalising for Beximco Pharmaceuticals

The Eq. (1) clearly represents that when x = max then Y = 1 and when x = min then Y = 0 thus the range stays within 0 to 1. Figure 4 shows the data after normalization.

Step 4. We split the data into train, test and validation set. In paper [9], the authors split data into 70% train, 15% test and 15% validation set. As our data set had a variable length, we modified the splitting boundary ratios for train, test and validation set. First 80% of the data was taken for training, next 10% was taken for testing and remaining 10% was taken for validation set. First a sequence length of 20 was used for each batch size.

(a) Use of Basic RNN cell
    Layers = [tf.contrib.rnn.BasicRNNCell(num_units = n_neurons,activation = tf. nn.elu) for layers in range(n_layers)
(b) User of Basic LSTM Cell
    Layers = [tf.contrib.rnn.BasicLSTMCell(num_units = n_neurons,activation = tf. nn.elu) for layers in range(n_layers)
(c) User of LSTM Cell with peephole
    Laers = [tf.contrib.rnn.BasicLSTMCell(num_units = n_neurons,activation = tf. nn.leaky_relu, use_peephples = True) for layers in range(n_layers)
(d) User of GRU Cell
    Layers = [tf.contrib.rnn.GRUCell(num_units = n_neurons,activation = tf.nn.lea- ky_relu) for layers in range(n_layers)

We have used a batch size of 50, 200 hidden neurons with four inputs, 2 layers and learning rate of 0.00001 for the training of neural networks. In all four of the architectures we used Adam optimizer [10] and Mean squared error for the loss function. Adam optimizer is an extended version of the stochastic gradient descent. It is designed to update the neighboring weights iteratively based in training data. Adam was introduced by Diederik from Open AI [10] and he listed that it was less memory intensive. Besides, it is very much suited for the non-stationary and sparse data [10]. Stock market data are highly volatile and noisy so choosing this optimization algorithm is favorable. According to Diederik [10], Adam uses the advantage of both the RMSprop and AdaGrad. RMsprop maintains a pre-parameter learning rate that are adapted based on the average of recent magnitude of the gradients for the weights. This refers to the fact that the algorithm does well on non-stationary data. Adagrad is also the same but it improves performance with sparse gradients. Adam realizes the advantages of both and it actually calculates an exponential moving average of the gradient and the squared gradient. For the first two neural networks we have used Exponential Linear Unit (elu) and for the latter two we have used Leaky Rectified linear unit as an activation function. The experiment ran for 100 epochs.

Step 5. This is the prediction part that measures the mean square error between the test_set and the predicted value to evaluate how close the predicted value is.

In the second methodology we have constructed the ARIMA model. In the first step we checked the stationary behavior of the data points. In the second step differencing is done to make sure the data is transformed into a stationary model. If the trend is increasing at a constant rate first differencing is done to make the data stationary if the trend is increasing at an increasing rate, then second differencing is done. Figure 5 depicts the correlation where Figs. 6 and 7 show the autocorrelation. Autocorrelation refers how strongly a data series is related to itself over time. Autocorrelations exits between −1 to +1 our correlogram shows a value which is close to 1 that means the data series is strongly related to each other with a lag of 1. We explore partial auto-correlation to further analyze the model and fit it into the ARIMA library from the statsmodel library [12]. Partial autocorrelations could determine the appropriate p values in AR(p) where ARIMA can be decomposed into AR and MA parameters. An AR model with only one parameter can be written as,

$$X(t) = A(1) * X(t-1) + E(t) \tag{2}$$

where X(t) = time series under investigation,
A(1) = the autoregressive parameter of order 1
X(t−1) = the time series lagged 1 period
E(t) = the error term of the model

The ARIMA model is differentiated into three patamaters p, d and q.

- AR (p): It captures the relations among current value and the values over a span of time.
- I(d): To make the time stationary integrated part makes a difference between current observation and previous observation.
- MA(q): It uses the dependency between an observation and a residual error from the lagged observations.

The general process of the ARIMA model is as the following:

- Visualize the time series data
- Make the time series data stationary
- Plot the Correlation and Autocorrelation Charts
- Construct the ARIMA model
- Use the model to make predictions

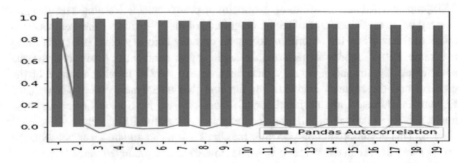

Fig. 5. Visual representation of the autocorrelation

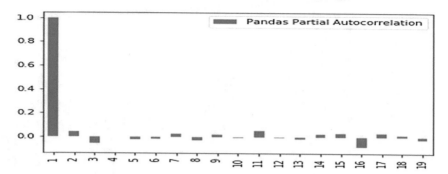

Fig. 6. Visual representation of partial autocorrelation

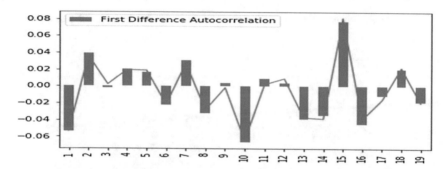

Fig. 7. Visual Representation of the first difference autocorrelation

We used statsmodels library of python to build the autocorrelation (acf), partial autocorrelation (pcf), the first difference autocorrelation and the ARIMA model. Following graphs are obtained using the statsmodels on the Grameenphone data set (Figs. 5, 6 and 7). In the last step of the prediction we feed the data into the ARIMA model and predicted the closing price. In the model we feed 90% of the data and predicted the 10% and then compared how well the model performed.

To further validate the experiment, we have also built a MACD indicator to see if the point of buying and selling can be cross checked and deep learning models can validate a better prediction for the decision making. MACD stands for moving average convergence divergence indicator. The standard methodology for this is that an exponential moving average (EMA) for 12 days and 26 days are calculated and subtracted. In the next step an exponential moving average of the MACD with a span of 9 days is calculated and plotted against the MACD chart. The common way to analyze the chart is that when the MACD line crosses the signal line it is an indication of selling the stock. Figure 8 shows the plotting of the MACD chart on Grameenphone data and the result is analyzed in the result section. In Fig. 9 we plot the MACD chart for last 100 days, where column A-F corresponds to trading date, closing price, 12 day EMA, 26 days EMA, difference between 12 day EMA and 26 days EMA, 9-day signal and finally MACD histogram respectively.

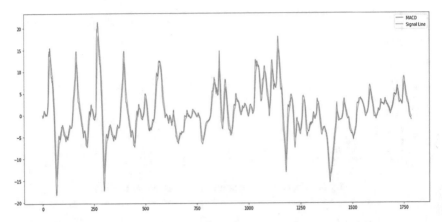

Fig. 8. Visualisation of MACD over Grameenphone dataset for 12 years

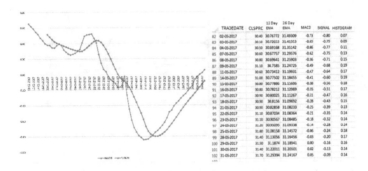

	TRADEDATE	CLSPRC	12 Day EMA	26 Day EMA	MACD	SIGNAL	HISTOGRAM
82	02-05-2017	30.40	30.76772	31.49309	-0.73	-0.80	0.07
83	03-05-2017	30.50	30.72653	31.41953	-0.69	-0.79	0.09
84	04-05-2017	30.50	30.69168	31.35142	-0.66	-0.77	0.11
85	07-05-2017	30.40	30.67757	31.29576	-0.62	-0.75	0.13
86	08-05-2017	30.60	30.69641	31.25903	-0.56	-0.71	0.15
87	09-05-2017	31.10	30.7585	31.24725	-0.49	-0.68	0.19
88	11-05-2017	30.60	30.73412	31.19931	-0.47	-0.64	0.17
89	14-05-2017	31.00	30.77502	31.18455	-0.41	-0.60	0.19
90	15-05-2017	30.80	30.77886	31.13606	-0.38	-0.56	0.18
91	16-05-2017	30.80	30.78212	31.12969	-0.35	-0.51	0.17
92	17-05-2017	30.90	30.80025	31.11267	-0.31	-0.47	0.16
93	18-05-2017	30.90	30.8156	31.09692	-0.28	-0.43	0.15
94	21-05-2017	30.90	30.82858	31.08233	-0.25	-0.39	0.13
95	22-05-2017	31.10	30.87034	31.08364	-0.21	-0.35	0.14
96	23-05-2017	31.10	30.90567	31.08485	-0.18	-0.32	0.14
97	24-05-2017	31.20	30.93090	31.09338	-0.14	-0.28	0.14
98	25-05-2017	31.80	31.08158	31.14572	-0.06	-0.24	0.18
99	28-05-2017	31.40	31.13056	31.16456	-0.03	-0.20	0.17
100	29-05-2017	31.50	31.1874	31.18941	0.00	-0.16	0.16
101	30-05-2017	31.40	31.22011	31.20501	0.02	-0.13	0.14
102	31-05-2017	31.70	31.29394	31.24167	0.05	-0.09	0.14

Fig. 9. Visualization of MACD on Grameenphone dataset for 100 days

4   Result and Analysis

All the neural network architectures outperformed the classical time series forecasting model ARIMA. The core reason why ARIMA was outperformed is that LSTM and other networks were able to take large dataset as an input whereas ARIMA could only take a short length of data to get the stationary time series. Besides the nonlinear behavior of market could be better captured by neural network models. Among the neural network architectures, the mean squared error were very close to each other. Table 1 depicts the Root Mean Squared Error (RMSE) of different techniques. For better visualization we plotted the result on a stacked bar chart in Fig. 10. It was clearly depicted that LSTM performed consistently well in all the datasets followed by GRU. Basic RNN was the one that fell short compared to the other architectures. LSTM worked well for dataset containing long term dependencies. This is also concluded in the paper [9]. In Fig. 8, the MACD indicator seems not be very helpful; the volatility and density of the data was not helping to make a proper buy and sell signal as the crossovers were not clearly identifiable. This is also one drawback of MACD reported in [11]. MACD have the lacking as it is a lag indicator. In the Fig. 9 we plotted the data for 100 recent days. The table shows a section of the data used to plot the graph. The crossing lines were indication the trade in and out days. As for only few days we get the signal, we make RMSE error for the whole set of data rather only for those few days.

Table 1. RMSE error in prediction

	GrameenPhone	Summit	Beximco	Apex	Brac
LSTM	0.011	0.0079	0.0085	0.00006	0.00989
LSTM-peephole	0.0268	0.0093	0.0083	0.000483	0.009
GRU	0.00271	0.024	0.00975	0.0093	0.0082
RNN	0.013	0.0084	0.05	0.0054	0.0069
ARIMA	1.994	0.49	3.67	6.67	1.78

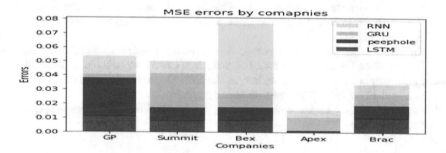

Fig. 10. MSE of different neural network architectures

5 Conclusion and Future Works

As per the experiment neural network provided some significant result on the forecasting of closing price of the stock. Stock data are highly unpredictable and the risk of return is still very high. This method offers more interesting and challenging work for the future. One of the broader aspect that this experiment is capable of addressing is portfolio optimization. This experiment requires more knowledge about finance as more parameters are actively involved in such decision making process. Furthermore, adding NLP and some numerical values after fundamental analysis could be beneficial for these neural network architectures.

References

1. Samuelson, P.: Proof that properly anticipated prices fluctuate randomly. Ind. Manag. Rev. **6**, 41–49 (1965)
2. Cootner, P.: The Random Character of Stock Market Prices. MIT Press, Cambridge (1964)
3. Ariyo, A., Adewumi, A., Ayo, C.: Stock price prediction using the ARIMA model. In: Proceedings - UKSim-AMSS 16th International Conference on Computer Modelling and Simulation, UKSim (2014). https://doi.org/10.1109/UKSim.2014.67

4. Fischer, T., Krauss, C.: Deep learning with long short-term memory networks for financial market predictions, FAU Discussion Papers in Economics, no. 11/2017. Friedrich-Alexander-Universitat Erlang-Nurberg, Institute for Economics, Erlang (2017)
5. Chen, K., Zhou, Y., Dai, F.: A LSTM-based method for stock returns prediction: a case study of China stock market. In: IEEE International Conference on Big Data, China (2015)
6. Yao, J., North, P., Tan, C.L.: Guidelines for financial forecasting with neural networks. In: Proceedings of International Conference on Neural Information Processing, Shanghai, China, 14–18 November, pp. 1–6 (2001)
7. Yoo, P.D., Kim, M.H., Jan, T.: Financial forecasting: advanced machine learning techniques in stock market analysis. In: 2005 Pakistan Section Multitopic Conference, Karachi, pp. 1–7 (2005). https://doi.org/10.1109/inmic.2005.334420
8. Korczak, J. Hernes, M.: Deep learning for financial time series forecasting in A-Trader system. In: Federated Conference on Computer Science and Information Systems (FedCSIS), pp. 905–912 (2017). https://doi.org/10.15439/2017f449
9. Samarawickrama, J., Fernando, T.G.I.: A recurrent neural network approach in predicting daily stock prices: an application to the sri lankan stock market. In: 2017 IEEE International Conference on Industrial and Information Systems (ICIIS). https://doi.org/10.1109/iciinfs.2017.8300345
10. Kingma, D., Ba, J.: Adam: a method for stochastic optimization. In: 3rd International Conference on Learning Representations, San Diego (2015). https://arxiv.org/abs/1412.6980
11. Limitations of MACD. http://oneminutestock.com/limitations-of-the-macd-indicator/
12. StatsModel Library. https://www.statsmodels.org/stable/index.html
13. Pandas Library. https://pandas.pydata.org/pandas-docs/version/0.23.4/generated/pandas.DataFrame.html

Hybrid Algorithm of Mobile Position-Trajectory Control

Gennady E. Veselov, Boris K. Lebedev, Oleg B. Lebedev$^{(\boxtimes)}$,
and Andrey I. Kostyuk

Southern Federal University, Rostov-on-Don, Russia
{gev,aikostyuk}@sfedu.ru, lebedev.b.k@gmail.com,
lebedev.ob@mail.ru

Abstract. The paper describes a hybrid algorithm of position-trajectory control of a moving object, based on the integration of the wave and ant algorithms. The process of tracing the trajectory is carried out step by step. At each step relative to the current position of the moving object, a zone is formed within which all the obstacles are localized with the help of the radar, after which a separate trajectory section is constructed which is a continuation of the previously constructed section. And the entire trajectory is a collection of individual sections. The time complexity of this algorithm depends on the ant colony lifetime l (the number of iterations), the number of vertices of the graph n, and the number of ants m, and is defined as $O(l \cdot n^2 \cdot m)$.

Keywords: Trajectory planning · Partial uncertainty ·
Two-dimensional space · Wave algorithm · Ant algorithm · Hybridization

1 Introduction

There are many classifications, statements and methods for solving problems of designing trajectories [1, 2]. But in any case, there are two classes of tasks. The tasks of the first class envisage the use of the project (trajectory) after it is fully designed. In problems of the second class, the processes of synthesis (tracing) of the trajectory and the movement of the moving object (MO) along it are connected, and the movement along the trajectory occurs almost simultaneously with its formation (tracing). This is due to the uncertainty of restrictions on the map of the area, preventing the tracing of trajectories. The presence of uncertainty is due to the fact that, as applied to MO, such obstacles are not localized, i.e. the control system previously has no information about their form, size or, especially, about the position. The identification of constraints that prevent the trajectory from the current position is detected after the trajectory reaches this position. Consistently at each step relative to the current position of the MO, a local zone of visibility (LZV) is formed. The coordinates of all obstacles are determined (localized) within of the LZV [3, 4]. After this, MO moves within the LZV along the shortest path from the current position to the new position, which is declared it is current.

The most popular for constructing a route on a plane received wave algorithms [5]. The map of the area with wave tracing is divided into squares (cells). At the first stage,

© Springer Nature Switzerland AG 2019
R. Silhavy (Ed.): CSOC 2019, AISC 985, pp. 287–295, 2019.
https://doi.org/10.1007/978-3-030-19810-7_28

in the process of wave propagation from the source to the target, the cells of the discrete working field (DWF) are assigned weighting estimates related to the accepted optimality criterion. At the second stage of the algorithm, the path is constructed. To do this, starting from the target cell, move in the direction opposite to the wave propagation direction, moving from a cell with a higher weight to an adjacent cell with a lower weight until the source cell is reached. The DWF cells selected during this process determine the desired optimal connection of the minimum length.

Wave algorithm provides the construction of the path of the minimum length. A distinctive feature of the wave algorithm is the presence of alternatives in the construction of the reverse trace of the minimum length. In this regard, the development of methods that allow the selection of the best option for the minimum path when constructing the reverse trace is relevant.

Recently, for solving various "complex" tasks, methods based on the use of random directed search methods are increasingly being used. Most of these algorithms are based on metaheuristics borrowed in nature [5, 6]. Such methods include the method of modeling annealing, methods of genetic search (evolutionary adaptation), methods of swarm intelligence, methods of alternative search adaptation based on probabilistic learning automata. In a number of papers [7–10], approaches based on various heuristic methods for solving this problem are described, which have not yet shown too good results. Analysis of methods for solving complex applied problems shows that the use of any one optimization algorithm (both classical and population) does not always lead to success. In hybrid (combined) algorithms that combine different or identical algorithms, but with different values of free parameters, the advantages of one algorithm can compensate for the disadvantages of another. Therefore, one of the main ways to increase the efficiency of solving optimization problems at present is the development of hybrid population algorithms [11].

2   Task Setting and Basic Algorithm Scheme

The terrain map is initially divided into a plurality of cells with a given discrete step and is represented as a discrete working field (DWF). A model of this kind is called receptor. Receptor model of the card can be represented as a graph. The vertices of the graph correspond to the cells. If two cells are adjacent, then the corresponding vertices of the graph are connected by an edge. The vertices of the graph represent all possible MO locations in the search space.

Obstacle data comes from the MO touch system. The coordinates of the obstacles are combined with the vertices of the graph. Vertices are marked as obstacles, and edges are removed around each vertex. In the space of the formed graph, the optimal path is searched.

The main indicators for planning a trajectory are: **Length (P_L), Safety Indicator (Sm), Task Lead Time (tm), Mission Success Rate (M)** [1, 4].

P_L - the length of the entire trajectory covered by the MO from the starting point to the target.

S_m - is the minimum distance between any MO sensor and any obstacle along the entire path. This indicator determines the maximum risk during the entire movement.

t_m - the time required to complete the movement.

V - the number of successful missions in non-deterministic environments with complex obstacles.

In the general case, the trajectory planning is done in accordance with the algorithm presented below.

1. $t = 0$.
2. Place the movable object (MO) in the current position (CP) $p(t)$.
3. Generate a local zone of visibility $LZV(t)$ relative to the point $p(t)$.
4. To fix motionless objects within $LZV(t)$.
5. Combine all generated LZV into one combined $OLZV(t)$. Assume that there are no obstacles on the field outside the $OLZV(t)$.
6. To build on the field the trajectory between the points $p(t)$ and the *target*.
7. If the point of the *target* is included in the $OLZV$, then the trajectory is fully formed and the transition to 10, otherwise the transition to item 8.
8. On the plot of the constructed trajectory included in the $LZV(t)$, the current position $p(t + 1)$ is selected, to which the MO moves.
9. $t = t + 1$. Transition to item 2.
10. The end of the algorithm.

The integral trajectory estimate: $W = k_1 P_L + k_2 S_m + k_3 t_m + k_4/V$.

3 Planning a Two-Dimensional Trajectory Under Partial Uncertainty Based on the Integration of Wave and Ant Algorithms

In the general case, the integration is based on the merging of the mechanisms of the wave and ant algorithms and is implemented when performing two basic steps. At the first stage, the search space R is formed by propagating the wave from the source to the target, representing the set of connected cells of the DWF reached by the wave with weights assigned to them. At the second stage, a route is laid using an ant algorithm based on the generated search space R.

A receptor model of the terrain map is formed, including a set of cells $E = \{e_i|$ $i = 1,2,...,n_e\}$. Each cell has coordinates (x_i, y_i). The e_f is marked as the initial position of the FP, into which the MO is placed, and the cell, which e_t is marked as the *target* position (*target*). Initially, all cells are considered free.

The trajectory planning process is iterative and includes five steps performed at each iteration t. At each iteration a separate section of the trajectory is built, which is a continuation of the previously constructed section. And the entire trajectory is a collection of separate sections, connecting the initial position of the MO with the target position.

At the first stage of the iteration t, according to the radar data located at the $CP(t)$ point, the local visibility zone $LZV(t)$ is formed. The coordinates of the $CP(t)$ point are determined at the previous iteration $(t-1)$. The $LZV(t)$ boundaries are defined on the terrain map model. Cells with obstacles contained in $LZV(t)$ are marked. The radar data about obstruction obtained at iteration t is combined with the data obtained at previous $(t-1)$ iterations.

At the second stage, by propagating the wave from the source (point $CP(t)$ to the *target* point, the trajectory search space $TSS(t)$ is formed. $TSS(t)$ is a set of connected DWF cells, reached by the wave, with weights assigned to them.

At the third stage, the trajectory $M(t)$ from the *target* point to the $CP(t)$ point is plotted by the ant algorithm on the $TSS(t)$. Note that the trajectory $M(t)$ consists of two parts. The second part of the trajectory $M2(t)$ passes through the zone free from obstacles, the first part $M1(t)$ passes through the a LZV with the obstacles contained in it.

At the fourth stage, in the first part of the trajectory $M1(t)$ selects its maximum weight cell e_k, which is the terminal part of $M1(t)$. The next e_n cell in the trajectory $M(t)$ behind the e_k cell already lies in the second part of the trajectory $M2(t)$.

At the fifth stage, an e_p cell is selected in $M1(t)$ with a smaller weight by δ compared to the e_k cell. The parameter δ is the control. The MO moves along the trajectory $M1(t)$ from the point $CP(t)$ to the cell e_p. The part of the trajectory $M1(t)$ between points $CP(t)$ and e_p is included in the planned trajectory. The e_p cell is then considered as the new current MO position at the next iteration $(t+1)$. Plot $M2(t)$ is deleted.

In Fig. 1 shows the formed map of the area with two points placed on it – the initial position of the MO and the *target*. The results of the implementation are reflected: the first stage – the a LZV was formed and the obstacles contained in it were revealed; the second stage – by propagating the wave from the source (point $CP(t)$) to the target point on the TSS, the search space of the path $TSS(t)$ is formed with weights assigned to the free cells; of the third stage – a trajectory $M(t)$ from the *target point* to the $CP(t)$ point is traced by the ant algorithm on the formed $TSS(t)$, the partitioning of $M(t)$ into two parts $M(t) = (e_f, e_k)$ and $M2(t) = (e_n, e_t)$; the fourth stage – fixation in $M1(t)$ of the terminal vertex e_k (cell with coordinates $(5; 4)$); the fifth stage – a new current position is selected (cell e_p with coordinates $(5; 3)$) on the segment $M1(t)$ to which ON. $e_f = (1; 1)$, $e_t = (10; 7)$, $e_k = (5; 4)$, $e_n = (5; 5)$, $e_p = (5; 3)$.

In Fig. 2 shows the results of the first stage at the second iteration: the LZV was formed, obstacles were identified, the route designed at the previous iteration was reflected. Point $TP(t)$ is placed in a cell with coordinates $(5, 3)$.

In Fig. 3 shows the results of the implementation: the second stage – by propagating the wave from the source (point $TP(t)$) to the target point to the TSS, the search space of the trajectory $TSS(t)$ is formed with weights assigned to the free cells; of the third stage – a trajectory $M(t)$ from the target point to the $TP(t)$ point, a split of $M(t)$ into two parts $M1(t)$ and $M2(t)$; the end vertex e_k (cell with coordinates $(8; 6)$ in $M1(t)$; the fifth stage – a new current position is selected (a cell with coordinates $(8; 5)$) on the segment $M1(t)$ to which the software moves.

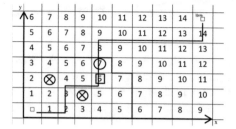

Fig. 1. Map of the area

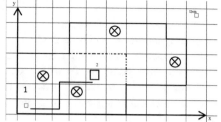

Fig. 2. The first stage at the 2nd iteration

In Fig. 4 shows the results of the implementation of the first stage at the third iteration: the LZV was formed, the obstacles contained in it, as well as the route, were identified.

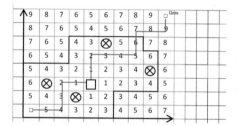

Fig. 3. Wave propagation

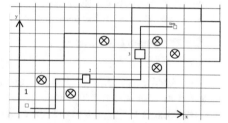

Fig. 4. Results of the first stage

Point $CP(t)$ is placed in a cell with coordinates (8; 5). Since the target point lies in sight, at the fourth stage of the third iteration, the ant algorithm lays out a plot of the planned trajectory $M(t)$ from the *target* point to the $CP(t)$ point (cell with coordinates (8; 5)).

4   Integration of the Ant and Wave Algorithms

In accordance with the hybrid algorithm, after propagation of a wave on a DWF from a source $CP(t)$ to the target point and weight assignment to DWF cells by an ant algorithm on the DWF, the route $M(t)$ is built from the *target* point to $CP(t)$ point.

Metaheuristics of the ant algorithm is based on a combination of two techniques: the general scheme is based on the basic method, which includes an embedded procedure. The built-in procedure is an independent algorithm for solving the same problem as the metaheuristic method as a whole. The basic method is to implement an iterative procedure for finding the best solution, based on the mechanisms of the adaptive behavior of an ant colony. The built-in procedure is a constructive algorithm for the ant to construct some specific interpretation of the solution [12].

Note that the search for a solution is carried out by a population of ants on a certain graph model of the search space.

Each iteration l of the base method includes three steps. At the first stage, each ant of the population finds a solution, at the second stage, each ant postpones the pheromone on some graph model of the search space, at the third stage, the pheromone evaporates. The work uses the cyclic (ant-cycle) method of ant systems [12], i.e. pheromone is deposited by each agent on the edges of the graph after the complete formation of the solution.

The basic idea of integration used in the hybrid algorithm discussed above is that the constructive algorithm for constructing an ant trajectory is based on the use of separate procedures of the wave algorithm. The ant is building a trajectory on the basis of the generated search space $TSS(t)$, after propagating the wave from the source to the target point to the DWF, assigning weights to the free cells. Like a wave algorithm, an ant builds a route on the $TSS(t)$, representing a collection of cells with a successively decreasing weight. All ants will build routes of the same minimum length, but differing configurations. In this case, the estimation of the route M_k, constructed by the agent a_k is the estimate of its configuration $F_k = \alpha f_{rk} + \beta f_{gk}$,

where α, β are control parameters that are selected experimentally,

f_{rk} - the number of turns of the route M_k,

f_{gk} is the number of obstacles containing cells adjacent to the cells of route M_k.

The smaller the f_{rk}, the smaller the indicator **Task completion time** - the time required to complete the movement.

The smaller the f_{gk}, the higher the **Safety Indicator**—the number of MO positions at the minimum distance between the MO and any obstacle along the entire path.

The goal of optimization is to minimize the value of F_k.

Note that the indicators P_L, S_m, t_m, V are not objects of optimization procedures. The optimization object is the index $F_k = \alpha f_{rk} + \beta f_{gk}$, the minimization of which based on the heuristics of the hybrid planning algorithm indirectly contributes to the minimization of the integral trajectory estimate $W = k_1 P_L + k_2 S_m + k_3 t_m + k_4 / V$.

The procedure for building a connection in a hybrid algorithm is represented as two algorithms:

A1- *Formation the trajectory search space $TSS(t)$.*

Based on the execution of the initial, first - fifth stages of the trajectory construction, the initial data necessary for the operation of the ant algorithm are formed: the target cell e_t and the source cell e_f and are determined; the area of the route R is formed; the weight $\rho(e_i)$ is assigned to all cells e_i of the trace area R.

A2- *Embedded procedure of ant algorithm.*

Agent a_k to build a route in trajectory search space $TSS(t)$ is placed at the initial vertex $v_0 = e_t$. Next, agent determines the set of vertices $E_k(t + 1) \subset E$ of candidates for inclusion in its route $M_k(t + 1)$. For each vertex $e_i \in E_k(t + 1)$, the probability P_i of its inclusion in the generated route $M_k(t + 1)$ is calculated. Randomly, in accordance with the calculated probabilities, the vertex $e_i^* \in E_k(t + 1)$ is selected, which is included in the route $M_k(t + 1)$.

5   Experimental Studies

To carry out the experiments, the procedure of synthesizing test examples with a known optimum was used by analogy with the well-known PEKU method (Placement Examples with Known Upper bounds on wirelength).

For the problem of planning a trajectory, four sets of examples were created with a known optimum along the trajectory length and the optimal value F_{opt} of estimating the configuration of the trajectory F_k. The following method was used to generate examples: a model of a local visibility zone containing obstacles and the coordinates of two connected points located on the LZV are input. Then, an example of tracing in LZV is generated, which has a previously known optimal value of F_{opt}. Optimization was carried out according to the criterion of the minimum length of the trajectory and the optimal value of the configuration estimate and was based on the methodology described in [1, 4].

For each example, the parameter N was first recorded—the population size and the value of the parameter I—the number of iterations was analyzed. A series of 10 experiments was carried out for each fixed set of parameters.

Based on the processing of experimental studies, an average dependence of the quality of solutions on the number of iterations was constructed (Fig. 5).

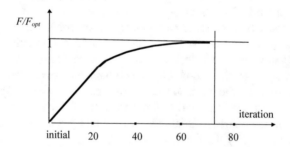

Fig. 5. Dependence of the quality of decisions on the number of iterations

Quality assessment is the value of F_k/F_{opt}, where F is the evaluation of the solution obtained. Studies have shown that the number of iterations at which the algorithm found the best solution lies within 68–80. From the graph it is clear that on average at the 76th iteration, the solution is close to the optimal one. The overall time complexity of the algorithm in any approach to hybridization lies within $O(n^2) - O(n^3)$.

Comparison of the criterion values obtained by the hybrid algorithm on test examples with a known optimum showed that 60% of the examples had the optimal solution, 15% of the solution examples were 5% worse, and 25% had no worse solution than 2%.

Comparative analysis was performed on a test set of scenes given in [1]. Comparing the developed hybrid algorithm with the known algorithms [9–12] using the integral estimation of the trajectory W showed that with a shorter operating time, the F_T values obtained with the help of the developed algorithm are less than 6% on average. Figure 6 shows examples of connections laid.

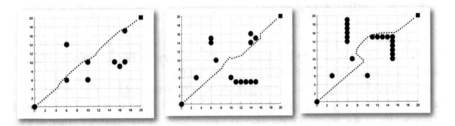

Fig. 6. Examples of scenes with built trajectories

6 Conclusion

Despite the rather large number of developed models and algorithms for controlling moving objects under uncertainty, researchers often face a number of problems, including the difficulty in justifying the quality of analysis results, taking into account the specifics of a specific task.

Among the promising trends are the development of hybrid methods. In hybrid (combined) algorithms that combine different or identical algorithms, but with different values of free parameters, the advantages of one algorithm can compensate for the disadvantages of another. The paper proposes a hybrid algorithm that allows you to build in real-time trajectories of the minimum length with simultaneous optimization of a number of other quality criteria for the constructed path. A distinctive feature is that the movement of a moving object along a trajectory occurs almost simultaneously with its formation (laying), in on-line mode. The key problem that was solved in this work is the technique of splicing the mechanisms of the ant and wave algorithms, which allowed to improve the issues related to the formation of the visibility zone and the choice (change) of the current initial positions, which accelerates the process of reaching the target object by the moving object.

Acknowledgements. This research is supported by grants of the Russian Foundation for Basic Research of the Russian Federation, the project №. 19-07-00645.

References

1. Pshihopov, V.Kh., et al.: Intellectual planning of the trajectories of moving objects in environments with obstacles, 345 p. (2014)
2. Pshihopov, V.Kh.., Medvedev, M.Yu.: Management of moving objects in certain and uncertain environments, 190 p. Science, Moscow (2011)
3. Guzik, V.F., Pereverzev, V.A., Pyavchenko, A.O., Saprykin, R.V.: Principles of constructing an extrapolating multidimensional neural network planner of an intelligent system of positional-trajectory control of moving objects. News SFU, Technical Science, vol. 2, no. 175, 67–80 (2016)
4. Pshihopov, V.Kh., Medvedev, M.Yu., Gurenko, B.V.: Algorithms of adaptive positional - trajectory control systems for moving objects. Probl. Control. **4**, 66–74 (2015)

5. Karpenko, A.P.: Modern search engine optimization algorithms. Algorithms inspired by nature: a tutorial, 448 p. Publishing House MSTU, Moscow (2014). N.E. Bauman
6. Caro, G.Di., Ducatelle, F., Gambardella, L.M.: AntHocNet: an adaptive nature inspired algorithm for routing in mobile ad hoc networks. Eur. Trans. Telecommun. **16**(5), 443–455 (2005)
7. Hoefler, T., Snir, M.: Generic topology mapping strategies for large-scale parallel architectures, pp. 75–85. University of Illinois at Urbana-Champaign Urbana (2011)
8. Neydorf, R.A., Polyakh, V.V., Chernogorov, I.V., Yarakhmedov, O.T.: The study of heuristic algorithms in the problems of laying and optimization of routes in an environment with obstacles. News of SFU, Technical Science, vol. 3, no. 176, pp. 127–143 (2016)
9. Fatemeh, K.P., Fardad, F., Reza, S.N.: Comparing the performance of genetic algorithm and ant colony optimization algorithm for mobile robot path planning in the dynamic environments with different complexities. J. Acad. Appl. Stud. **3**(2), 29–44 (2013)
10. Chen, S., Eshaghian, M.M.: A fast recursive mapping algorithm. Department of Computer and Information Science, New Jersey, USA, pp. 219–227 (2013)
11. Raidl, G.R.: A unified view on hybrid metaheuristics. In: Lecture Notes in Computer Science, pp. 1–12. Springer (2006)
12. Lebedev, O.B.: Models of adaptive behavior of the ant colony in design problems. Monograph. Publishing house of SFU, Taganrog, 199 p. (2013)

Awareness of Information and Communication Technology Induced Climate Change and the Developing Countries

Ramadile Moletsane[✉]

Vaal University of Technology, Private Bag X021, Vanderbijlpark, South Africa
Ramadilem@vut.ac.za

Abstract. Information and Communication Technology (ICT) is central to every aspect of our lives and it has influenced peoples' lives positively and negatively. The objective of this study is three fold: To explore how ICT contributes to carbon footprint generation. To determine how carbon footprint influence climate change. Thirdly suggest ways that can be used to reduce carbon footprint caused by ICT disposal. In this narrative review paper the online databases were searched for literature including grey literature. Papers included in this narrative reviews were written in English only. Hand searches of the references of retrieved literature was also conducted. ICT contributes to the production of carbon footprint. The amount of produced carbon footprint traps the heat in the atmosphere thereby raising the global temperature. Most people in the developing countries are not aware of how ICT contributes to global warming and climate change. Good side of ICT is that it can also act as "reducing agent" of carbon footprint generated itself and by other industries. Global warming and cause of climate change, is due to human made activities. For example, hazardous emissions of e-waste to the environment are man-made. People in developing countries are unaware about climate change, global warming and harmful effects of e-waste to the environment. To minimise carbon footprint generated by e-waste during disposal, this paper suggests proper management strategies. Formal recycling is identified as safer management strategy and friendly to the environment relative to informal recycling.

Keywords: Climate change · Greenhouse gases · Global warming ·
Carbon footprint · Information and communication technology ·
Electronic waste · Developing countries

1 Introduction

The objectives of this study is threefold. Firstly, it explores how information and communication technology (ICT) contributes to carbon footprint, then determine how carbon footprint influence climate change. Lastly recommend ways to combat climate change by reducing the carbon footprint generated by ICT devices when obsolete. One of the reasons that motivates this study is the occurrence of severe disasters that are induced by climate change [1]. Climate change induce global phenomena such as flooding, extreme snow, cyclones and wild fires to mention few. Secondly the

© Springer Nature Switzerland AG 2019
R. Silhavy (Ed.): CSOC 2019, AISC 985, pp. 296–306, 2019.
https://doi.org/10.1007/978-3-030-19810-7_29

perceived lack of knowledge and awareness about how climate change affect our planet today and maybe in the future. According to Shahid and Piracha [2] absence of awareness especially in the developing countries is a barrier to adjust to climate change. The climate change research has always been on scientific analysis and its implications and ignored how climate change affect the developing countries [2].

Earlier this study touched a little on ICT without defining it. There are numerous definitions of ICT that exist out there. For example, ICT can comprehensively be characterized as the tools and equipment that furnishes the required environment with the physical foundation and the services for the production, transmission, processing, saving and disseminating of data in all forms including voice, text, information, graphics, and video [3]. According to Afolabi and Abidoye [4] ICT is "the usage of electronic devices such as computers, telephones, internet and satellite systems to store, retrieve and disseminate information in the form of data, text image and others [4]." What is central to both definitions is dissemination, store, retrieve information or data in the form of text, image and others. Therefore, in the context of this study the following definition will be used: ICT is any electronic equipment used to store, retrieve, and disseminate information or data in the form of text, image, voice and others.

Nowadays ICT is central to every aspect of our lives and it has influenced peoples' lives positively and negatively [5, 6]. Positive aspects of ICT are prevalent in education, healthcare, timely dissemination of information through the media—anywhere and anytime, online shopping, provide new job opportunities for our people, online banking and many more. In any case, this is just a single side of the coin to the ICT innovation in our lives; the more splendid side that is. The darker and increasingly dismal side of the ICT industry is its exponentially increasing energy consumption [7]. As our dependence on ICT gadgets and services develops quickly, so does our requirement for electricity to manufacture and produce, and power these gadgets [7]. The production of this much-needed electricity to make and operate all the ICT equipment available today is a huge contributing reason towards the generation of carbon dioxide, a main Greenhouse Gas (GHG), and also other global warming pollutants [8].

It is clear that production, disposal and use of ICT products contributes carbon dioxide (CO_2) to carbon footprint [9]. According to Park, Hoerning [10] if ICT equipment that has reached its end-of-life not properly managed it becomes hazardous to the environment. The emissions from the ICT gadgets and consequently their environmental effect originates from electricity utilization used both, in manufacturing these gadgets, and additionally running them. In addition, digging for earth metals utilized in the manufacturing of ICT gadgets and waste disposal are extra contributors to the aggregate ICT industry CO_2 footprint [8]. Therefore, every life cycle of ICT product contributes to carbon footprint. The ICT devices that reached their end-of-life are referred to as electronic waste [6]. The disposal methods such as burning in the open or incineration release toxic fumes and materials to the atmosphere and the environment thereby affecting climate change [10].

It should be made clear at this stage that the focus of the study is carbon footprint generated by ICT equipment and specifically when they have reached their end-of-life. Although the carbon footprint produced by ICT is insignificant compared to the one generated by sectors in other industries, it is equally important to deal with it for the benefit of the planet [8]. However, a closer look reveals that energy consumption of ICT equipment excluding smartphones account to 8% of total worldwide electricity consumption, and it estimated to get to 14% by the year 2020 [11]. Green Car Congress [12] assert that by 2020 smartphone will be the most damaging ICT device to the environment. In contrast, data centres due to their number in growth and size, will be the most contributor to carbon footprint. Their [data centres] contribution will rise fivefold from the year 2002 to 2020 businesses and governments and add more data centres to be more efficient.

Although smartphones consume less energy to operate, 85% of their emissions comes from their production [12]. A smartphone printed circuit board and chip requires most amount of electricity as they are made from precious raw materials mined at high cost [12]. Worldwide greenhouse emissions data demonstrates that the big contributors to worldwide emissions by economic sector in 2015 were energy (electricity) production 29%, transportation 27%, industry at 21%, pursued by business and residential at 12% and 9% for farming [13]. It is estimated that ICT alone contributes about 2 to 2.5% of the total carbon footprint or GHG yearly [14]. Out of total ICT contribution to GHG, operational impacts of personal computing are projected to account for almost 60% of carbon footprint, with the remaining 40% attributed to manufacturing [15]. The bulk of emission is due to energy consumption of personal computers and their screens at 40%, followed by data centres at 23%. A further 24% of emission is contributed by fixed and mobile telecommunications [16]. The global share of ICT emissions is significant and rising [15].

The remainder of this paper is organised as follows: Sect. 2 presents the methods used for this study. Section 3 present the results of the study. Under the results, Sect. 3.1 present greenhouse gases and carbon footprint. In Sect. 3.3 climate change and the developing countries is presented. Sections 3.4 and 3.5 present developing countries with high climate change venerability and climate change and information and communication technology respectively. Finally, the study concludes in Sect. 4 with discussions.

2   Methods

In this narrative review paper the following online databases were searched for literature: Elsevier, PubMed and IEEE Explorer. The literature search was restricted to English only research articles published from 2001 to 2018. The keywords that was used to get information on climate change and the developing countries were 'climate change', 'climate change and the developing countries'.

A second literature search included grey literature, IEEE Explorer and Elsevier aiming at climate change and ICT. Combination of the terms: 'Electronic waste and the environment', 'disposal of ICT equipment', and 'ICT and climate change'.

Third literature search that included databases: IEEE Explorer and Elsevier was conducted in order to present global warming, greenhouse gases and carbon footprint. The following keywords were used: 'global warming', 'carbon footprint', and 'greenhouse gases'. The search was further conducted using combination of this keywords. The keywords were combined using "+" operator.

In this study an assessment on climate change and the developing world, and the countries with high venerability of climate change were presented. Furthermore, this paper then presented an extensive review of ICT negative and positive impacts on climate change.

3   Results

3.1   Greenhouse Gases and Carbon Footprint

What actually is GHG? According to [17] GHG is any gaseous compound in the atmosphere that is capable of absorbing infrared radiation, thereby trapping heat in the atmosphere. Any gas that is able to trap heat in the atmosphere is called a greenhouse gas. examples of heat trapping gases include nitrous acid, methane, ozone, water vapour and carbon dioxide [18]. Trapping of heat in the atmosphere causes temperature to rise and hence global warming [19]. Many GHG occurs naturally, but human activities add more to greenhouse effect that is contributing to the rising temperature referred to as global warming [20]. Greenhouse effect is a natural process that keeps the earth warmer than otherwise it would be [19].

According to Adedeji, Reuben [21] the these are some of the human made GHG includes the chlorofluorocarbons (CFCs), hydrofluorocarbons (HFCs) and Perfluoro-carbons (PFCs), as well as sulphur hexafluoride (SF6). The list include carbon dioxide (almost three quarters of warming impact), nitrous acid (around 8%), and methane at around 14% and fluorinated gases are at 1.0% [22]. The most abundant GHG is believed to be carbon dioxide, but not water vapour. Carbon dioxide is the second GHG gas found in abundance after water vapour [23] Carbon dioxide (CO_2) is second at 0.04%, but increasing steady due to human activities [22]. In contrast, the 2015 statistics on GHG emissions in the European Union and the world, recorded in 2015 revealed CO_2 as the most emitted GHG [20]. The numbers were the following, CO_2 at 81%, followed by methane at 10.6%, nitrous acid at 5.5% and hydrofluorocarbons at 2.5% [20].

A persistent heat in the atmosphere caused by these heat trapping compounds such as CO_2 and nitrous oxide (N_2O) causes adverse effects such as global warming over time [17]. Therefore, since ICT products emits CO_2 among other substances when heated or in use ICT equipment then contribute to GHG. According to farming con-tributes 60% of global methane emissions and 40% of global nitrous acid emissions related to human activities [23]. Livestock in total contributes about 14.5% to GHG emissions [24].

On the other hand, carbon footprint represents a total of CO_2 and other GHG emissions that are directly or indirectly caused by human activities [25]. It is a tool used that serves to manage and reduce GHG emissions [26]. This also include the emissions accumulated in all life stages of a product or service expressed in terms of CO_2 [27]. CO_2 has the lion's share of all other GHG [26]. Most people abuse the term carbon footprint [28]. with 'footprint" part of the word, we mean the impact something has while carbon is a shorthand for all greenhouse gases that contribute to global warming [28].

Carbon footprint is an indicator for assessing the impact of human activities (directly or indirectly) to global warming and climate change [29]. Examples of direct activities that contributes to carbon footprint includes the use of fertilisers lead to emissions of nitrogen oxide, some chemicals when manufactured produces methane. The GHG emissions of an equipment over its entire life time is measured through carbon footprint [25].

3.2 Climate Change and Global Warming

Most people use climate change interchangeably with global warming and assume that global warming will bring uniform warming of the earth [21]. In reality this is not true, it implies that some areas will be become warmer while others become cold. This is due to changes of circulation in the atmosphere [21]. There is more to climate change than warming trend, and hence inappropriate to use the terms (global warming and climate change) synonymously [21]. According to Radu, Scrieciu [26] global warming is one of the examples of adverse results of GHG.

Global warming is the long term raising global temperature, whereas climate change is a variety of global phenomena predominantly caused by natural fuels. According to Rahman [30] climate change refers to changes in today's climate predominantly caused by humans. On the other hand, global warming is the effect of GHG [26]. It is clear that human beings are central and main sources of climate change problem. Again it will take humans to reverse the effects of climate change by being responsible to their activities. According to Statista [31] whether global warming is a man-made phenomenon, or cycles of temperature fluctuations occurring in nature it is debatable and hard to prove. Next the study presents how climate change impact the developing countries.

3.3 Climate Change and the Developing Countries

Climate change has devastating effects to economy, social—including human and livestock health, and the environment when left unattended [32]. Climate change when unchecked poses a devastating effects such as threat to poverty elevation and undoing of developments that took decades to build [21]. Some of the future predictions of climate change impacts, given that no action is taken by the nations of this world are the following: by the year 2100, the temperature would have risen by between 1.1 °C and 6.4 °C. Other parts of the world will receive more rain, while becomes drier [21].

Drought or dry land and more rain are not good for agriculture, given that developing countries rely on agriculture as a source of livelihood [33]. Essays [34] noted that communities in the agriculture for occupation are poorer compared to those working in different sectors of the economy. Given their current economic status of these communities, effects of climate change will be disastrous in particularly to them. Climate change could be barrier to millennium goal I, that is to eradicate extreme poverty [35].

In South Africa climate change was blamed to be behind the water shortage crisis that struck Western Cape Province last year (2017). Between 2015 and 2017 Western Cape province has experienced third of its lowest rainfall years on the record [36].

Although Philippines have been a victim to tropical cyclones, recently it has been struck by severe storms to the extent that scientist blame it to climate change [37].

Another concern is how electronic waste (e-waste) is managed in the developing countries. Literature posits that e-waste is not properly managed in the developing countries. The reasons include lack of awareness about e-waste negative impact on the environment and health, absence of regulation, corrupt custom officials, poverty and unemployment, illegal trade in e-waste from the developed countries, lack of state-of the- art facilities to recycle e-waste much safer, labour without protective gears, etc. [6, 9, 10]. When e-waste is managed improperly it becomes danger to the environment, livestock and humans. Sadly, it also contributes to the GHG emissions that eventually causes climate changes.

3.4   Developing Countries with High Climate Change Venerability

According to Shahid and Piracha [2] Pakistan is one of the developing countries with a high venerability to climate change. In 2016 Zimbabwe, Haiti and Fuji were the most affected regions and for the period 1997 to 2016 Haiti, Honduras and Myanmar ranked the highest [38]. Philippines is a country worth mentioning in the list of the most venerable regions to the climate change [39]. Some refer to Philippines as most disaster prone place on Earth. Beside this, the Philippines are in the race against climate change whereby they are involved in an expensive project (US$14 billion) in which they are developing a city that is said to be resistant and sustainable to climate change effects. The metropolis city is larger than the United States Manhattan [39].

Sadly, the most venerable countries are the least contributors to carbon footprint. Man-made climate change will hit the hardest on African states, costal states, low-lying islands, Asian mega deltas and the polar areas [40]. United States has withdrawn from Paris pact under Donald Trump administration without explanation [39]. It is worth mentioning to also mention big contributors towards GHG. In the past, United States was the biggest contributor of carbon dioxide to GHG.

Recently China has surpassed the United States. China emissions are more than United States and European Union combined [41]. The order of the five countries that recorded the highest to the world's total CO_2 emissions in 2017 are: China (27.1%), United States (14.58%), India (6.82%) and Russia (4.68%) and Japan (3.33%). The said news is that these big contributors won't be affected as much as the least contributing countries. Climate change will affect the most the developing countries of which are not big contributors of manmade greenhouse gases.

3.5   Climate Change and Information and Communication Technology

In this section, this paper will explain how each cycle of ICT devices impact negatively climate change by contributing to greenhouse gases. Then move to discuss more on e-waste effects on climate change. Lastly suggest how emissions of e-waste can be reduced to relieve climate change impact. According to Hilty, Hischier [42] ICT equipment has the following life stages: production, use and end- of-life. With reference to Berkhout and Hertin [43] ICT direct negative impact originate from the production, use and disposal of ICT equipment. Berkhout and Hertin [43] identified three major effects of ICT on the environment. ICT has both positive and negative impacts on the environment and these impacts are classified as first order effects, second order and third order effects [43]. Table 1 shows the ICT impacts (both positive and negative) on the environment.

Table 1. ICT impacts on the environment

	Positive impacts	Negative impacts
First order effects	The use of ICT applications for the environment, for example remote sensing technologies and platforms for *environmental monitoring* and *disaster management*	Environmental impacts related to the use, production and disposal of ICT equipment. For example, *e-waste*
Second order effects	Dematerialization and structural changes, for example *electronic directories*	Incomplete substitution e.g. '*white vans*' in addition to *private shopping trips*
Third order effects	Life style changes, for example consuming environmentally friendly products	'Rebound effect', e.g. *growth for long distance travel*

In this study as indicated earlier, the focus is ICT equipment contribution to carbon footprint. Therefore, the study looks at negative impacts of ICT on the environment under the first order effects [43]. Environmental impacts related to production of ICT equipment: the energy (electricity) needed to mine raw materials (e.g. gold, silver and copper) needed to produce ICT components is derived from fossil energy. Energy is consumed when 'turning' the mined raw substances into desirable ICT products. The energy used, and emissions from raw metals in this process contributes to GHG. The production of ICT equipment releases a huge amounts of carbon dioxide. In one of the studies it was found that for every ton of cathode ray tube screens produced, about 2.9 metric tons of carbon dioxide were emitted [44]. Environmental impact related to use of ICT equipment: when ICT devices are in operational, need energy (electricity). When ICT products left unattended in operational mode cause further consumption of electricity that contributes to carbon footprint.

Environmental impacts related to the disposal of ICT equipment is referred to as e-waste: Most of e-waste generated globally is buried and not recycled [9, 45]. Failure to properly recycle e-waste put stress on the demand of raw materials and energy. If we had recycled almost every e-waste we would have saved energy and raw materials. Again when e-waste is recycled or treated under unsafe conditions it emits relatively higher amount of GHG, that are also harmful to the recycle workers and the surrounding communities from where recycling is done [9]. Practices of burning e-waste, for example burning cables in the open to retrieve copper are dangerous to the environment. This practice allows potential GHG to the atmosphere and harmful consequences to human health. Furthermore, e-waste substances are non-degradable and examples includes mercury, beryllium, etc. The impacts of ICT on the environment can be both direct and indirect. Examples for direct impacts include e-waste and energy consumption. On the other hand, the indirect impacts examples include impacts of ICT applications, buildings and smart grids, and intelligent transport systems [46].

This paper has identified and explained some of the man-made activities that are unfriendly to the environment and ultimately contribute to climate change that threatens our survival. Furthermore, we also noted how ICT equipment contributes to carbon footprint. Having said that, developing countries fail to manage e-waste properly [6, 10, 47]. Good news with ICT is that it can be used to reduce carbon footprint from other industries [8]. It is today estimated that by the year 2020, ICT applications could help slash global carbon footprint by 15% [48].

4 Discussion

Global warming and climate change are the reality of today. We have seen and noted its devastating outcomes primarily in the developing countries. In this study it is found that e-waste contributes to global warming, although at a lower rate relative to other industries. ICT equipment that has reached their end of life (e-waste), if not managed properly contributes to carbon footprint. ICT life cycle contribute to carbon footprint (production cycle, usage cycle and disposal cycle). Global warming and climate change are believed to emanate from human made activities.

Therefore, humans have power to put halt on global warming and subsequent consequences. Though ICT is found to contribute to carbon footprint, it can also act as a 'reducing agent 'of carbon footprint in ICT and other industries. People are not aware of global warming and climate change especially in the developing countries. Absence of awareness is also noted with e-waste scavengers especially in the developing countries. To reduce the carbon footprint generated from ICT waste, it is important to dispose of e-waste using method friendly to the environment. Therefore, it is important to make awareness a priority. Formal recycling is found to be environment friendly compared to informal recycling.

References

1. Kramers, A., Hojer, M., Lovehagen, M., Wangel, J.A., et al.: Information and communication technology and climate change adaptation: evidence from selected mining companies in South Africa. J. Disaster Risk Stud. **8**(3), 1–9 (2014)
2. Shahid, Z., Piracha, A.: Awareness of climate change impacts and adaptation at local level in Punjab, Pakistan. In: Maheshwari, B., Singh, V.P., Thoradeniya, B. (eds.) Balanced Urban Development: Options and Strategies for Liveable Cities, pp. 409–428. Springer, Switzerland (2016)
3. Asabere, N.Y., Enguah, S.E.: Use of information & communication technology (ICT) in tertiary education in Ghana: a case study of electronic learning (E-learning). Int. J. Inf. Commun. Technol. Res. **2**(1), 62–68 (2012)

4. Afolabi, A.F., Abidoye, J.A.: Integration of information and communication technology in library operations towards effective library services. In: 1st International Proceedings on International Technology, Education and Environment Conference African Society for Scientific Research, pp. 620–628. Human Resource Management Academic Research Society (2011)

5. de Wet, W., Koekemoer, E., Nel, J.A.: Exploring the impact of information and communication technology on employees' work and personal lives. SA J. Ind. Psychol. **42**(1), 1–11 (2016)

6. Moletsane, R., Venter, C.: The 21st century learning environment tools as electronic waste. In: World Congress on Engineering and Computer Science, pp. 188–192. IAENG, USA (2018)

7. Gelenbe, E., Caseau, Y.: The impact of information technology on energy consumption and carbon emissions. Ubiquity, pp. 1–15 (2015)

8. Belkhir, L., Elmeligi, A.: Assessing ICT global emissions footprint: trends to 2040 & recommendations. J. Clean. Prod. **177**, 448–463 (2018)

9. Perkins, D.N., Brune, D., Nxele, M.-N., Sly, T.D.: E-waste: a global hazard. Ann. Global Health **80**(4), 286–295 (2014)

10. Park, J.K., Hoerning, L., Watry, S., Burgett, T., Matthias, S.: Effects of electronic waste on developing countries. Adv. Recycl. Waste Manag. **2**(2), 1–6 (2017)

11. Pickavet, M., Vereecken, W., Demeyer, S., Audenaert, P., Vermeulen, B., Develder, C., Colle, D., Dhoedt, B., Demmester, P.: Worldwide energy needs for ICT: the rise of power-aware networking. In 2nd International Symposium on Advanced Networks and Telecommunication Systems, pp. 1–3. IEEE, USA (2008)

12. Green Car Congress: Study projects global carbon footprint from ICT will be equivalent to half of transportation's current level by 2040. https://www.greencarcongress.com/2018/03/20180306-mcmaster.html. Accessed 01 Jan 2019

13. United States Environmental Protection Agency: Greenhouse gas emissions. https://www.epa.gov/ghgemissions/sources-greenhousegas-emissions. Accessed 30 Dec 2018

14. International Telecommunications Union: ICTs and Energy Efficiency. https://www.itu.int/en/action/climate/Pages/energy_efficiency.aspx. Accessed 01 Jan 2019

15. Teehan, P., Kandlikar, M.: Comparing embodied greenhouse gas emissions of modern computing and electronics products. Environ. Sci. Technol. **47**(9), 3997–4003 (2013)

16. Mugisa, J.: Design of an interface between the smart grid control plane and the telecommunications network control plane/data center control plane. In: Barcelona School of Informatics (FIB) pp. 1–49. University Polytechnic of Catalunya, Barcelona, Spain (2014)

17. Zein, A.L.E., Chehayeb, N.A.: The effect of greenhouse gases on earth's temperature. Int. J. Environ. Monit. Anal. **3**(2), 74–79 (2015)

18. Forabosco, F., Chitchyan, Zh, Mantovani, R.: Methane, nitrous oxide emissions and mitigation strategies for livestock in developing countries: a review. South African J. Anim. Sci. **47**(3), 268–280 (2017)

19. Lallanila: Greenhouse Effect. https://www.livescience.com/37743-greenhouse-effect.html. Accessed 29 Dec 2018

20. European Parliament: Greenhouse gas emissions by country and sector. http://www.europarl.europa.eu/news/en/headlines/society/20180301STO98928/greenhouse-gas-emissions-by-country-and-sector-infographic. Accessed 28 Dec 2018

21. Adedeji, O., Reuben, O., Olatoye, O.: Global climate change. J. Geosci. Environ. Prot. **2**(2), 114–122 (2014)

22. The Guardian: What are the main man-made greenhouse gases? https://www.theguardian.com/environment/2011/feb/04/man-made-greenhouse-gases. Accessed 20 Dec 2018

23. U.S Energy Information Administration: What are greenhouse gases and how do they affect the climate? https://www.eia.gov/tools/faqs/faq.php?id=81&t=11. Accessed 19 Dec 2018
24. Rojas-Downing, M.M., Nejadhashemi, P., Harrigan, T., Woznick, S.: Climate change and livestock: impacts, adaptation, and mitigation. Clim. Risk Manag. **16**, 145–163 (2017)
25. Gao, T., Liu, Q., Wang, J.: A comparative study of carbon footprint and assessment standards. Int. J. Low-Carbon Technol. **9**(3), 237–243 (2014)
26. Radu, A.L., Scrieciu, M.A., Caracota, D.M.: Carbon footprint analysis: towards a projects evaluation model for promoting sustainable development. Procedia Econ. Finance **6**(2013), 353–363 (2013)
27. Ologun, O.O., Wara, S.T.: Carbon footprint evaluation and reduction as a climate change mitigation tool. Int. J. Renew. Energy Res. **4**(1), 176–182 (2014)
28. The Guardian: What is a carbon footprint? https://www.theguardian.com/environment/blog/2010/jun/04/carbon-footprint-definition. Accessed 23 Dec 2018
29. Angelakoglou, K., Gaidajis, G., Lymperopoulos, K., Botsaris, P.N.: Carbon footprint analysis of municipalities – evidence from Greece. J. Eng. Sci. Technol. Rev. **8**(4), 15–23 (2015)
30. Rahman, M.I.: Climate change: a theoretical review. Interdisc. Description Complex Syst. **11**(1), 1–13 (2013)
31. Statista: Largest producers of territorial fossil fuel CO2 emissions worldwide in 2017, based on their share of global CO2 emissions. https://www.statista.com/statistics/271748/the-largest-emitters-of-co2-in-the-world/. Accessed 02 Jan 2019
32. Aleke, B.I., Nhamo, G.: Information and communication technology and climate change adaptation: evidence from selected mining companies in South Africa. J. Disaster Risk Stud. **8**(3), 1–9 (2016)
33. Acharya, S.S.: Sustainable agriculture and rural livelihoods. Agric. Econ. Res. Rev. **19**, 205–217 (2006)
34. Essays: The Importance of Agriculture in Developing Countries Economics Essay. https://www.ukessays.com/essays/economics/the-importance-of-agriculture-in-developing-countries-economics-essay.php. Accessed 02 Jan 2019
35. Kumar, S., Kumar, N., Vivekadhish, S.: Millennium development goals (MDGs) to sustainable development goals (SDGs): addressing unfinished agenda and strengthening sustainable development and partnership. Indian J. Commun. Med. **41**(4), 1–4 (2016)
36. New, M., Otto, F., Piotr Wolski, P.: Global warming has raised risk of more severe droughts in Cape Town. https://www.iol.co.za/news/opinion/global-warming-has-raised-risk-of-more-severe-droughts-in-cape-town-18656511. Accessed 02 Jan 2019
37. EcoWatch: How is Climate Change Affecting the Philippines? https://www.ecowatch.com/how-is-climate-change-affecting-the-philippines-1882156625.html. Accessed 02 Jan 2019
38. Eckstein, D., Künzel, V., Schäfer, L.: Global climate risk index 2018: who suffers most from extreme weather events. https://germanwatch.org/en/node/14987. Accessed 01 Jan 2019
39. Placido, D.: BS-CBN news, climate change a 'day-to-day problem': Duterte. https://news.abs-cbn.com/news/07/25/18/climate-change-a-day-to-day-problem-duterte. Accessed 02 Dec 2018
40. Huq, S., Ayers, J.: Critical list: the 100 nations most vulnerable to climate change. Institute of developmental studies (2007). http://www.eldis.org/document/A34452. Accessed 02 Jan 2019
41. Rapier, R.: China emits more carbon dioxide than the U.S. and EU combined. https://www.forbes.com/sites/rrapier/2018/07/01/china-emits-more-carbon-dioxide-than-the-u-s-and-eu-combined/#49c912f628c2. Accessed 01 Jan 2019

42. Hilty, L., Hischier, R., Ruddy, T.F., Som, C.: Informatics and the life cycle of products. In: 4th International Congress on Environmental Modelling and Software, pp. 1602–1611. Empa, Switzerland (2008)
43. Berkhout, F., Hertin, J.: Impacts of Information and Communication Technologies on Environmental Sustainability: Speculations and Evidence. University of Sussex, Brighton (2001)
44. Climate Institute: Adams, R.: E-waste and How to Reduce It. http://climate.org/e-waste-and-how-to-reduce-it/. Accessed 03 Jan 2019
45. Xu, Y., Park, H., Baek, Y.: A new approach toward DS: an activity focused on writing self-efficacy in a virtual learning environment. Educ. Technol. Soc. **14**(4), 181–191 (2011)
46. Houghton, J.W.: ICT and the environment in developing countries: a review of opportunities and developments. In: What Kind of Information Society? IFIP International Conference on Human Choice and Computers, pp. 236–247. Springer, Heidelberg (2010)
47. Kitila, A.W.: Electronic waste management in educational institutions of Ambo Town, Ethiopia. East Africa. Int. J. Sci. **24**(4), 319–331 (2015)
48. Finlay, A.: Global Information Society Watch: Focus on ICTs and Environmental Sustainability. APC and Hivos (2010)

Bot Detection on Online Social Networks Using Deep Forest

Kheir Eddine Daouadi[1(✉)], Rim Zghal Rebaï[2], and Ikram Amous[2]

[1] MIRACL-FSEGS, Sfax University, 3000 Sfax, Tunisia
khairi.informatique@gmail.com
[2] MIRACL-ISIMS, Sfax University, Tunis Road Km 10, 3021 Sfax, Tunisia
rim.zghal@gmail.com,
ikram.amous@enetcom.usf.tn

Abstract. Nowadays, social networks are widely used not only by humans, but also by bots (automated agents). Recent studies have reported that bot accounts are used across different dimensions and in different granularities (e.g. performing terrorist propaganda, spreading misinformation and polluting content). Bot detection has become a significant challenge, especially on online social networks. Today, researchers through Twitter are attempting to propose approaches for bot detection. However, they are confronted with certain challenges owing to the problems inherent to text and the use of language-dependent features. Therefore, we defined a set of statistical features which proved their importance in our work. Our proposed features are based on how much the others interact with the posts of the user, when the user interacts and how much the user interacts. We demonstrated that the mere use of information from the metadata of user profiles and the metadata of posts with a Deep Forest algorithm is sufficient in order to detect bot accounts accurately. In fact, this yielded an Accuracy result of 97.55%.

Keywords: Social network analysis · Twitter user classification ·
Bot detection · Metadata of user profiles · Metadata of posts, deep forest

1 Introduction

Nowadays, several parties in the community are attracted by the so-called social networks like Twitter, Facebook etc. This gain of interest is due to the increasing applications targeted by researchers, organizations, political parties and even governmental institutions etc. The use of social networks has become an important part of our everyday life; millions of users exchange a huge amount of data on such social networks. Twitter is one of the leading social networks, which allows users to post the so-called 'tweets'. Users can generate three main types of tweets: a 'tweet', which is an original post; a 'retweet', which is a reposting of another post; and a 'reply', which is a comment on another post. This free-microblogging has been very widely used not only by humans to create, publish and share content, but also by bots to spread misinformation, pollute contents etc. Bot detection on Twitter has become one of the significant problems, these days. This is underscored by the challenge recently organized by the

© Springer Nature Switzerland AG 2019
R. Silhavy (Ed.): CSOC 2019, AISC 985, pp. 307–315, 2019.
https://doi.org/10.1007/978-3-030-19810-7_30

Defense Advanced Research Projects Agency (DARPA) to detect bot activities [1]. A recent study reported that between 9% and 15% of the active Twitter accounts were bots [2]. Today, researchers through Twitter are attempting to propose approaches for bot detection. However, they are confronted with certain challenges owing to the diversity of the types of data propagated throughout Twitter. In addition, they use language-dependent features. Moreover, they are confronted with problems inherent to the text. In this paper, a new statistical approach for bot detection on Twitter is proposed. The suggested approach aims at detecting bot accounts by exploiting the metadata of user profiles and the metadata of posts collected from the user profiles. We demonstrated the importance of using parameters from the metadata of user profiles and the metadata of posts in order to recognize the type of user account accurately, quickly and without considering the users' language.

The rest of this paper is organized as follows: In the second Section, we will discuss the related works. In Section Three, we will describe our proposed approach. In Sect. 4, we will discuss the results of our experiments. In Sect. 5, the benefits from our proposed approach will be discussed. Finally, we will summarize the main contributions of our proposed method.

2 Related Works

Classifying Twitter account types is considered as a type of latent attribute inferences in which the main purpose is to infer various properties of online accounts. Several works have been done on a single and specific perspective such as bot detection [1–7], age prediction [8], organization detection [9] etc. Three main types of approaches for Twitter bot detection were used in the literature [10]. The first type is the crowd-sourcing approach, also named blacklist approach [11]. When using this type of approach, "expert" workers are still needed to accurately detect fake accounts. The second is the social graph-based features approach: this type of approach is very expensive and more complex than other types of approaches. The third type is the feature-based approach: Different classes of features are commonly employed to capture the orthogonal dimensions of the users' behaviors (e.g. content features, temporal features, network features etc.). We will discuss the literature of the feature-based bot detection approach.

In [2], the authors used more than one thousand features. These can be classified into six types, namely user-based features, friend features, network features, temporal features, content and language features, and finally sentiment features. This yielded a reasonable AUC of 95%. In [12], the authors exploited digital DNA using both supervised and unsupervised approaches. They used a sequence of types of posts, and a sequence of feature-based content; then they defined a similarity measure in order to characterize both bot and human accounts. In [13], the authors proposed an approach based on the textual content solely. They adopted the Latent Dirichlet Allocation (LDA) to obtain a topic representation of each user. This yielded an F1-measure of 76.55% and 14.69% using the Lybia Honeypot dataset and the Libya dataset, respectively. In [14], the authors proposed an approach based on the content of publications and metadata of user profile features. This yielded an F1-measure from 77% to

91% using Random Forest algorithm. In [15], the authors proposed an unsupervised approach based on topological, behavioral, content and overlapping community structure features. This yielded an F1-measure of 88%. In [16], the authors proposed an approach based on the post frequency in order to distinguish between human, bots and cyborg users (a profile which is managed by humans and automatically). This yielded an F1-measure of 88% using a supervised learning approach.

In summary, current approaches for bot detection are not sufficient to detect bot accounts accurately, quickly and without considering the users' language. In addition, the messages from social networks are short, imprecise and may be written in different languages and dialects. To address the limits previously mentioned, a novel feature-based approach for bot detection was proposed.

3 Proposed Method

Following the limits mentioned in our related works, we proposed a novel feature-based approach for bot detection. Our proposed approach consists of three main phases: data acquisition, feature extraction and classification (Table 1).

Table 1. Corpus description.

Dataset label	Active		Suspended, removed or private	
	Human	Bot	Human	Bot
D1 [1]	1388	720	359	106
D2 [17]	11404	12931	7872	9292

3.1 Data Acquisition

This phase aims at collecting data from the user accounts. We used Twitter timeline (API), which allows collecting the K most recent posts of such users. The posts come with information from the metadata of the user profile (User), the metadata of the post (Post), the time and date of the post (Time). However, we focused on two datasets which were published by [1] and [17]. The first one was collected by a manual-labeling method, and the second one was collected by the honeypot method. These datasets contain a user_id with a label as 0 or 1 (Human or Bot). We used the user_id with the Twitter timeline (API) in order to collect the 200 most recent posts of each labelled user. We observed the users' accounts' statuses via the statuses/user_Timeline API endpoint. These statuses can take on one of four values: active account, suspended account, removed account and private account. Since we could not access the posts of user accounts which are private, removed or suspended. In this work, we used only active users.

3.2 Feature Extraction

This phase aims at building out the feature vector from the user account based on our proposed features. These include metadata of user profiles and metadata of posts.

First, we used 13 parameters from the (User) attribute. The suggested features include binary and numerical parameters as described in Table 2. These parameters have a demonstrated utility in classifying user accounts by our cited related works.

Table 2. Metadata of user profile features.

Label	Description
Geo enabled	Whether a user has enabled the possibility of geo-tagging his/her posts
URL	Whether the user has provided a URL in association with his/her profile
Default	Whether the profile has the default settings (default profile)
Location	Whether the user has provided information on his/her location
Verified	Whether the user has a certified profile
Description	Whether the user has provided a description in association with his/her profile
Followers	The number of followers that the user has
Following	The number of followings that the user has
Ratio	The ratio of the number followers to the number of followings
Lists	The number of public lists that the user is a member of
Favorites	The number of posts liked by the user
Post	The number of posts issued by the user
Average post	The number of posts per day (during the lifecycle of the user profile)

Second, we used parameters from the metadata of posts. Our proposed metadata of posts features are based on how the others interact with the posts of the user, when the user interact and how much the user interacts. We built on the hypothesis that the metadata of posts parameters are a decisive factor in order to perform Bot vs. Human classification task. Table 2 shows our proposed metadata of posts features. These include the number of different types of posts in the hours of the day, in the days of the week, the average number of posts per day and per week, the minimum and maximum interval between each type of posts, the number of unique sources used by the user, the average number of hashtags per post, the number of unique hashtags used by the user, the number of posts that contain a URL, etc. (see Table 3).

To filter the most relevant features, one feature selection method was used. The chosen method is information gain, which outputs each feature from 0 to 1 in importance. Fortunately, there wasn't even one feature that obtained 0 from the information gain method. All the proposed features could not be ignored and had relevant results in the classification process.

Table 3. Proposed metadata of posts features.

Label	Description
T, R, P	The number of tweets, retweets, and replies
TR, TP, RP	Ratio of T to R, ratio of T to P, ratio of R to P
AD	Average number of T, R, P, T+R, T+P, R+P and T+R+P per day
AW	Average number of T, R, P, and posts per week
NH	Number of T, R, P, and posts per hour of day
ND	Number of T, R, P and posts per day of week
Min	Minimum interval between T, R, P and T+R+P
Max	Maximum interval between T, R, P and T+R+P
AI	Average interval between T, R, P and T+R+P
Ret	Total number of retweets of T, R, and P
Fav	Total number of favorites of T, R, and P
ARet	Average number of retweets (T, R, P, T+R+P)/number of followers
AFav	Average number of favorites (T, R, P, T+R+P)/number of followers
URL	The number of posts that contain a URL
Source	The number of unique sources used by the user in order to post a post
Hash1	Average number of hashtags per post
Hash2	Number of unique hashtags used by the user
Quoted	Number of retweets which were reposted with a comment
Mention	Number of posts which mention another user
GEO	Number of geo-localized posts
place	Number of posts that were posted in association with a place

3.3 Classification

The third phase of our proposed approach is the classification task. We used a supervised learning approach; a set of algorithms were tested using our proposed features. Then a comparison of the results of each algorithm was performed. We evaluated the performance of our predictive model and the impact of the post metadata features in improving class distinction. For each datasets, we built three models: one based on the metadata of user profile features alone, one based on the post metadata features alone and one based on both metadata of user profile and metadata of posts features. These three separate models were used in order to evaluate the marginal impact of adding metadata of post features in the bots vs. human classification. Our final predictive model achieved high performance metrics. This yielded an Accuracy result of 97.55%.

4 Experiment and Evaluation

Once the proposed features had been constructed, we tested over 30 traditional supervised learning algorithms such as Bagging (B), Multi-Layer Perceptron (MLP), AdaBoost (AB), Random Forest (RF), Simple Logistic (SL), etc. Moreover, we used the Synthetic Minority Oversampling Technique (SMOTE) [18] in order to enrich our datasets with more examples of bot and human samples. In addition, we tested the Deep Forest (DF) algorithm, which is a powerful method that has seen successfully in many domains [19]. We used each algorithm with their standard parameter in order to ensure the comparability between them. The tests were implemented using the Waikato Environment for Knowledge Analysis (WEKA), and a code developed in Python programming language to run the Deep Forest algorithm. In order to evaluate the performance of our proposed approach, we used five performance metrics, namely Precision, Recall, F-measure, Accuracy and AUC. The 10-fold cross validation approach was used in order to calculate the performance metrics.

First, we tested the traditional supervised learning algorithms. As shown in Table 4, the RF algorithm outperformed the other traditional algorithms using all datasets in terms of AUC measure. In order to compare the results of our proposed approach with the baselines, we evaluated the AUC of the corresponding models:

Table 4. AUC results of the traditional supervised learning algorithms using each dataset.

Dataset label	SL	AB	B	RF	MLP
D1	0,94	0,92	0,93	**0,94**	0,92
D2	0,96	0,95	0,98	**0,98**	0,95
D1 + D2	0,96	0,94	0,98	**0,98**	0,94

- D1: We used the manual-labeling datasets that were used by [2] for training and testing. This yielded an AUC result of 94.4% (Vs. 89.0% [2] and 89.1% [7]), a reasonable accuracy considering that the dataset contains more sophisticated and recent bots.
- D 2: We used the honeypot datasets that were used by [2] for training and testing. This yielded an AUC result of 98.3% (Vs. 95% [2], 94.6% [7]).
- D1+D2: We used both the manual-labeling and honeypot datasets for training and testing. This yielded 98.1% AUC (vs 94.0% [2], 93.9% [7]) (Table 5).

Table 5. Random Forest 10-fold cross validation measures. (U: metadata of user profile, P = metadata of posts,).

		Precision	Recall	F1-measure	AUC	Accuracy
D1	U	80.3	80.6	80.3	87.4	80,6
	P	84.5	84.7	84.5	94.2	84.7
	U+P	**85.2**	**85.8**	**85.6**	**94.4**	**85.7**
D2	U	88.4	88.4	88.4	94.9	88.4
	P	92.0	92.0	92.0	97.7	91.9
	U+P	**93.2**	**93.2**	**93.2**	**98.3**	**93.1**
D3	U	87.4	87.3	87.3	94.1	87.3
	P	91.3	91.3	91.3	97.5	91.2
	U+P	**92.4**	**92.4**	**92.4**	**98.1**	**92.3**

Second, to test our hypothesis, stating that the metadata of posts are a decisive factor in the Bot Vs. Human classification task, nine models were built as previously mentioned. Table 4 shows the performance metrics of each model. Several insights were observed as follows:

- D1: 7% and 5.3% were gains where we added our proposed metadata of posts features in the AUC and F1-measure, respectively.
- D2: 3.4% and 4.8% were gains where we added our proposed metadata of posts features in the AUC and F1-measure, respectively.
- D1+D2: 4% and 5.1% were gains where we added our proposed metadata of posts features in the AUC and F1-measure, respectively (Table 6).

Table 6. Accuracy results using D1+D2.

Algorithm	Accuracy%
Random Forest (RF)	92.31
(RF)+SMOTE	95.65
Deep Forest (DF)	94.26
(DF)+SMOTE	97.55

Third, we used SMOTE in order to enhance our datasets with more examples of bot and human samples. The main advantage of using the SMOTE algorithm is to enhance previously labeled datasets based on the feature vector of the labelled users. However, it effectively forces the decision region of each class by generating a new sample based on the feature vectors of each class. This occurs without any very expensive labelling step and avoiding the over-fitting problem. Significant gains were observed when using SMOTE; this yielded an accuracy result of 95.65%.

Fourth, we tested the Deep Forest algorithm with the standard parameters [19]. The main advantage of using the Deep Forest algorithm is to use deep model with a non-differentiable module. Significant gains were observed in the accuracy result when using our proposed features with the SMOTE and the DF algorithms. This yielded Accuracy result of 97.55% (Vs 94.26% using (DF), 95.65% using (RF+SMOTE) and 92.31% using (RF)).

In summary, we developed a highly accurate model that classifies user accounts belonging to humans or bots. We illustrated the impact of using the metadata of posts as features for the Bot vs. Human classification task. Our proposed approach outperformed the three baselines which were proposed in [2, 7] and using only the metadata of user profiles. The suggested approach aims at exploiting parameters from the metadata of user profiles and metadata of posts; these were used by the Deep forest algorithm in order to classify user accounts as being accounts of bots or humans.

5 Conclusion and Future Work

Nowadays, social networks are widely used not only by humans, but also by bots (automated agents). Recent studies have reported that bot accounts are used across different dimensions and in different granularities (e.g. spreading misinformation, polluting content, and even performing terrorist propaganda). Today, researchers through Twitter are attempting to propose approaches for bot detection. However, they are confronted with certain challenges owing to the problems inherent to texts and the use of language-dependent features. In addition, the current approaches for bot detection are not sufficient to detect bot accounts accurately. The ability of distinguishing between the two account types is needed for developing many applications such as consumer-product-opinion-mining tools, information-dissemination platforms, etc. In this paper, an efficient feature-based approach for bot detection is proposed by exploiting statistical parameters from the metadata of user profiles and the metadata of posts. Our proposed framework outperforms previous ones by achieving a higher performance of classification. In addition, we illustrated the main advantage of using the SMOTE algorithm and its contribution in order to avoid the over-fitting problem and to enhance the overall accuracy results. Moreover, we exploited the Deep Forest algorithm, which outperformed the other traditional supervised-learning algorithms. Our final predictive model achieved high Accuracy result of 97.55%. As future work, we will extend our work in order to be able to extract the main topics discussed by the user.

References

1. Subrahmanian, V.S., Azaria, A., Durst, S., Kagan, V., Galstyan, A., Lerman, K., Stevens, A.: The DARPA Twitter bot challenge. arXiv preprint arXiv:1601.05140 (2016)
2. Varol, O., Ferrara, E., Davis, C., Menczer, F., Flammini, A.: Online human-bot interactions: detection, estimation, and characterization. In: 11th International AAAI Conference on Web and Social Media, pp. 1–10. AAAI, Canada (2017)
3. Kantepe, M., Ganiz, M.C.: Preprocessing framework for Twitter bot detection. In: Computer Science and Engineering (UBMK), pp. 630–634. IEEE, Turkey (2017)
4. Pozzana, I., Ferrara, E.: Measuring bot and human behavioral dynamics. arXiv preprint arXiv:1802.04286 (2018)
5. Chavoshi, N., Hamooni, H., Mueen, A.: On-demand bot detection and archival system. In: Proceedings of the 26th International Conference on World Wide Web Companion, pp. 183–187. ACM, Australia (2017)
6. Cai, C., Li, L., Zengi, D.: Behavior enhanced deep bot detection in social media. In: International Conference on Intelligence and Security Informatics (ISI), pp. 128–130. IEEE, China (2017)
7. Ferrara, E.: Disinformation and social bot operations in the run up to the 2017 French presidential election. First Monday, 22(8) (2017)
8. Guimaraes, R.G., Rosa, R.L., De Gaetano, D., Rodriguez, D.Z., Bressan, G.: Age groups classification in social network using deep learning. IEEE Access 5, 10805–10816 (2017)
9. Kim, S.M., Paris, C., Power, R., Wan, S.: Distinguishing individuals from organisations on Twitter. In: Proceedings of the 26th International Conference on World Wide Web Companion, pp. 805–806. ACM, Australia (2017)

10. Ferrara, E., Varol, O., Davis, C., Menczer, F., Flammini, A.: The rise of social bots. Commun. ACM **59**(7), 96–104 (2016)
11. Wu, T., Wen, S., Liu, S., Zhang, J., Xiang, Y., Alrubaian, M., Hassan, M.M.: Detecting spamming activities in twitter based on deep-learning technique. Concurr. Comput. : Pract. Exp. **29**(19), e4209 (2017)
12. Cresci, S., Di Pietro, R., Petrocchi, M., Spognardi, A., Tesconi, M.: Social fingerprinting: detection of spambot groups through DNA-inspired behavioral modeling. IEEE Trans. Dependable Secure Comput. **15**(4), 561–576 (2018)
13. Morstatter, F., Wu, L., Nazer, T.H., Carley, K.M., Liu, H.: A new approach to bot detection: striking the balance between precision and recall. In: Proceedings of the 2016 IEEE/ACM International Conference on Advances in Social Networks Analysis and Mining, pp. 533–540. IEEE Press, USA (2016)
14. Gilani, Z., Kochmar, E., Crowcroft, J.: Classification of twitter accounts into automated agents and human users. In: Proceedings of the 2017 IEEE/ACM International Conference on Advances in Social Networks Analysis and Mining, pp. 489–496. ACM, Australia (2017)
15. Bindu, P.V., Mishra, R., Thilagam, P.S.: Discovering spammer communities in Twitter. J. Intell. Inf. Syst., 1–25 (2018)
16. Tavares, G.M., Mastelini, S.M., Barbon Jr., S.: User classification on online social networks by post frequency. CEP **86057**, 970 (2017)
17. Lee, K., Eoff, B.D., Caverlee, J.: Seven months with the devils: a long-term study of content polluters on Twitter. In: ICWSM, pp. 185–192 (2011)
18. Chawla, N.V., Bowyer, K.W., Hall, L.O., Kegelmeyer, W.P.: SMOTE: synthetic minority over-sampling technique. J. Artif. Intell. Res. **16**, 321–357 (2002)
19. Zhou, Z.H., Feng, J.: Deep forest: towards an alternative to deep neural networks. arXiv preprint arXiv:1702.08835 (2017)

Theoretical and Experimental Evaluation of PSO-K-Means Algorithm for MRI Images Segmentation Using Drift Theorem

Samer El-Khatib[1], Yuri Skobtsov[2(✉)], Sergey Rodzin[1],
and Semyon Potryasaev[3]

[1] Southern Federal University, Rostov-on-Don, Russia
samer_elkhatib@mail.ru, srodzin@yandex.ru
[2] St. Petersburg State University of Aerospace Instrumentation,
Saint Petersburg, Russia
ya_skobtsov@list.ru
[3] St. Petersburg Institute of Informatics and Automation,
Russian Academy of Sciences (SPIIRAS), Saint Petersburg, Russia
spotryasaev@gmail.com

Abstract. Image segmentation is the process of subdividing an image into regions that are consistent and homogeneous in some characteristics. An important factor in the recognition of magnetic resonance images is not only the accuracy, but also the speed of the segmentation procedure. Modified Exponential Particle Swarm Optimization algorithm is proposed in paper. The time complexity of proposed algorithm is investigated using consequences from Drift theorem. It is established that the proposed algorithm has a polynomial estimation of complexity. Images from the Ossirix image dataset and real medical images were used for testing.

Keywords: MRI image segmentation · Particle swarm optimization · K-means · Swarm intelligence · Segmentation · Bio-inspired methods

1 Introduction

One of the most difficult tasks in the processing of medical images is segmentation. All the subsequent image-processing steps - classification, identification and feature extraction are directly dependent on the segmentation results. This problem has been studied in many papers and a large number of image segmentation algorithms has been proposed [1]. Segmentation - process of the image fragmentation into disjoint parts characterized by the definite homogeneous features. This procedure is used to solve a wide range of tasks: finding objects in satellite images (mountains, rivers, etc.), face recognition, medical image processing – (MRI, Computer Tomography, X-Ray), etc. At computer processing and recognition of images the wide range of problems is solved. One of the main stages of recognition is the process of dividing the image into non-overlapping areas (segments) that cover the entire image and are homogeneous by some criteria. Segmentation simplifies the analysis of homogeneous areas of the image, as well as brightness and geometric characteristics. The segmentation is implemented using special methods.

© Springer Nature Switzerland AG 2019
R. Silhavy (Ed.): CSOC 2019, AISC 985, pp. 316–323, 2019.
https://doi.org/10.1007/978-3-030-19810-7_31

Their goal is to separate the analyzed object, structure or area of interest from the surrounding background. Segmentation based on MRI data is very important but at the same time it is a time-consuming task if it is performed manually by medical specialists. Therefore, there is a need for computer analysis of image to facilitate diagnosis.

The given article is an expansion of the previous articles [3, 4, 8], where there was proposed a modification of algorithm based on the ant-colony optimization image segmentation technique. To evaluate the practical and theoretical time complexity of the developed algorithm for segmentation of MRI images, a technique was proposed using the consequences of Drift's theorem [10].

This article is dedicated to development and evaluation for hyper-heuristic particle swarm segmentation method for MRI images.

2   Hyper-heuristic Particle Swarm Method for Image Segmentation

Hyper-heuristic is a search procedure aimed to automate the process of selection, combination and adaptation of several simple heuristics to effectively solve a problem. The main idea of the proposed hyper-heuristic method is the application of several heuristics, each of which has its weak and strong places, and then their use depending on the current state of the solution.

Formulation of the problem is similar to the statement of the segmentation problem using the ant colony optimization algorithm [3, 4] and consists of the following: for the given source images in the form of a set of pixels with visual properties such as brightness, color, texture, and also a certain size, level of noise, contrast and quality, it is necessary, within the available time resources, to find the markup of image onto K segments, which provides acceptable accuracy and quality of image recognition.

Hyper-heuristic particle swarm segmentation method is a management system, in the subordination of which there are three bio-inspired heuristics: PSO-K-means [6], Modified Exponential PSO [5], Elitist Exponential PSO [7]. Each of the hyper-heuristics is applied depending on the quality of the source images: good quality (no noise and other artifacts); with the presence of noise; contrast images, blurred images. The control scheme of the method is shown on Fig. 1.

In this paper we introduce investigation for PSO-K-means algorithm.

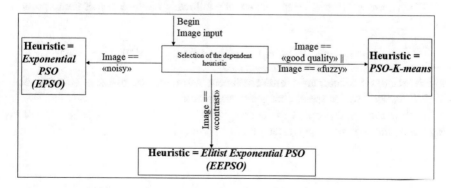

Fig. 1. Dependent hyper heuristics management scheme

2.1   PSO-K-Means Segmentation Algorithm

When implementing the PSO-K-means heuristic, the n-dimensional search space is populated by a swarm of m particles (solution population) [2]. The coordinate of the i-th particle ($i \in [1: m]$) is described by the vector $x_i = (x_i^1, x_i^2, \ldots, x_i^n)$, that defines a set of optimization parameters: velocity $v_i(t)$ and position $x_i(t)$ at the moment of the time t. The position of the particle changes according to the following:

$$x_i(t+1) = x_i(t) + v_i(t+1) \tag{1}$$

where $x_i(0) \sim U(x_{min}, x_{max})$.

For each position of the n-dimensional search space which the i-th particle has visited, the value of the fitness-function $f(x_i)$ is evaluated. In this case, the best value of the fitness-function is stored, as well as the position coordinates in the n-dimensional space corresponding to this value of the fitness-function.

Correction of each j-th coordinate of the velocity vector ($j \in [1 : n]$) of i-th particle is performed according to the formula:

$$v_{ij}(t+1) = v_{ij}(t) + c_1 r_{1j}(t)[y_{ij}(t) - x_{ij}(t)] + c_2 r_{2j}(t)[\hat{y}_j(t) - x_{ij}(t)] \tag{2}$$

where $v_{ij}(t)$ – j-th velocity component ($j = 1, .., n_x$) of the i-th particle in the moment of time t; $x_{ij}(t)$ – j-th coordinate of the particle i; c_1, c_2 – constant acceleration coefficients, which define the effectiveness of PSO-K-means heuristic; $r_{1j}(t), r_{2j}(t) \sim U(0,1)$ – random values in the range of [0, 1]; $y_{ij}(t)$ and $\hat{y}_j(t)$ – cognitive (the best position of i-th particle on coordinate j (gbest)) и and social components of the swarm.

The best position (gbest) at the moment of time $(t + 1)$ is calculated as:

$$y_i(t+1) = \begin{cases} y_i(t) & \text{if } f(x_i(t+1)) \geq f(y_i(t)) \\ x_i(t+1) & \text{if } f(x_i(t+1)) < f(y_i(t)) \end{cases} \tag{3}$$

The social component of the swarm $\hat{y}_j(t)$ at the moment of the time t is calculated as

$$\hat{y}(t) \in \{y_0(t), \ldots, y_{n_s}(t)\} | f(y(t)) = \min\{f(y_0(t), \ldots, f(y_{n_s}(t)))\} \tag{4}$$

where n_s – the common amount of the particles.

When correcting the velocity vector v_i the following modification of the formula (2) used:

$$v_{ij}(t+1) = \omega v_{ij}(t) + c_1 r_{1j}(t)[y_{ij}(t) - x_{ij}(t)] + c_2 r_{2j}(t)[\hat{y}_j(t) - x_{ij}(t)] \tag{5}$$

which includes the factor ω – inertia coefficient before j-th coordinate of velocity vector for i-th particle, so the speed changes more smoothly.

Each particle x_i represents the K centers as $x_i = (m_{i1}, \ldots, m_{ij}, \ldots, m_{iK})$, where m_{ij} represents the center of the cluster j for the particle i.

The fitness-function for each set of clusters is calculated as:

$$f(x_i, Z_i) = \omega_1 \overline{d}_{\max}(Z_i, x_i) + \omega_2(z_{\max} - d_{\min}(x_i)) \tag{6}$$

where $Z_{\max} = 2^s - 1$ for s-bit image; Z – the connection matrix, which represents connectivity between pixel and cluster for the particle i. Each element of the matrix z_{ip} represents, if the pixels z_p belongs to the cluster C_{ij} for the particle i. Constants ω_1 and ω_2 are user defined; $\overline{d}_{\max}$ – the maximum mean Euclidean distance from the particles to the associated clusters:

$$\overline{d}_{\max}(Z_i, x_i) = \max_{j=1..K} \left\{ \sum_{\forall Z_p \in C_{ij}} d(Z_p, m_{ij}) / |C_{ij}| \right\}, \tag{7}$$

$d_{\min}(x_i)$ – minimal Euclidean distance between pairs of cluster centers:

$$d_{\min}(x_i) = \min_{\forall j_1, j_2, j_1 \neq j_2} \left\{ d(m_{ij_1}, m_{ij_2}) \right\} \tag{8}$$

During the search, the particles exchange information among themselves, on the basis of which they change their location and speed of movement, using global and local solutions at the present moment. The global best solution is known for all particles, and in case of finding the best value, it is corrected [9].

The input data for PSO-K-means heuristic are: the number of clusters K; the number of the particles m, that directly perform the segmentation; the maximum number of iterations of the method n_{t0} to find a solution; acceleration coefficients c_1 and c_2 – personal and global components of the final particle velocity.

3 Theoretical and Experimental Evaluation of PSO-K-Means Algorithm Using Drift Analysis

Drift analysis is a modern tool to estimate the complexity of bio-stochastic methods [10]. The method for the estimation of the time complexity using drift theorem is as follows.

In the finite state space S there is a certain function $f(x)$, $x \in S$. Need to find

$$max\{f(x); x \in S\} \tag{9}$$

Let x^* be the state with the maximum value of the function $f_{max} = f(x^*)$. An abstract bio-inspired method for solving the optimization task includes the following steps:

1. Initialization of the solutions population (randomly or heuristically) $\xi_0 = (x_1, \ldots, x_{2n})$ which contains 2n individuals (n is integer). Let be $k = 0$. For each population ξ_k need to determine $\xi_k = max \{f(x_i): x_i \in \xi_k \}$.
2. *Generation* of the population $\xi_{k+1/2}$ using evolutionary operators.

3. *Selection* and *reproduction* of 2n individuals from populations $\xi_{k+1/2}$ and ξ_k and deriving a new population ξ_{k+S}.
4. If $f(\xi_{k+S}) = f_{max}$, then *stop*, otherwise $\xi_{k+1} = \xi_{k+S}$, $k = k + 1$ and go back to the step 2.

Let x^* be the optimum point. Lets denote $d(x, x^*)$ as distance between points x and x^*. If there are many optimums S^*, then $d(x, S^*) = min\{d(x, x^*): x^* \in S^*)\}$ is the distance between point x and the set S^*. We denote this distance by $d(x)$. Then $d(x^*) = 0$, $d(x) > 0$ for any $x \notin S^*$.

Considering that the population $X = min\{x_1, \ldots, x_{2n}\}$, we set

$$d(X) = min\{d(x) : x \in X\} \tag{10}$$

Formula (10) serves to measure the distance between the population and the optimal solution.

The sequence $\{d(\xi_k); k = 0, 1, 2, \ldots\}$, generated by bio-inspired method is random sequence and it can be modeled using homogeneous Markov chain.

Then *drift* of the random sequence in the time moment k can be defined as

$$\Delta(d(\xi_k)) = d(\xi_{k+1}) - d(\xi_k) \tag{11}$$

The stopping time of the method can be estimated as $\tau = min\{k : d(\xi_k) = 0\}$. Need to study the relationship between the time τ and the dimension of the problem n. At what drift values $\Delta(d(\xi_k))$ can we estimate the mathematical expectation $E[\tau]$? Can the presented method find optimal solution in polynomial time or it will require exponential time?

The idea of drift analysis is quite simple. The key issue here is the evaluation of the relation between d and Δ.

A bio-inspired method can solve an optimization problem in the polynomial average time under the following drift conditions:

- If polynomial $h_0(n) > 0$ (n – the dimension of problem) exists such as

$$d(X) \leq h_0(n) \tag{12}$$

for any given population X, i.e. the distance from any population to the optimal solution is a polynomial function of the dimension of the problem;

- At any moment $k \geq 0$, if $d(\xi_k) > 0$, then polynomial $h_1(n) > 0$ exists such as

$$E[d(\xi_k) - d(\xi_{k+1}) | d(\xi_k) > 0] \geq 1/h_1(n) \tag{13}$$

i.e. drift of the random sequence $\{d(\xi_k); k = 0, 1, 2, \ldots\}$ in relation to optimal solution always positive and limited by reverse polynomial.

The consequence of the drift analysis results is in the fact that the estimation of the drift value is converted into an estimate of the operation time of the method, and the local property (drift of one step) is transformed into a global property (the operation time of the method until the optimum is found). This is a new result in estimation of the time complexity of bio-inspired methods obtained with the drift analyzing. Drift can be evaluated easier. With the help of the drift analysis there were determined conditions that guarantee the solution of some problems on average in polynomial time, as well as the conditions under which the method requires an average exponential time for solving the problem.

As it can be seen from above with respect to developed hybrid PSO-k-means segmentation method the use of drift analysis method for time complexity estimation is of interest. It is necessary to estimate the drift values based on image data during the segmentation process and to perform a polynomial time complexity satisfactory test[8]. Let's estimate the fulfilment of these conditions for specific values of the objective functions for the developed image segmentation method.

Table 1 represents the values of the objective function of the PSO-K-means algorithm with the number of particles $m = 5$, number of the clusters $k = 5$, image size is 109×106. Objective functions f_1 are presented on iteration k as in formulas (6).

Table 1. Objective function of PSO-k-means algorithm at m = 5.

$k = 1$	$k = 2$	$k = 3$	$k = 4$	$k = 5$	$k = 6$	$k = 7$	$k = 8$	$k = 9$	$k = 10$
216,22	219,93	252,71	252,71	252,71	258,19	258,19	258,19	258,19	258,19

Then values $f_1 - f_3$ in formula (12) for objective function are as follows:

$$d(X) \leq \sum_{\substack{i=1..width* \\ height}} \sum_{\substack{j=1..width* \\ height, j \neq i}} \sqrt{(X_i - X_j)^2 + (Y_i - Y_j)^2} \leq$$
$$width * height * \sqrt{width^2 + height^2}$$

где *width* – width of the image, *height* – height of the image.

The formula (13) for the mathematical expectation of drift can be represented as a combination $E[d(\xi_k) - d(\xi_{k+1})|d(\xi_k) > 0] \geq min(d(X))$

For the Table 1 above, the mathematical expectation of drift[8] has the form shown in Table 2.

Table 2. Drift of the objective function of the PSO-k-means segmentation algorithm at $m = 5$.

	$k = 2$	$k = 3$	$k = 4$	$k = 5$	$k = 6$	$k = 7$	$k = 8$	$k = 9$	$k = 10$
E	3,71	32,78	0	0	5.48	0	0	-	-

As it can be seen from the results, formulas (10) and (11) are valid when real data values are substituted, therefore, according to [10] the hybrid PSO-k-means method solves the segmentation problem in polynomial time.

4 Conclusion

In the presented paper, there has been introduced the combined hybrid PSO-k-means segmentation algorithm and proposed investigation of its theoretical and practical time complexity using Drift theorem. The presented algorithm based on the particle population. Each particle makes its own decision. Experimental results showed that PSO-k-means segmentation method can solve segmentation task in polynomial time.

The proposed algorithm can be enhanced by means of the additional research, especially of pseudo-random heuristic coefficients, as well as their impact on convergence and the final result of the processing.

Acknowledgements. The research described in this paper is partially supported by the Russian Foundation for Basic Research according to the research project № 19-07-00570 "Bio-inspired models of problem-oriented systems and methods of their application for clustering, classification, filtering and optimization problems, including big data" (Southern Federal University).

The research described in this paper is partially supported by the Russian Foundation for Basic Research (grants **16-29-09482-ofi-i, 17-08-00797, 17-06-00108, 17-01-00139, 17-20-01214, 17-29-07073-ofi-i, 18-07-01272, 18-08-01505, 19-08-00989**) state order of the Ministry of Education and Science of the Russian Federation № **2.3135.2017/4.6**, state research **0073–2019–0004**, International project ERASMUS +, Capacity building in higher education, № 73751-EPP-1-2016-1-DE-EPPKA2-CBHE-JP, Innovative teaching and learning strategies in open modelling and simulation environment for student-centered engineering education.

References

1. Gonzalez, R.C., Woods, R.E.: Digital Image Processing, 3rd edn. Prentice-Hall, Englewood Cliffs (2008)
2. Kennedy, J, Eberhart, R.C.: Particle swarm intelligence. In: Proceedings of the IEEE International Joint Conference on Neural Networks, pp. 1942–1948 (1995)
3. El-Khatib, S., Rodzin, S., Skobtcov, Y.: Investigation of optimal heuristical parameters for mixed ACO-k-means segmentation algorithm for MRI images. In: Proceedings of III International Scientific Conference on Information Technologies in Science, Management, Social Sphere and Medicine (ITSMSSM 2016). Part of series Advances in Computer Science Research, vol. 51 pp. 216–221. Atlantis Press (2016). https://doi.org/10.2991/itsmssm-16.2016.72. ISBN: 978-94-6252-196-4
4. El-Khatib, S., Skobtsov, Y., Rodzin, S.: Theoretical and experimental evaluation of hybrid ACO-k-means image segmentation algorithm for MRI images using drift-analysis. In: Proceedings of XIII International Symposium « ntelligent Systems – 2018» (INTELS 2018), St. Petersburg, Russia, 22–24 Oct 2018, pp. 1–9. Procedia Computer Science
5. El-Khatib, S.: Modified exponential particle swarm optimization algorithm for medical image segmentation. In: Proceedings of XIX International Conference on Soft Computing and Measurements (SCM 2016), St. Petersburg, 25–27 May 2016, vol. 1, pp. 513–516
6. Saatchi, S., Hung, C.C.: Swarm intelligence and image segmentation swarm intelligence. ARS J (2007)
7. Das, S., Abraham, A., Konar, A.: Automatic kernel clustering with a multi-elitist particle swarm optimization algorithm. Pattern Recogn. Lett. **29**, 688–699 (2008). Science direct

8. El-Khatib, S., Skobtsov, Y., Rodzin, S., Zelentsov, V.: Hyper-heuristical particle swarm method for MR images segmentation. In: Silhavy, R. (eds) Artificial Intelligence and Algorithms in Intelligent Systems. CSOC2018. Advances in Intelligent Systems and Computing, vol. 764, pp. 256–264. Springer, Berlin (2018)
9. Skobtsov, Y.A., Speransky, D.V.: Evolutionary computation: hand book. The National Open University "INTUIT", Moscow, 331 p. (2015). (in Russian)
10. He, J., Yao, X.: Drift analysis and average time complexity of evolutionary algorithms. In.: Artificial Intelligence, vol. 127, no. 1, pp. 57–85 (2001)

Proposal of Data Pre-processing for Purpose of Analysis in Accordance with the Concept Industry 4.0

Veronika Grigelova, Jela Abasova[✉], and Pavol Tanuska

Faculty of Materials Science and Technology in Trnava,
Slovak University of Technology in Bratislava, Bratislava, Slovakia
{veronika.simoncicova, jela.abasova,
pavol.tanuska}@stuba.sk

Abstract. The paper deals with a process of lowering amount of errors on automated screwing mechanism. Research focuses on analysis and individual evaluation of problematic robot screwing heads. The anticipated result is finding a problematic point of the whole process, what should lead to lowering the amount of errors in the process, shortening errors duration, and to improving effectiveness of predictive maintenance. Many errors occur in the screwing process due to not meeting the criteria of key parameters – adherence pressure, axial torque, screwing depth. There are additional parameters that may affect the process, like quality and chemical composition of material, its thickness, placing of the screw, etc. The paper focuses on identification of useful data sources, joining and pre-processing the downloaded data, and performing basic analysis thus preparing for analysis via DM methods. We will use the methods of KDD and Big Data, with respect to Industry 4.0 and CRISP-DM methodology.

Keywords: Screwing head · Big data · Pre-processing · CRISP-DM · Industry 4.0

1 Introduction

Companies today have to respond flexibly to rapid changes in fields of knowledge and technology. Globalization and growing demands on quality and quantity of production lead to the forming of Industry 4.0, or the fourth industrial revolution. It is characterized by complexity, digitalisation, and by gradational merging of biological, physical and digital realms. Industry 4.0 includes various technologies and methodologies, e.g. robotics, artificial intelligence, internet of things, biotechnologies, nanotechnologies, 3D print, and many others. In the field of data science, Industry 4.0 is manifested i. a. by methods of Big Data, Knowledge Discovery in Databases (KDD), and Data Mining (DM) [9].

For purposes of multi-dimensional analysis, Big Data, KDD and DM are preferred over traditional statistical methods. The concepts were developed for identifying patterns in large data sets. Traditional statistical methods are not well suited to evaluating the probability of coincidental patterns in high-dimensional data sets, that is why there was need for designing specific methods mentioned above [6].

© Springer Nature Switzerland AG 2019
R. Silhavy (Ed.): CSOC 2019, AISC 985, pp. 324–331, 2019.
https://doi.org/10.1007/978-3-030-19810-7_32

Big Data is a name for modern technologies aimed for operating huge data volumes (HDV). It is interchangeable, to a certain extent, with KDD, what is a complex process of getting potentially useful knowledge from HDV [8].

KDD has five phases (the phases can slightly differ depending on an expert). The first phase, selection, means examining available data sources, choosing the ones that meet criteria of the project, and downloading the data. Next comes pre-processing: identifying and selecting of useful data in every data set, choosing the common primary key, and merging the data sets. After that follows transformation, that are adjustments (textual, numerical etc.), cleaning, sorting of the data etc. The next phase is data mining, the process of getting potentially useful knowledge from the data sets and identifying patterns, that are not visible on first sight. The last part is interpretation/ evaluation: using the acquired data in the context of the problem, thus acquiring necessary knowledge about the object [1].

KDD and DM are a starting point of a standardized methodology approach – CRISP-DM. The methodology is based on practical experience of various data scientists with KDD and DM methods. This approach is standardized for all industries regardless of the origin of the data [10].

CRISP-DM is a robust methodology for operating HDV. It includes and describes six phases: business understanding, data understanding, data preparation, modeling, evaluation, and deployment [2].

Crisp-DM in Praxis. We dealt with a real-world company problem of lowering amount of errors on automated screwing mechanism. Many errors occur in the screwing process due to not meeting the criteria of key parameters – adherence pressure, axial torque, screwing depth. There are additional parameters that may affect the process, like quality and chemical composition of material, its thickness, placing of the screw, etc. The anticipated result was finding a problematic point of the whole process, what should lead to lowering the amount of errors in the process, shortening errors duration, and to improving effectiveness of predictive maintenance.

In the first phase of the project it was necessary to define the direction of the project and the problem that was going to be addressed. At this point we had to cooperate with experts to determine which aspects of the process should we focus on. In collaboration with the robot maintenance department and empirical experience of workers, we identified a specific problematic screwing head. We have added one more screwing head that is opposite to the previous one and performs the same operation and has a lower error rate, so that we can possibly compare waveforms. The analysis helped us determine the lengths and the number of errors in the screwing head.

Phase of business understanding, data understanding phase and initial preprocessing data partially overlap in time, especially with regard to the first analysis, which serve only to determine the faulty screwing head. After defining the business concepts, we focused on the phase of exploring the data.

The data comprehension phase includes a detailed view of the data available for datamining. This part of the analysis is critical to avoid unexpected problems in the next phase of the process.

2   Data Sources

The painting process is a complex one, with inputs from and outputs into various processes within the company. In the research phase we gathered great amount of information about various data sources. Where it was possible, we obtained a small sample of data; if the data seemed useful for the purpose of our project, we started downloading and/or get historical data from the source.

By the end of the data-gathering phase we had data from four sources for the purpose of further analysis:

A. Screwing heads errors:
The data source consists of automatically recorded values from the robots (screwing heads) of the whole production hall. It contains errors of the robots and their process status reports. We focus on the technical problems dealing with robot errors. When an error occurs, it makes a robot stop, and it will not start again until the alarm is cancelled by operator. That means seconds or event minutes after the error occurred, due to the distance of the operator. The solution also depends on severity of the error. The file contains information about time of the error (resp. occurring of error and cancelling of its alarm), duration, location/name of the robot and station, and note (further description). Exports can be made for a maximum of 6 months.

B. Errors of the connecting technology attached to the screwing heads:
The data source does not depend on the previous one. They are not time-synchronised, but if an error on the screwing head occurs, it will be recorded in both files. This data source includes further information about the error, resp. about IO ("In Order") and NiO ("Not in Order") processes that run on specific connecting technology (the screwing one). Data are stored in the following attributes:

- ID (program) – number/name of program used for screwing a specific contact (depends on the type of a product)
- ID (product) – number/name of product on which the screwing is executed
- Moment [Nm] – force by which the screw-bolt is screwed
- Angle – amount of turns made by screwing head during the process
- Analog depth – how deep the screw-bolt is withdrawn (negative value meaning the screw-bolt is above level of material, positive value means below the level)
- IO/NiO – whether the process ran correctly or not, IO = correct, NiO = error/alarm
- Duration – duration of screwing
- Cycle – amount of cycles from the last exchange of the screwing head
- Date – date and time of the record

C. Measure points:
The data source contains values of measure points on the products. The values are important for meeting the criteria of construction process. The source includes ID of a product and circa 200 measure points with exact name and description of location, and limiting values (trend borders for ideal values, alarm borders for acceptable values).

There is an additional attribute for every measure point, explaining whether the point is of big importance for the construction or not so important.

D. Product moves:
The data source consists of three parameters recording, when the product (characterised and recognizable by its ID) passes through the sensor. This is an additive data set, as it helps with joining/merging data another sets, and it also can be used for finding out which product was operated when the error occurred.

3   Data Pre-processing

Data had to be cleared from missing values and erroneous and inconsistent data, the formats of some parameters had to be modified and the correct time interval and the correct data sample had to be selected.

That is followed by a very important step of integrating data sets, which requires analytical tools and methods. This process cannot be performed without prior knowledge of data sets. Incorrect integration of data sets could lead to a degradation of data sources and could affect the overall quality of results and solutions.

Many data sources did not have a common unambiguous identifier. This problem was solved by integrating data sources using the available attributes and other available data resources. In order to link the available data samples, there was a need for another aid kit that determines the movement of the product along the production hall to accurately identify where and, most importantly, which product was in the robot where the error occurred. We used a script for this. The additional file contains only the time information and the unique identifier of the product.

I. Data transformation
In the process of data transformation, we mainly adjusted textual data, put numerical attributes into correct format and generated new attributes where necessary.

We executed several text adjustments, e.g. added names of units to the names of attributes, so that their values can be only numerical ones. Names of attributes had been written in three languages, so they were translated, renamed and unified.

Attributes that should be numerical were put into correct format by parsing numbers, guessing their type and by replacing decimal comma (used in two of the local languages as a decimal mark) by decimal point.

We decided to generate some new attributes. Units of values measured on the same robot with different (sometimes even with the same) programs may differ. Attributes "Type1", "Type2", and "Type3" are units of these measured values that help to identify such cases. Attributes "Robot", "Station", and "Hall" were got by division of the original "Location" attribute into three parts, what offers flexibility in further analysis. We also generated date/time attributes representing time units as month, week, day of week, hour etc.

II. Data cleaning

Before the data could be used for analysis, we had to clean them by removing missing and redundant values. Missing values may be connected to the problems with screwing heads that sometimes do not recorded at all. Redundancy usually originates in the fact the data are exported from various sources, but it may be due to the problems with recording as well. We removed also values from non-working days, because there is no production in the company during weekends and holidays. Still, some of the errors stay open during these days, what may lead to misrepresentation of the data. The last category that had to be deleted were supplementary attributes used to generating another attributes in previous phases of the process. Lastly, the attributes were sorted and organized to secure clarity.

4   Results

After the pre-processing phase we decided to perform basic analysis so we could choose the course of further analysis. They also serve as a verification of the previous phases. We designed a data mining model using analytic tool Rapid Miner, as can be seen in the figure (Fig. 1).

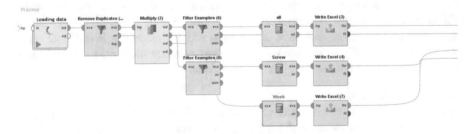

Fig. 1. Datamining model

We executed the basic analysis of the pre-processed data. In the figure (Fig. 2) you can see the amount of errors on individual screwing heads that occurred during six months (which is also the maximum time period for storing the data from screwing heads errors – source A). We analysed appearance of various error types during individual weeks.

We realised that the major problem of the screwing process is the screwing technique, so after that we focused mostly on the data source B, because it contains more detailed description of the attributes evaluated during the process. We chose to use 10 robots with the biggest number of errors. In the figure (Fig. 3) you can see overall duration of errors that occurred on them.

KDD process was applied into praxis in a real company problem solution. The process consisted of following steps:

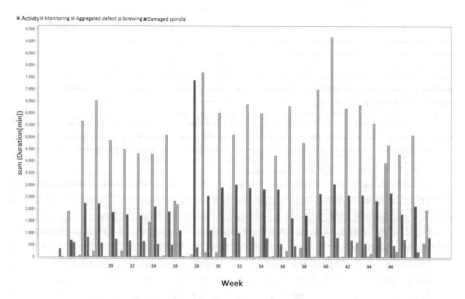

Fig. 2. Error analysis in accordance with type and time

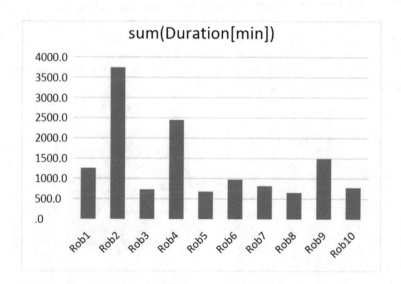

Fig. 3. Duration of errors for individual screwing heads

It begins with a phase of business understanding, where we defined an aim of the work. We chose the painting process in a real-world company. That is a complex process including various subprocesses, and connected to many other processes in the company.

Then it was necessary to get adequate data, what is a long and complicated part of the process. There are many data sources in the company, where data with different structure and quality are stored. We get available information about those sources, what includes obtaining sample of each data set. We decided to use 4 data sources: screwing heads errors, errors of connecting technology, measure points, and product moves. After deciding which data sets are useful for purposes of the project, we obtained historical data from every of the chosen sets. The longer history is available, the more precise analysis can be done.

Data of each set underwent pre-processing, which consisted of merging data while using a unique identifier, data transformation (text adjustments, putting numbers into correct format, generating new attributes), and data cleaning (filtering missing values, redundant values and non-working days).

In the next step we performed basic analysis on the merged, transformed and cleaned data. This analysis served as a verification of the previous phases. It produced statistics and charts, that provide a base for next steps of the project.

Presentation of results is an important phase, as it validates selection of the datasets and the data obtained from them. In this phase we confronted our results (represented by basic analysis of pre-processed data) with management and experts of the company. Specialised department evaluated the obtained results and provided us further advice. We also consulted our previous steps with the management and developed an idea of the next direction of the project.

The next step is application of results into praxis by performing analysis using various DM methods (e.g. neural networks, decision trees, etc.) then using the results (connections and patterns) to improve predictive maintenance, in searching for the sources of errors, and for advancing managerial decisions.

Based on these steps, we designed a process diagram that describes individual phases of adjusting data (Fig. 4):

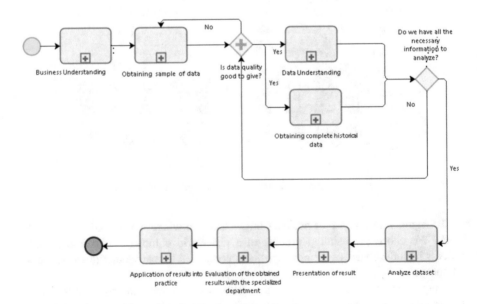

Fig. 4. Data pre-processing diagram

5   Conclusion

The paper dealt with the data pre-processing of four different data sources, what were very important steps before integrating data sets and there was applied KDD process into praxis in a real company. In the future we plan to use the verified data in further analysis via various DM methods (decision trees, neural networks, linear regression, polynomial regression etc.). Connections and patterns acquired during the process should be used for purpose of predictive maintenance and can serve as a basis for managerial decisions.

Acknowledgement. This publication is the result of implementation of the project VEGA 1/0272/18: "Holistic approach of knowledge discovery from production data in compliance with Industry 4.0 concept" supported by the VEGA.

References

1. Grimes, S.: Naming & classifying: text analysis vs. text analytics (2019). http://www.huffingtonpost.com/seth-grimes/naming-classifying-text-a_b_4556621.html6. Accessed 10 Jan 2019
2. The CRISP-DM model (2019). https://mineracaodedados.files.wordpress.com/2012/04/the-crisp-dm-model-the-new-blueprint-for-data-mining-shearer-colin.pdf. Accessed 10 Jan 2019
3. Gurusamy, V.: Preprocessing techniques for text mining (2019). https://www.researchgate.net/publication/273127322_Preprocessing_Techniques_for_Text_Mining. Accessed 10 Jan 2019
4. Gupta, G., Malhotra, S.: Text documents tokenization for word frequency count using rapid miner (Taking resume as an example). Int. J. Comput. Appl. ISSN 0975-8887 (2015). International Conference on Advancement in Engineering and Technology (ICAET 2015)
5. Akthar, F., Hahne, C.: RapidMiner 5 operator reference (2019). https://rapidminer.com/wp-content/uploads/2013/10/RapidMiner_OperatorReference_en.pdf. Accessed 10 Jan 2019
6. Harrison Jr., J.H.: Introduction to the mining of clinical data. Clin. Lab. Med. **28**, 1–7 (2008)
7. RapidMiner text mining extension (2019). http://www.predictiveanalyticstoday.com/rapidminer-text-mining. Accessed 10 Jan 2019
8. TECHOPEDIA INC. 2017. Knowledge Discovery in Databases (KDD). https://www.techopedia.com/definition/25827/knowledge-discovery-in-databases-kdd. Accessed 10 Jan 2019
9. Industry4.sk. (2019). https://www.industry4.sk. Accessed 10 Jan 2019
10. the-modeling-agency-com (2019). https://www.the-modeling-agency.com/crisp-dm.pdf. Accessed 10 Jan 2019

Predicting Regional Credit Ratings Using Ensemble Classification with MetaCost

Evelyn Toseafa[✉] and Petr Hajek

Institute of System Engineering and Informatics,
Faculty of Economics and Administration, University of Pardubice,
Studentska 84, 532 10 Pardubice, Czech Republic
evelyn.toseafa@student.upce.cz, petr.hajek@upce.cz

Abstract. Ensemble classifiers are learning algorithms that combine sets of base classifiers in order to increase their diversity and, thus, decrease variance and achieve better predictive performance compared to single classifiers. Previous research has shown that ensemble classifiers are more accurate than single classifiers in predicting credit ratings. Here we deal with highly imbalanced multi-class data of regional entities. To overcome these problems, we propose a novel hybrid model combining data oversampling and cost-sensitive ensemble classification. This paper demonstrates that the use of the SMOTE technique to balance the multi-class data solves the imbalance problem effectively. Different misclassification cost assigned in cost matrix solves the problem of ordered classes. This approach is combined with ensemble classification within the MetaCost framework. We show that more accurate prediction can be achieved using this approach in terms of average cost and area under ROC. This paper provides empirical evidence on the dataset of 451 regions classified into 8 rating classes, as obtained from the Moody's rating agency. The results show that Random Forest combined with MetaCost outperforms the rest of the base classifiers, as well as other benchmark methods.

Keywords: MetaCost · Random Forest · Ensemble learning · Regions · Credit rating

1 Introduction

Over the years, credit rating has been given considerable attention in the global financial markets since it has become the most typical way of measuring credit risk. Credit rating can be conducted at the level of firms, banks, regions, countries, etc. Usually the rating is given to a specific entity based on a thorough report of its financial history of creditworthiness realized by a rating agency. A rating class from a rating scale denotes the credit risk. Higher credit rating indicates a lower risk of the entity's default, thus giving investors the assurance about the safety of their loans and investments. This gives investors ideas about the credibility of the entity, and the risk factors attached to particular entity thereby enabling investors decide as to whether to invest or not. Rating agencies regularly review the ratings given to particular entities to enable present investors keep their loans or sell them [1]. Credit ratings are significant

© Springer Nature Switzerland AG 2019
R. Silhavy (Ed.): CSOC 2019, AISC 985, pp. 332–342, 2019.
https://doi.org/10.1007/978-3-030-19810-7_33

in the financial market especially when it comes to the distribution of funds by firms, government and other institutions.

A region's or country's credit rating is the evaluation of a government's ability and readiness to meet its payable obligations. Those ratings assess the possibilities of a borrower's failure to pay its obligations on time. Generally, governments inquire credit ratings to ease their own access and the access of other issuers domiciled within their borders to international capital markets, where lots of investors, mainly U. S. investors, choose rated securities over unrated securities of apparent similar credit risk. Again, these credit ratings give investors a deeper understanding of the level of risk involved by investing in a particular region or country. It is also important because it affects the domestic market operations that serve as the gateway for the flow of funds into the region or country. These rating plays a key role for banks to even cross boarders for funding to boost a country or region's economy. Besides, they are very important because some of the largest issuers in the international capital markets are regional or national governments which affect the ratings assigned to borrowers of the same nationality. The global financial market has brought about the interconnection between regions/countries and the sectors within them, making them more dependent on one another [2]. The key variable that affects a region's or country's position in the capital markets is generally the rating issued by the major international credit rating agencies, such as Moody's, Fitch, and standard and Poor's [3].

The predictive models of credit rating aim to model the rating process performed by the rating agencies to save time and provide a decision-support tool that generates a more robust and objective solution. The models can be generally divided into statistical methods such as logistic regression and artificial intelligence methods [4]. Lots of researchers have recently analysed the technique of combining the predictions of multiple classifiers to produce a single prediction model. The ensemble methods combine the predictions of several base machine learning algorithms in order to improve accuracy and robustness over single algorithms. The resulting classifier known as the ensemble is mostly more accurate than any of the single classifiers making up the ensemble. Both theoretical and empirical research established that a good ensemble is one where each classifier in the ensemble is both accurate and makes its errors on different parts of the input space [5]. This is why ensemble classifiers have become one of the most popular credit rating prediction methods [6, 7].

Previous research has been narrowed to the prediction of corporate and sovereign credit ratings, while neglecting the regional credit ratings. These represent a challenging problem due to many financial and economic factors affecting their assignment. Moreover, the rating classes are distributed in an imbalanced manner and different misclassification cost is associated with different wrong classification. Previous research has not approached these problems adequately and, therefore, here we aim to fill the gap by employing the MetaCost classifier combined with ensemble learning on the oversampled training dataset. Regional credit ratings of the Moody's rating agency from the year 2015 to 2016 are used to empirically verify our approach. To predict them, we used a dataset of financial and economic variables for 451 regions from 2015 as the input variables and 2016 as output variables. We hypothesize that ensemble classification with MetaCost classifier outperforms the previously used statistical methods. Hence, the central focus of this paper is to employ MetaCost classifier and ensemble classifiers to accurately predict the credit rating of regional entities. A cost matrix and area under ROC curve are used to compare the different models.

The remaining part of this paper is organized as follows. In the next section, the related literature is reviewed. Section 3 describes the dataset and presents the research methodology. Section 4 presents the empirical results and Sect. 5 concludes the paper.

2 Predicting Regional Credit Ratings – A Literature Review

A region's rating is more often the measure of each government's ability and readiness of rated governments to pay commercial debt obligations in full and on time [8, 9]. It is vital since it also affects the local market which is made up of companies.

Regional credit ratings give investors a deeper understanding of the level of risk involved in investing in a particular region. It has direct effect on corporate and financial ratings [10]. This is mostly destructive when a sovereign credit rating is downgraded below investment grade since this downgrade stimulates liquidation and deep price falls or cliff effects. Regional credit ratings play an important role in acquiring funds internally and externally on better terms to boost a country or region's economy. Any change in the credit rating is likely to affect the interest rates of the assets in another region due to economic and financial connections amongst these regions. Therefore, the credit ratings have complex impact on both within region and across regions.

Bennell et al. [11] used a neural network to predict the credit ratings of countries. They found that the neural network model predicts country credit ratings much more accurate than statistical methods. Several studies have examined the potential of ensemble classification that combines multiple base models [6].

Assessment of regional ratings is very paramount not merely for the sovereign country, but for other issuers in the country such as banks, corporations and public sector entities, this is as a result of the sovereign ceiling above which is impossible for other borrowers situated in the country. With developmental projects on the minds of governments to meet the needs of their citizenry, as well as score political points in the midst of limited access to internal funds, countries resort to international borrowing from international institutions. Such institutions should therefore have a reason of giving out credits and this can effectively be carried out with country and regional ratings. A study by [12] showed that sovereign and regional defaults are likely to be preceded by falls in private credit in countries and regions where banks are expected to hold a major portion of their assets in government bonds and financial institutions that are more developed. As a result of a rating downgrade, there has been a significant increase in wholesale debt and deposit funding costs because a portion of the increment is as a result of investors demanding for high compensation for taking on the credit risk.

Country and regional ratings are usually the assessment of each government's ability and willingness to service its debts in full and on time. Rating is a forward-looking evaluation of the default probability. Country and regional ratings address the credit risks of national (regional) governments but not the precise default risk of other issuers. Ratings given to other private and public sector entities in each country can, and often do vary. Ratings of some issuers could be the same as the country's, while others are a bit lower [13].

The important set of research places emphases on the determinants of credit ratings. The existing research on these subjects has been analysed in order to address the issues relating to the methodology applied, the data collection, the construction of indicators and variables best suitable for the research goals, and the understanding of the results [14, 15]. Looking over the years, Panetta et al. [14] present a unique comparison of country ratings and find that Standard and Poor's and Moody's ratings turned out to be slightly more accurate than those of Fitch.

3   Research Methodology

To evaluate the performance of the proposed prediction system, we acquired a collection of dataset of regional (sub-sovereign) units that has been collected over the period of 2015 and 2016, extracted from Moody's rating from 451 regions categorized into 8 rating classes. The prediction model was based on several categories of variables, including economic, financial, debt, unemployment etc. the study was designed to maximize prediction accuracy with ensemble classifiers.

Our research methodology was as follows:

Step 1: Input and output variables were extracted from Moody's financial reports from 2015 and 2016. Variables from 2015 represent the inputs of the models, while variable from 2016 represents the output of the models. The prediction model was based on several groups of variables, including economic, financial and debt indicators (Table 1). This is the set of variables used by the Moody's rating agency and represents the determinants of ratings used also in earlier research [8]. In Fig. 1, it can be seen that the regions were classified into 8 classes, ranked from Aaa to Ca. In addition, the histogram also shows that this was a highly imbalanced dataset.

Table 1. The list of input variables for credit rating prediction.

Variable	Mean	Variable	Mean	Variable	Mean
Population	2848.5	Debt/OR	73.07	Fin. surplus/TR	−4.61
GDP	100,037	Net debt/TR	68.19	Cash surplus/TR	−1.83
GDP per capita	104.10	Debt before swap	9.56	Borrowing need/TR	10.28
GDP/Nat.Avg.	27,056	Debt after swap	7.82	TE per capita	4,057
GDP (PPP)	27,881	Short-term debt/Debt	15.79	TE/GDP	14.71
Real GDP	1.96	Long-term debt/Debt	50.56	OB/OR	7.39
Unempl. rate	8.527	Maturity debt	8.43	Gross OB/OR	5.04
Nat. unempl. rate	8.95	Own revenue/OR	40.03	Net OB/OR	−1.63
Total debt	12,426	Govern. transfers/OR	40.37	Self-financing ratio	0.92
Net debt	10,895	Earmarked revenue/OR	25.81	Capital spending/TE	17.03
Net debt per capita	3,263	Interests/OR	2.52	TR-TE	−0.04
Debt/GDP	11.86	Debt service/TR	8.17	NWC/TE	14.7

Notes: GDP – gross domestic product, OR – operating revenue, TR – total revenue, PPP – purchasing power parity, TE – total expenditures, OB – operating balance, NWC – net working capital.

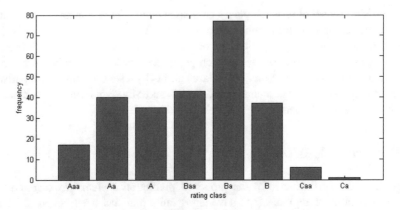

Fig. 1. Histogram of the rating classes

Step 2: 10-fold cross-validation was performed to obtain training and testing data.
Step 3: Oversampling was performed on training data using SMOTE technique to solve the imbalance problem. It is used to acquire a synthetically class-balanced or nearly class-balanced training set, which is then used to train the classifier [16]. In other words, after using SMOTE, all classes had the same frequency as the Ba class in the training data. Note that testing data were not used in this step and, thus, were highly imbalanced.
Step 4: Cost matrix was proposed to consider the ordinal character of the output class.
Step 5: MetaCost classifier takes into consideration the cost matrix to ensure better prediction. Here, we combined MetaCost with several base classifiers, including ensemble classifiers. The experiments were carried out in WEKA 3.8.
Step 6: Statistical tests were performed to detect significant differences between the ROC and average cost of the models.

Here we also present the brief description of the methods used for the prediction. In this paper, we propose to use MetaCost, a principled method for making an arbitrary classifier cost-sensitive by wrapping a cost-minimizing technique around it. This technique treats the fundamental classifier as a black box, requiring no knowledge of its functioning or change to it. MetaCost is applicable to any number of classes and to arbitrary cost matrices. Realistic trials on a large group of target databases show that MetaCost almost always yields large cost cutback compared to the cost-blind classifier used. The MetaCost operator makes its base classifier cost-sensitive by using the cost matrix specified. MetaCost relies on an internal cost-sensitive classifier in order to relabel classes of training examples [17]. The MetaCost operator is a nested operator and has a sub-process. The sub-process must have a learner as the operator that expects an example set and generates a model. This operator tries to build a better model using the learner provided in its sub-process. MetaCost [17] is based on the Bayes optimal prediction that reduces the expected cost $R(j|x)$ [18]:

$$R(j|x) = \sum_i^I p((j|x)cost(i,j)), \qquad (1)$$

where $P(i|x)$ is the probability of class i given example x and cost (i, j) is the cost of misclassifying a class i example as class j. The Bayes optimal prediction rule implies a division of the example space into I classes, such that class i is the minimum expected cost prediction in class i. If misclassifying class i becomes more expensive relative to misclassifying others, then parts of the former non-class i regions shall be re-allocated as class i since it is now the minimum expected cost prediction.

Table 2 illustrates the 8 rating classes of the rating agencies used as the cost matrix to determine the cost sensitivity. Note that for the sake of simplicity, the cost was linearly increasing with the number of misclassified classes.

Table 2. Cost matrix for credit ratings.

	Rating class			
	Aaa	Aa	...	C
Aaa	0	1	...	8
Aa	1	0	...	7
...	...	...	...	...
C	8	7	...	0

As the base classifiers in MetaCost, we used the following methods: AdaBoost, LogitBoost, Bagging, Voting, Decorate, Random Forest, Naïve Bayes, Bayes networks, Support Vector Machine (SVM), Logistic Regression and J48 decision tree. These methods represent a set of benchmark algorithms used previously in the related credit rating literature [19–23].

Lately, AdaBoost is tagged to as discrete and used for classification, rather than regression. It can be used to improve the performance of any machine learning algorithm such as decision trees. These trees are very short and contain only one decision for classification and are therefore called decision stumps. Each instance in the training dataset is weighted. The initial weight is set to: $weight(x) = 1/n$, where x is the training example and n is the number of training examples. AdaBoost is best used to boost the performance of decision trees on binary classification problems. LogitBoost performs additive logistic regression and, thus, can effectively handle multi-class problems.

Bagging, also called Bootstrap Aggregating, reduces the variances of forecasts by creating supplementary dataset for training novel dataset using combinations with recurrences to produce multisets of the same size as the novel data. It is used to generate multiple forms of prediction. When the size of the training set is augmented, the predictive force cannot be improved, but when the variance is reduced, it barely turns the prediction to the expected result, and then use the new dataset to combine predictor. The Decorate ensemble method represents an alternative approach by constructing artificial training data and, thus, increasing diversity of base classifiers. Therefore, it is suitable for smaller datasets in particular.

Voting ensembles can be used for classification and regression problems. It selects multiple sub-models and allows another model to specify and learn how best to combine the predictions from the sub-models. It uses a Meta model for the best combination of the predictions of sub-models, this technique is sometimes called blending, as it blends its predictions together. The build classifier selects a classifier from the set of classifiers by error on the data, instances serve as training data. Training data are used to generate the boost classifier.

Random Forest is one of the most popular and powerful machine learning algorithms. Its algorithm makes a small twist to Bagging and turns it in to a very powerful classifier. It is also ensemble learning method for classification, regression and other tasks that operate by building a multitude of decision tree training time and outputting the class that is of the mode or the mean prediction of the individual tree.

Naïve Bayes is a classification method based on Bayes' Theorem with an expectation of individuality among predictors. Naive Bayes classifier assumes that the presence of a particular variable in a class is not similar to the presence of any other variable. Even if these variables influence each other upon the existence of the other features, all of these properties independently contribute to the probability. Besides its simplicity, Naive Bayes is known to outperform even highly sophisticated classification methods. A Bayesian network (Bayes Net) model is a probabilistic graphical model which is a kind of statistical model that denotes a set of random variables and their conditional dependencies via a directed acyclic graph.

The J48 classification algorithm is the process of building models of class from a set of records that contain class labels. The decision tree algorithm is used to find out the way the variables vector behaves for a number of examples, and also on the basis of the training examples the classes newly generated examples.

The Sequential Minimal Optimization (SMO) was used to train SVM. SMO is an iterative algorithm used for solving quadratic programming (QP) problems. One example where QP problems are relevant is during the training process of SVM.

4   Experimental Results

In this section, the description of all the settings of the experiments are outlined and the results are presented in terms of two prediction measures: average cost and area under ROC (receiver operating characteristics). Average cost was calculated as the sum of misclassification cost from Table 2 divided by the number of examples. 10-fold cross-validation procedure was used to avoid overfitting of the machine learning methods. The training data were first pre-processed with the SMOTE technique to solve the minority classes and oversampling of the examples. The MetaCost operator makes its base classifier cost-sensitive by using the cost matrix specified in Table 2.

To train MetaCost, 10 iterations were used in the learning process in all the base ensemble classifiers. The settings of the base classifiers were as follows. The size of each bag in Bagging was 100. AdaBoost M1 was performed with 10 iterations and decision stump as base learner; and 100 random trees were used in random forest with the learning rate of bagging and random subspace was also performed in 10 iterations. LogitBoost was also trained with decision stump as base learner and the Z max threshold for responses was set to 3. The J48 algorithm was used as base learner in the Decorate ensemble method, the desired number of classifiers in the ensemble was 15 and the ensemble diversity was set to 1.0. In Random Forest, maximum depth of trees was unlimited, the number of trees to be generated was 100 and the number of variables randomly sampled as candidates at each split was calculated as $\log_2$(#predictors) + 1. For Voting, the combination of ensemble classifiers was Bayes Net, Naïve Bayes, SVM, Random Forest and J48. The J48 algorithm was trained with the confidence factor of 0.25 and the minimum number of examples per leaf of 2. The K2 algorithm was used to explore the search space of Bayes Net with a simple estimator. SVM was trained using polynomial kernel function and complexity, ranging from $C = 1$ to $C = 128$ using the grid search procedure.

The results of the experiments are summarized in Table 3. The results presented show the average results and their standard deviations over 10 validation experiments. A high value of area under ROC (close to 1) indicates that the algorithm performed well on imbalanced classes. The results in Table 3 indicate that MetaCost with Random Forest base classifier outperformed the other classifiers with an average cost value of 0.492 and ROC value of 0.886, while the Decorate ensemble classifier was the second best with an average cost value of 0.584 and ROC value of 0.856. Bagging also attained an average cost of 0.635 and an ROC values of 0.823. Besides the above classifiers, Logistic Regression and Bayes Net performed better than Naïve Bayes and J48 classifiers. AdaBoost M1 had a high average cost of 1.152 and a low value of 0.594 for the ROC. This can be attributed to the poor performance of Decision Stump used as its base classifier. Overall, ensemble classifiers utilizing bagging performed better than those using boosting or simple voting procedures. This can be attributed to the variety of regions used in the dataset. In fact, the regions come from many countries in the world, such as Mexico, Italy, Canada or Russia. To handle this variety, it is apparently more effective to learn the base classifiers on the smaller samples of the regions and only then merge those classifiers into a Meta framework. The empirical hypotheses stated in the introduction signifies from the results of the experiment that MetaCost classifier combined Random Forest performed significantly better than the other ensemble classifiers at $P < 0.05$ (using Wilcoxon signed rank test).

To compare the best model with a benchmark, we chose the Random Forest trained in a cost-sensitive setting. This can be considered to be a state-of-the-art method used in this domain [18–20]. Random Forest trained in this mode also performed well with an average cost value of 0.515 and an ROC value of 0.885, thus was not outperformed by MetaCost + Random Forest significantly. However, substantial reduction in cost (almost 5%) achieved by MetaCost + Random Forest compared with the benchmark method can be important for investors. Overall, we can conclude that the model proposed in this study can help predict credit ratings of region more accurately in terms of cost and area under ROC.

Table 3. Results of the experiments – misclassification cost and area under ROC curve.

Method	Cost	Area under ROC
MetaCost + AdaBoost M1	1.635 ± 0.213	0.594 ± 0.045
MetaCost + Bagging	0.635 ± 0.082	0.823 ± 0.031
MetaCost + Voting	0.650 ± 0.080	0.838 ± 0.033
MetaCost + LogitBoost	0.643 ± 0.107	0.837 ± 0.018
MetaCost + Decorate	0.584 ± 0.067	0.856 ± 0.021
MetaCost + Random Forest	**0.492 ± 0.063**	**0.886 ± 0.021**
MetaCost + SVM	0.843 ± 0.119	0.611 ± 0.022
MetaCost + Bayes Net	0.775 ± 0.070	0.792 ± 0.033
MetaCost + Naïve Bayes	0.804 ± 0.087	0.743 ± 0.047
MetaCost + Logistic Regression	0.641 ± 0.077	0.836 ± 0.029
MetaCost + J48	0.708 ± 0.145	0.733 ± 0.047
Cost-sensitive Random Forest	0.515 ± 0.081*	0.885 ± 0.019*

*performs significantly similar as the best at $P < 0.05$ (Wilcoxon signed rank test)

5 Conclusion

The results from the experiments showed that ensemble learning algorithms with MetaCost classifier combined with Random Forest performed best among the competing methods. The results indicate that MetaCost with Random Forest outperformed the other single and ensemble classifiers significantly. On the other hand, Random Forest used as a single classifier achieved better results than the other ensemble classifiers trained in MetaCost framework. Therefore, RandomForest was confirmed to be the benchmark method in this domain. Moreover, when combined with MetaCost, it can provide substantial savings in misclassification cost. This can have huge impact on bond investors and affect the interest rates associated with those bonds.

The good performance on this imbalanced dataset can be attributed to the SMOTE oversampling method that reduced the risk of overfitting. It is therefore recommended to combine oversampling techniques with cost-sensitive learning to achieve good performance for all classes. In future, the performance should be tested for a long-term rating prediction. Another limitation of this study was the linear cost matrix. This can be adapted to better consider the actual cost (interest rates and default costs) of rating misclassification. We recommend this for future research. Alternatively, the cost matrix can defined by the group of experts.

Acknowledgments. This article was supported by the grant No. SGS_2019_17 of the Student Grant Competition.

References

1. Zhong, H., Miao, C., Shen, Z., Feng, Y.: Comparing the learning effectiveness of BP, ELM, I-ELM, and SVM for corporate credit ratings. Neurocomputing **128**, 285–295 (2014)
2. Reusens, P., Croux, C.: Sovereign credit rating determinants: a comparison before and after the european debt crisis. J. Banking Finance **77**, 108–121 (2017)
3. Gaillard, N.: The determinants of Moody's sub-sovereign ratings. Int. Res. J. Finance Econ. **31**(1), 194–209 (2009)
4. Zhu, Y., Xie, C., Wang, G.J., Yan, X.G.: Comparison of individual, ensemble and integrated ensemble machine learning methods to predict China's SME credit risk in supply chain finance. Neural Comput. Appl. **28**(1), 41–50 (2017)
5. Opitz, D., Maclin, R.: Popular ensemble methods: an empirical study. J. Artif. Intell. Res. **11**, 169–198 (1999)
6. Finlay, S.: Multiple classifier architectures and their application to credit risk assessment. Eur. J. Oper. Res. **210**(2), 368–378 (2011)
7. Toseafa, E.: Using meta learning methods to forecast sub-sovereign credit ratings. J. East. Eur. Res. Bus. Econ. **2018**, 1–12 (2018)
8. Gaillard, N.: Determinants of Moody's and S&P's Subsovereign Credit Ratings. World Bank, Fondation Nationale Des Sciences Politiques, France (2006)
9. Bhatia, M.A.V.: Sovereign Credit Ratings Methodology: An Evaluation. International Monetary Fund, New York (2002)
10. Beck, R., Ferrucci, G., Hantzsche, A., Rau-Goehring, M.: Determinants of sub-sovereign bond yield spreads – the role of fiscal fundamentals and federal bailout expectations. J. Int. Money Finance **79**, 72–98 (2017)
11. Bennell, J.A., Crabbe, D., Thomas, S., Ap Gwilym, O.: Modelling sovereign credit ratings: neural networks versus ordered probit. Expert Syst. Appl. **30**(3), 415–425 (2006)
12. Borensztein, M.E., Valenzuela, P., Cowan, K.: Sovereign Ceilings Lite? The Impact of Sovereign Ratings on Corporate Ratings in Emerging Market Economies (EPub). International Monetary Fund (2007)
13. Eijffinger, S.C.W.: Rating agencies: role and influence of their sovereign credit risk assessment in the eurozone. J. Common Mark. Stud. **50**(6), 912–921 (2012)
14. Panetta, F., Correa, R., Davies, M., Di Cesare, A., Marques, J.M., Nadal de Simone, F., Wieland, M.: The Impact of Sovereign Credit Risk on Bank Funding. MPRA Paper (2011)
15. Flandreau, M., Flores, J.H.: Bonds and brands: foundations of sovereign debt markets, 1820–1830. J. Econ. Hist. **69**(3), 646–684 (2009)
16. Chawla, N.V., Bowyer, K.W., Hall, L.O., Kegelmeyer, W.P.: SMOTE: synthetic minority over-sampling technique. J. Artif. Intell. Res. **16**, 321–357 (2002)
17. Domingos, P.: MetaCost: a general method for making classifiers costsensitive. In: Proceedings of the Fifth International Conference on Knowledge Discovery and Data Mining, pp. 155–164. ACM Press (1999)
18. Michie, D., Spiegelhalter, D.J., Taylor, C.C.: Machine Learning, Neural and Statistical Classification. Prentice Hall, Upper Saddle River (1994)
19. Hajek, P., Michalak, K.: Feature selection in corporate credit rating prediction. Knowl.-Based Syst. **51**, 72–84 (2013)
20. Abellan, J., Mantas, C.J., Castellano, J.G.: A random forest approach using imprecise probabilities. Knowl.-Based Syst. **134**, 72–84 (2017)

21. Zhu, Y., Xie, C., Wang, G.J., Yan, X.G.: Comparison of individual, ensemble and integrated ensemble machine learning methods to predict China's SME credit risk in supply chain finance. Neural Comput. Appl. **28**(1), 41–50 (2017)
22. Gmehling, P., La Mura, P.: A Bayesian inference model for the credit rating scale. J. Risk Finance **17**(4), 390–404 (2016)
23. Jones, S., Johnstone, D., Wilson, R.: An empirical evaluation of the performance of binary classifiers in the prediction of credit ratings changes. J. Banking Finance **56**, 72–85 (2015)

Human Activity Identification Using Novel Feature Extraction and Ensemble-Based Learning for Accuracy

Abdul Lateef Haroon P.S[✉] and U. Eranna

Department of Electronics and Communication Engineering,
Ballari Institute of Technology and Management, Ballari, Karnataka, India
abdulbitm@gmail.com

Abstract. The area of human activity recognition is gaining momentum with the rise of smart appliances towards tracking and monitoring human behavior system. Till last decade, there have been various works being carried out towards building such a robust system that has led its way to commercial products too. However, after an in-depth investigation, it was found there is a far way to go in order to build up a true and dependable recognition system. Therefore, the proposed system introduces a novel framework meant for human activity recognition system with the sole target to enhance the precision factor in the identification process. A simplified feature extraction process has been introduced in this work that after being subjected to ensemble-training approach is found to improve the identification performance significantly. The study outcome shows better accuracy as well as good system performance.

Keywords: Human activity recognition · Activity identification system · Ensemble learning · Feature extraction

1 Introduction

With the increasing usage of smart appliances, tracking and monitoring operations have become quite inevitable in almost all the pervasive devices [1, 2]. One such operation is called as human activity recognition system which is about tracking any specific mobility pattern [3]. A different application of it is already on the commercial markets, e.g. health care, video surveillance, and interaction between human and smart appliance (computing system) [4]. With the upgradation of modern imaging technologies, there is also an improvement in the process of capturing the feed of human-based activities. However, there are various challenges associated with such activity recognition system e.g. variations of intra-classes and similarity of inter-classes, interaction of multiple subjects, background complexity, group activity, etc. [5] There have been various attempts towards usage of machine learning-based approach for solving classification problems in activity-recognition system, e.g. neural network, deep learning, support vector machine, etc. [6, 10]. However, very less work has considered accuracy as the prominent performance parameters. Apart from this, accuracy in the activity recognition system is not an easy task to be accomplished as various parameters control

© Springer Nature Switzerland AG 2019
R. Silhavy (Ed.): CSOC 2019, AISC 985, pp. 343–352, 2019.
https://doi.org/10.1007/978-3-030-19810-7_34

accuracy, and one prominent parameter in this regard is a possible feature. One of the significant factors that are ignored in the majority of the existing approaches is associated with the identification of the index attribute in the human body that is associated with mobility and displacement. If such index attributes are identified, then it offers enriched information towards feature extraction.

Therefore, the prime objective of the proposed research work is to offer simplified modeling of human activity recognition system where the core emphasis is given to the feature extraction process and accuracy. Existing research-based solutions towards activity recognition are briefed in Sect. 2 followed by highlighting of the research problem in Sect. 3. Section 4 introduces about the adopted methodology followed by Sect. 5 discussing the algorithm implementation. Discussion of results obtained is carried out in Sect. 6 while the conclusion is briefed in Sect. 7.

2   Related Work

Our prior review work has offered a brief idea about the research work towards an activity recognition system [11]. The most recent work of Chen et al. [12] has used a convolution neural network considering the distance between joints of inter-frames for performing activity recognition system over the new dataset. The neural network was also adopted by Chen et al. [13] for extracting more information from the event called as knowledge distillation. However, it has been seen that dataset offers importance towards recognition system. Adoption of the sensory dataset is found more frequent Franco et al. [14]. There are also good numbers of research where recognition of gesture and activity is captured from Kinect Gavrilova et al. [15]. The work of Hegde et al. [16] has presented an automatic mechanism of daily activity recognition using sensory data. A similar line of research is also carried out by Hsu et al. [17] where principal component analysis is used for dimensional reduction. Khalifa et al. [18] have introduced a unique study where recognition is carried out from energy harvesting data. Khan et al. [19] have presented a correlational factor as a feature from the activity dataset captured using sensors using body area network. Usage of Hidden Markov Model has been used for constructing a framework for activity recognition as seen in the work of Li et al. [20] along with usage of k-means clustering for feature extraction. Plotz and Guan [21] have discussed the significance of deep learning for activity recognition system associated with the mobile computing system. Recognition system over online data has been reported by Qi et al. [22] considering a dictionary learning mechanism. Hidden Markov model and a supervised classification technique are also reported to be used in the activity recognition system Sok et al. [23]. It was also reported in the existing system that in-depth learning-based training approach could be used for image-based activity recognition system Tan et al. [24], Wang et al. [25]. Activity recognition on the basis of smart-based data is discussed by Wang et al. [26]. Usage of the network, e.g., WiFi signal and radio-frequencies were also reported to be used in human activity recognition Wang et al. [27] and Wenyuan et al. [28]. Unique recurrent modeling is presented using dilation operation for activity recognition system Xi et al. [29]. Other unique research implementations are also witnessed in the work of Yang et al. [30] and Yao et al. [31]. The next section discusses the research problem.

3 Problem Description

A closer look into the existing approaches will show that the majority of the existing research approaches towards activity recognition system is carried out over sensory data and not much on images. Although, this mechanism offers simplified approaches, they were never accessed for their accuracy rate. Apart from this, the test environment is only restricted to the static dataset of activity recognition, and no much analysis has been carried out for online dynamic real-time feeds of events. Usage of the neural network is too much abundant and not much flexibility is offered towards classification problems associated with human activities. Therefore, the classifying the real-time feed is one challenging aspect in activity recognition.

4 Proposed Methodology

The proposed work is an extension of our prior work [32] towards an activity recognition system where the current work completely emphasizes improving the accuracy of the identified activity. The proposed system identifies all the activities on the basis of the limb movement of the human body, and hence better accuracy can be maintained. Therefore, the proposed system hypothesizes that if the major limb movements can be identified and recognized than any form of application-oriented human activity can be developed commercially.

Figure 1 highlights the mechanism of taking the event feed as an input which undergoes an extraction process of its index points of mobility along with depth map. All the significant mobility points associated with the subject is retained in a matrix from where temporal and spatial information has been used for extracting features. The extracted features are then subjected to ensemble learning mechanism in order to perform activity classification.

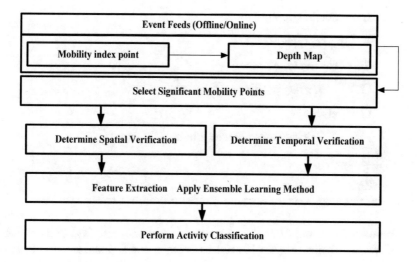

Fig. 1. Schematic architecture of proposed system

5 Algorithm Implementation

The algorithm of the proposed system is designed for precise identification of the human activity with the aid of a simple ensemble learning mechanism. The idea of using ensemble learning mechanism is targeted for offering accuracy in the identification process. In order to incorporate a greater degree of accuracy in the process of the activity recognition system, it is necessary that every smaller piece of information associated with the event should be recognized. Therefore, the proposed system implements a mechanism where the features are built up from spatial and temporal attributes connected to the real-time events in the proposed system. The algorithm takes the input of real-time event video in the form of I_s (input sample) which after processing yields an outcome of s_{feat}(sampled feature). The step involved in the proposed algorithm is as follows:

Algorithm for activity recognition using ensemble learning

Input: I_s (input sample)

Output:s_{feat}(sampled feature)

Start

1. init I_s,

2. **For**i=1:length(s_{db})

3. $I_{dep} \rightarrow s_{db}(i)$ //image depth

4.   Obtain L_{cor}, $R_{cor} \rightarrow (x,y,z)$

5. $DA \rightarrow \sqrt{(\sum(L_i-R_i)^2)}$, where i=x,y,z

6. $a \rightarrow f(P1r, rP2)$

7. $A \rightarrow a.180/pi$

8.   obtain $s_{feat} \rightarrow [A]$

9. $[p_{op} \ score] \rightarrow f_2(s_{feat})$

10.   **If** score(p_{op})>-*thres*

11. $R \rightarrow$ flag positive recognition

12.   **else**

13. $R \rightarrow$ flag negative recognition

13. End

End

Algorithm Step Discussion: For simpler processing of real-time feed of an event, the proposed system digitizes the feed in order to obtain it as I_s (input sample) which has both spatial and temporal information (Line-1). The next part of the algorithm is to obtain significant features from all the spatial data (s_{db}), which is further used for extracting the image depth I_{dep} (Line-3). In this stage, both the colored version of the core skeleton as well as image depth is obtained. The colored image S is obtained from the data structure of s_{db} while image depth I_{dep} is obtained from Line-3. The next part of the algorithm is focused on obtaining various potential features associated with the ankle and wrist as they are a prominent indicator of any significant activity. The proposed algorithm obtains the three-dimensional coordinates system from both left and right ankle (Line-4). A Euclidean distance is applied for computing spatial the differences between the left and right ankle (Line-5) considering all the three dimensional coordinates. Similar algorithmic steps, i.e. Line-4and Line-5 is used for computing the Euclidean distance between the left and the right wrist too. Although, the logic focuses on the identifying the spatial attribute of the displacement considering the significant limbs, e.g. ankle and wrist, however, the temporal attribute of the displacement can be known if the orientation measure is considered for the event captured. Therefore, the algorithm performs the computation of the orientation measure of another limb, i.e. both the elbow. The algorithm computes the effective displacement of both the elbows viz. P1r and rP2 and applies a function $f(x)$ on this to obtain the actual orientation factor associated with the elbow (Line-). The function $f(x)$ computes inverse tangent of four quadrants over two variable viz. (i) variable-1 for normalized cross product of P1r and rP2 and (ii) dot product of P1r and rP2. Finally, the orientation score is obtained A (Line-7). This mechanism is performed for both elbow and legs orientation to finally obtain all the features sfeat from A (Line-8). The algorithm applies another function $f_2(x)$ over the obtained features to extract the final set of features (Line-9). However, the obtained feature is just for feature identification that directly assists in the activity recognition system. The operation carried out till this stage ensure 70% of accuracy owing to the adoption of spatial and temporal based features; therefore, the proposed system upgrade the function $f_2(x)$ using ensemble learning based approach.

There are various justifications of using ensemble-based learning approach. Adoption of this approach allows the implementation of multiple forms of machine learning approaches for the purpose of efficient predictive recognition of human activity. The proposed system performs prediction of data with respect to the data history, and hence all the factors (spatial as well as temporal aspects) are considered while performing the training operation. When this function is applied on a set of features, the training yields predictive operator p_{op} and its respective score *score* (Line-9). After obtaining the value of the score, it is compared with the cut-off value *thres* where the positive value will ensure perfect recognition R of the activity (Line-11) of the event while the negative value will call for negative recognition R (Line-13). A closer look into this step of operation will show that this is meant for improving the accuracy of the recognition capability. Although the process is bit iterative in this regards, interestingly it offers the capability to reduce down the error-prone identifications. According to the algorithm, if the process does find a perfect match with its trained dataset with respect to the spatial and temporal factor, it will mean that the subject display a very different movement and hence the recognition system captures that the subject is different and distinct.

6   Results Discussion

The prime strategy to analyze the outcome is to assess the mobility of the subject in order to determine their precise activity identification. The study considers approximately 20 skeletal joints. Using the similar test environment of our prior experiment [32], the proposed system has captured a 20 × 20 matrix associated with all the sequences of the individual video. In this system, all the cells in the matrix depict the sectors that are occupied by histograms. Different numbers of subjects are considered for this experiment where different video sequences have been captured and stored. The scripting of the proposed logic was carried out in MATLAB, and it offers supportability of both live as well as offline analysis. After the capturing of the event has been carried out, the proposed system computes all the essential mobility aspects associated with the ankle, elbow, wrist, leg, etc. Figure 2 highlights the computation of the ankle and wrist displacement.

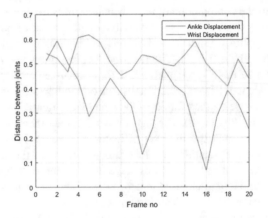

Fig. 2. Wrist and ankle displacement calculation

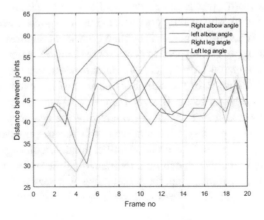

Fig. 3. Leg and elbow calculation

Figures 2 and 3 shows that the proposed system is capable of tracking all the minor and insignificant movement of the subject too. The system also tracks information associated with ankle, wrist, elbows, and leg as shown in Fig. 4.

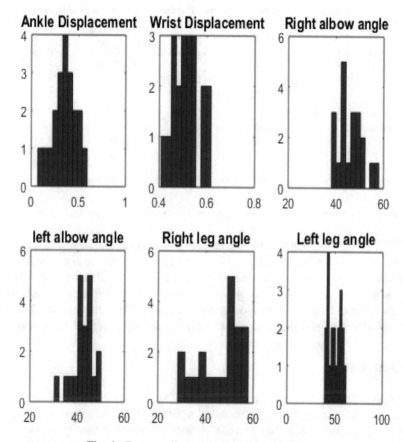

Fig. 4. Event attribute detection using histogram

The proposed system also estimates the mean matrix as well as stream map as exhibited in Fig. 5 that offers the comprehensive evidence of accuracy obtained for all the activity identification.

Fig. 5. Matrix representations for different histograms

The idea of the proposed system is assessed with respect to the existing approaches of activity recognition system, e.g. Zdraveski et al. [33], Nakagawa et al. [34], and Henpraserttae et al. [35]. The summarized outcome is highlighted in Table 1.

Table 1. Comparative analysis of activity recognition system

Approaches	Practicality	Computational complexity	Accuracy
Zdraveski et al. [33]	Not accessed	Not accessed	92–95%
Nakagawa et al. [34]	Highly dependent on massive data for identification	Not accessed	75.122%
Henpraserttae et al. [35]	Doesnt consider data variability	Not Accessed	90%
Proposed system	Applicable to both live and offline analysis	Memory consumption: 10–15% Time consumption: 0.5341 s	98%

It should be noted that all the existing system are experimented and analyzed on the basis of the standard dataset. Although, it is least important about the format of the data, however, analysis over static dataset doesn't prove the practicality of the existing solution. Moreover, the existing concepts towards activity recognition are quite hypothetical as they were never assessed for their computational complexity. However, the proposed system bridges this gap and proves that it maintains a good balance between accuracy and its other associated performance. The performance of the result is the same for both real-time captures of subject mobility as well as static frames too.

7 Conclusion

This paper discusses the novel and simple modeling of the activity recognition system considering the case study of human activity. The study outcomes prove the applicability of the model toward the real-life scenario. The study also presents a unique extraction of features that always ensures that all the information associated with the spatial as well as temporal factors is considered in the analysis of the activity identification system. The study outcome shows that the proposed system performs much better as compared to the existing approaches of the activity recognition system.

References

1. Pham, C.: MobiRAR: real-time human activity recognition using mobile devices. In: 2015 Seventh International Conference on Knowledge and Systems Engineering (KSE), Ho Chi Minh City, pp. 144–149 (2015)
2. Fortin-Simard, D., Bilodeau, J., Gaboury, S., Bouchard, B., Bouzouane, A.: Human activity recognition in smart homes: combining passive RFID and load signatures of electrical devices. In: 2014 IEEE Symposium on Intelligent Agents (IA), Orlando, FL, pp. 22–29 (2014)

3. Fu, Y.: Human Activity Recognition and Prediction. Springer, Heidelberg (2015)
4. Guesgen, H.W.: Human Behavior Recognition Technologies: Intelligent Applications for Monitoring and Security: Intelligent Applications for Monitoring and Security. IGI Global, Hershey (2013)
5. Zhang, S., Wei, Z., Nie, J., Huang, L., Wang, S., Li, Z.: A review on human activity recognition using vision-based method. Hindawi J. Healthcare Eng. **2017**, 31 (2017)
6. Cheng, L., Guan, Y., Zhu, K., Li, Y.: Recognition of human activities using machine learning methods with wearable sensors. In: 2017 IEEE 7th Annual Computing and Communication Workshop and Conference (CCWC), Las Vegas, NV, pp. 1–7 (2017)
7. Marinho, L.B., de Souza Júnior, A.H., Rebouças Filho, P.P.: A new approach to human activity recognition using machine learning techniques. In: International Conference on Intelligent Systems Design and Applications (ISDA), At Porto, Portugal (2016)
8. Lara, O.D., Labrador, M.A.: A survey on human activity recognition using wearable sensors. IEEE Commun. Surv. Tutorials **15**(3), 1192–1209 (2013)
9. Vrigkas, M., Nikou, C., Kakadiaris, I.: A review of human activity recognition methods. Front. Robot. Artif. Intell. **2** (2015). https://doi.org/10.3389/frobt.2015.00028
10. Ann, O.C., Theng, L.B.: Human activity recognition: a review. In: 2014 IEEE International Conference on Control System, Computing and Engineering (ICCSCE 2014), Batu Ferringhi, pp. 389–393 (2014)
11. Lateef Haroon P.S, A., Eranna, U.: Insights on research-based approaches in human activity recognition system. Commun. Appl. Electron. **7**(16), 23–31 (2018)
12. Chen, Y., Yu, L., Ota, K., Dong, M.: Robust activity recognition for aging society. IEEE J. Biomed. Health Inform. **22**(6), 1754–1764 (2018)
13. Chen, Z., Zhang, L., Cao, Z., Guo, J.: Distilling the knowledge from handcrafted features for human activity recognition. IEEE Trans. Ind. Inform. **14**(10), 4334–4342 (2018)
14. De-La-Hoz-Franco, E., Ariza-Colpas, P., Quero, J.M., Espinilla, M.: Sensor-based datasets for human activity recognition – a systematic review of literature. IEEE Access **6**, 59192–59210 (2018)
15. Gavrilova, M.L., Wang, Y., Ahmed, F., Polash Paul, P.: Kinect sensor gesture and activity recognition: new applications for consumer cognitive systems. IEEE Consumer Electron. Mag. **7**(1), 88–94 (2018)
16. Hegde, N., Bries, M., Swibas, T., Melanson, E., Sazonov, E.: Automatic recognition of activities of daily living utilizing insole-based and wrist-worn wearable sensors. IEEE J. Biomed. Health Inform. **22**(4), 979–988 (2018)
17. Hsu, Y., Yang, S., Chang, H., Lai, H.: Human daily and sport activity recognition using a wearable inertial sensor network. IEEE Access **6**, 31715–31728 (2018)
18. Khalifa, S., Lan, G., Hassan, M., Seneviratne, A., Das, S.K.: HARKE: human activity recognition from kinetic energy harvesting data in wearable devices. IEEE Trans. Mob. Comput. **17**(6), 1353–1368 (2018)
19. Khan, M.U.S., et al.: On the correlation of sensor location and human activity recognition in body area networks (BANs). IEEE Syst. J. **12**(1), 82–91 (2018)
20. Li, W., Tan, B., Xu, Y., Piechocki, R.J.: Log-likelihood clustering-enabled passive RF sensing for residential activity recognition. IEEE Sens. J. **18**(13), 5413–5421 (2018)
21. Plötz, T., Guan, Y.: Deep learning for human activity recognition in mobile computing. Computer **51**(5), 50–59 (2018)
22. Qi, J., Wang, Z., Lin, X., Li, C.: Learning complex spatio-temporal configurations of body joints for online activity recognition. IEEE Trans. Hum.-Mach. Syst. **48**(6), 637–647 (2018)
23. Sok, P., Xiao, T., Azeze, Y., Jayaraman, A., Albert, M.V.: Activity recognition for incomplete spinal cord injury subjects using hidden Markov models. IEEE Sens. J. **18**(15), 6369–6374 (2018)

24. Tan, T., Gochoo, M., Huang, S., Liu, Y., Liu, S., Huang, Y.: Multi-resident activity recognition in a smart home using RGB activity image and DCNN. IEEE Sens. J. **18**(23), 9718–9727 (2018)
25. Wang, J., Zhang, X., Gao, Q., Yue, H., Wang, H.: Device-free wireless localization and activity recognition: a deep learning approach. IEEE Trans. Veh. Technol. **66**(7), 6258–6267 (2017)
26. Wang, C., Xu, Y., Liang, H., Huang, W., Zhang, L.: WOODY: a post-process method for smartphone-based activity recognition. IEEE Access **6**, 49611–49625 (2018)
27. Wang, C., Chen, S., Yang, Y., Hu, F., Liu, F., Wu, J.: Literature review on wireless sensing-Wi-Fi signal-based recognition of human activities. Tsinghua Sci. Technol. **23**(2), 203–222 (2018)
28. Wenyuan, L., Siyang, W., Lin, W., Binbin, L., Xing, S., Nan, J.: From lens to prism: device-free modeling and recognition of multi-part activities. IEEE Access **6**, 36271–36282 (2018)
29. Xi, R., et al.: Deep dilation on multimodality time series for human activity recognition. IEEE Access **6**, 53381–53396 (2018)
30. Yang, Z., Raymond, O.I., Zhang, C., Wan, Y., Long, J.: DFTerNet: towards 2-bit dynamic fusion networks for accurate human activity recognition. IEEE Access **6**, 56750–56764 (2018)
31. Yao, L., et al.: Compressive representation for device-free activity recognition with passive RFID signal strength. IEEE Trans. Mob. Comput. **17**(2), 293–306 (2018)
32. Lateef Haroon P.S, A., Eranna, U.: An efficient activity detection system based on skeleton joints identification. Int. J. Electr. Comput. Eng. (IJECE) **8**(6), 4995–5003 (2018)
33. Zdravevski, E., et al.: Improving activity recognition accuracy in ambient-assisted living systems by automated feature engineering. IEEE Access **5**, 5262–5280 (2017)
34. Nakagawa, E., Moriya, K., Suwa, H., Fujimoto, M., Arakawa, Y., Yasumoto, K.: Toward real-time in-home activity recognition using indoor positioning sensor and power meters. In: 2017 IEEE International Conference on Pervasive Computing and Communications Workshops (PerCom Workshops), Kona, HI, pp. 539–544 (2017)
35. Henpraserttae, A., Thiemjarus, S., Marukatat, S.: Accurate activity recognition using a mobile phone regardless of device orientation and location. In: 2011 International Conference on Body Sensor Networks, Dallas, TX, pp. 41–46 (2011)

Neural Network Comparison for Paint Errors Classification for Automotive Industry in Compliance with Industry 4.0 Concept

Michal Kebisek[1(✉)], Lukas Spendla[1], Pavol Tanuska[1],
Gabriel Gaspar[1], and Lukas Hrcka[2]

[1] Faculty of Materials Science and Technology,
Slovak University of Technology, Trnava, Slovakia
{michal.kebisek,lukas.spendla,pavol.tanuska,
gabriel.gaspar}@stuba.sk
[2] PredictiveDataScience, s. r. o., Bratislava, Slovakia
lukas.hrcka@predictivedatascience.sk

Abstract. The proposed paper focuses on utilization of neural networks for paint errors classification in the area of automotive industry. The paper utilizes hypothesis, that outdoor weather has significant impact on the number of paint errors, as a basis for comparison of neural network algorithms. For the neural network algorithms comparison we used real production data from the paint shop process. The paper deals also with definition of classification classes and attributes selection, as well as the data integration process itself that utilizes Hadoop platform as an intermediate data storage.

Keywords: Automotive · Error classification · Neural network · Industry 4.0

1 Introduction

Current modern trend in manufacturing focus on the improvement of flexibility, cost reduction and continuous improvement of quality. These areas are heavily supported by the utilization of the Industry 4.0 concept, which introduces modern technologies as a Big Data and Internet of Things [1–3]. These have a high impact in the area of quality improvement, since they allow industries to meet the wide data requirements needed for analysis of various events across the manufacturing process.

Through time, each production process has defined hypotheses, which are verified in practice. However this verification is usually based on the user experiences or process observation, without any proof or evidence. Therefore, with the availability of the required data, through the introduction of modern technologies, it is interesting for companies to verify these hypotheses and use the results to improve the quality of the production processes.

© Springer Nature Switzerland AG 2019
R. Silhavy (Ed.): CSOC 2019, AISC 985, pp. 353–359, 2019.
https://doi.org/10.1007/978-3-030-19810-7_35

2 Methods

Our objective is to confirm or rebut a hypothesis based on the experience of paint shop employees that says "Painting quality is dependent on external influences, mainly weather, that are not directly related to the defined parameters of the process".

In order to verify the given hypothesis, it was necessary to obtain information about:

- Painting errors that can be identified during the paint checking.
- Path of the bodywork to identify the movement through the painting process.
- Weather data.

For analysis purposes, it was necessary to modify and transform the data obtained from the data sources so that they could be used as an inputs for classification using neural networks. This is captured as the BPMN (Business Process Model and Notation) process on Fig. 1.

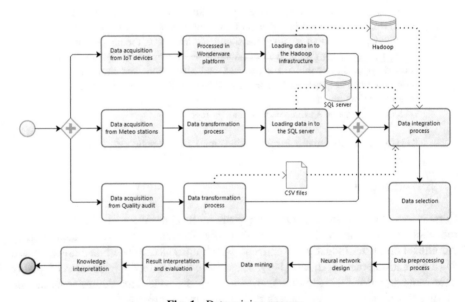

Fig. 1. Data mining process

Data are stored in multiple data sources. The painting process quality data are stored in the Quality Audit system. This system stores all the data from the quality control that is performed after the painting process. Individual parts of the bodywork are checked for the quality of the paint structure, scratches, dried droplets, chromaticity and other defects. The system also contains additional quality information that we do not take into account in our analysis.

Bodywork movement through the painting process information comes from the Wonderware MES (Manufacturing Execution System) system [4], which contains all the process data that are collected throughout the painting process. The number of collected parameters exceeds 1500 process tags. Due to the number of workplaces that are duplicated in the painting process, the painted bodywork may pass through several of these workplaces. Therefore it was necessary to precisely map the path and movement of bodywork through the paint shop.

For the sake of our analysis we also use IoT (Internet of Things) devices that identify the movement of the bodywork during the painting process. Through them, it is possible to precisely locate the position of the bodywork in the individual painting phases.

Meteorological weather data, which according to the hypothesis, should have a significant impact on the painting process, are collected from several stations located directly in the company's vicinity and near the paint shop and are stored in the Meteo Station system. It provides an overview of temperature, pressure, precipitation and humidity that are collected every 5 min. Due to the requirement of historical data, we have only obtained aggregated daily values (from the last 12 months).

Since the data sources mentioned above store data from the non-interconnected systems, integration of these data was necessary. The standard total time that bodywork spent in the paint shop was between 10 h to 14 h, but there were cases where this time exceeded 30 h (mainly due the repairs). For this reason, the path of the painted bodywork had to be investigated in detail. We used an additional data set that contained information about the bodywork movement in the paint shop. Based on the time stamps, we were able to assign bodywork data to the individual positions in the painting process, and subsequently link them with the corresponding weather data.

It should be noted, that the painting process data do not contain the final quality evaluation of the paint structure. Therefore it has been necessary to integrate these data with the Quality Audit data set, which contained the final paint quality evaluation.

Subsequently, it was possible to select a suitable data subset and preprocessed it. Preprocessed data served as an input data set in the data mining process. Obtained results were transformed to the appropriate knowledge for verifying the given hypothesis.

3   Results

To verify the defined hypothesis, it was necessary to choose a specific neural networks algorithm. In our paper, we used 3 different neural network algorithms that Rapid Miner software platform offers [5]:

- Deep Learning – based on a multi-layer feed-forward artificial neural network that is trained with stochastic gradient descent using back-propagation.
- Neural Net – learns a model by means of a feed-forward neural network trained by a back propagation algorithm (multi-layer perceptron).
- Auto MLP – simple algorithm for both learning rate and size adjustment of neural networks during training that combines ideas from genetic algorithms and stochastic optimization.

As a target attribute for neural network classification, the average number of paint errors per bodywork was determined. We defined 5 classification classes for this attribute, with the exact ranges listed in Table 1. Column "Range" defines the number of paint errors that were identified on one painted bodywork. The ranges were set up empirically, based on paint shop employee's knowledge, due to the fact, that standard methods as discretize by binning or frequency achieved unreasonable results in number of paint error in individual classification classes.

Table 1. Defined classification classes for paint errors.

Nominal value	Range	Absolute count	Fraction
Minimum error	−∞–2.075	3	0.027
Less than average error	2.075–2.575	54	0.491
Average error	2.575–3.075	17	0.155
More than average error	3.075–3.575	3	0.027
Maximum error	3.575–∞	33	0.300

Another important step was to define key attributes for a set of input parameters for neural networks. These were selected based on the weights that we obtained by evaluating the result of the Rapid Miner operator "Weight by Information Gain". This operator determined the relevance of attributes based on the information gain. For comparing neural network algorithms, we selected five input attributes that obtained the highest weight value shown in Table 2.

Table 2. Weights of available attributes in data set.

Attribute	Weight
Week day	0.512
Average humidity	0.311
Maximal temperature	0.230
Average air pressure	0.172
Average outdoor temperature	0.151
Day in month	0.141
Minimal temperature	0.130
Average precipitation	0.115
Average wind	0.032

After classification classes and key input parameters identification we designed a data mining model suitable for applying selected neural network algorithms. Part of the proposed model is shown on Fig. 2. The input data set consists of pre-processed integrated data based on the process described in the previous section. Due to one of the basic requirements of neural networks, it was necessary to modify input parameters to be represented by numerical values.

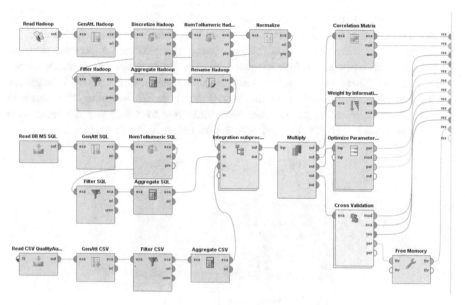

Fig. 2. Part of the proposed data mining model

For this reason, we used "Discretize" and "Nominal to Numerical" operators to correctly convert input parameter values. These values were then divided into 70% to 30% training and test data set. For this purpose, the "Cross validation" Rapid Miner operator was used.

The results obtained from the proposed data mining model for individual neural network algorithms are processed and shown in Table 3.

Table 3. Neural network results evaluation.

Neural network type	Accuracy	Kappa	Correlation
Neural Net	82.73%	0.652	0.860
Auto MLP	80.00%	0.601	0.841
Deep Learning	81.82%	0.639	0.851

The highest accuracy of 82.73% was achieved by the Neural Net algorithm, and it also reached the highest value of the Kappa parameter, which represents the actual improvement in the classification rate over the chance rate divided by the maximum possible improvement over the chance rate [6]. The correlation parameter represents the degree of association between real and predicted target attributes and produce a weights vector based on these correlations. Achieved correlations in all algorithms were very similar. Neural Net algorithm has also the highest correlation. Based on these results, we chose the Neural Net algorithm as the most appropriate for verifying the specified hypothesis.

The number of input neurons of the selected algorithm was 5+1 neurons. One neuron formed the so-called threshold node, with the other neurons representing weekdays and weather information (average humidity, maximal temperature, average air pressure and average outdoor temperature). The optimal number of hidden layer neurons was obtained experimentally using an optimization operator. He selected the appropriate number of neurons in the hidden layer as 10, including the threshold neuron. The number of output neurons was predefined to 5 due to a defined output value as polynomial. Selected neural network topology is shown on Fig. 3.

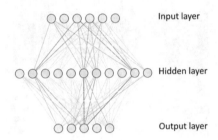

Fig. 3. Selected neural network topology

Based on the neural network evaluation results, we have selected the Neural Net algorithm, which achieved the classification accuracy of 82.73%, to be the most suitable for classification of our data set. The Table 4 shows sample of production data classification results, with achieved confidence for each classification class. The classes in Table 4 correspond with the classification classes defined in Table 1.

Table 4. Neural network class predictions

No. of errors	Prediction [No. of errors]	Confidence [Class 1]	Confidence [Class 2]	Confidence [Class 3]	Confidence [Class 4]	Confidence [Class 5]
Class 3	Class 3	0.019	0.086	0.821	0.051	0.023
Class 1	Class 1	0.873	0.045	0.029	0.024	0.029
Class 5	Class 5	0.076	0.057	0.027	0.124	0.716
Class 3	Class 4	0.087	0.085	0.347	0.357	0.124
Class 2	Class 2	0.086	0.758	0.084	0.057	0.015
Class 4	Class 4	0.042	0.063	0.078	0.805	0.012
Class 1	Class 1	0.883	0.028	0.012	0.038	0.039
Class 5	Class 5	0.028	0.039	0.085	0.046	0.802
Class 2	Class 1	0.421	0.389	0.124	0.042	0.024
Class 4	Class 4	0.028	0.035	0.074	0.812	0.051
...	...	...	...	...	...	...

From the results of the production data classification, it was found that the quality of the paint process was most affected by day of the week parameter, with Saturday and Sunday being the days with the highest error rate. The defined hypothesis was not fully confirmed, because the weather effect on the quality of the paint is notable but less significant than the day of the week influence. It follows that, it would be appropriate to include into analysis also other parameters from the painting process and external influences that could have a significant impact on the obtained results.

In order to verify the obtained results other neural network algorithms, as well as other data mining algorithms such as decision trees, genetic algorithms could also be applied.

4   Conclusion

In our paper we have compared neural network algorithms for classification of paint errors in the area of automotive industry. As a baseline for this comparison served the hypothesis, that the outdoor weather has significant impact on the number of paint shop errors. Therefore we used real production data from the paint shop process to evaluate the selected neural network algorithms, namely Neural Net, Auto MLP and Deep Learning.

From the results of our neural network comparison, presented in this paper, we can conclude, that all considered neural network algorithms for classification are suitable for this task. For our data sets, the Neural Net algorithm reached the highest accuracy and also had the highest kappa parameter. Based on the neural network target attribute classification analysis we can assume that this algorithm is suitable for the classification of paint errors.

The underlying hypothesis, which served as a basis for our comparison, can be evaluated as not fully confirmed. Outdoor weather parameters have impact on the number of paint errors, according to the production data used for classification. However generally accessible parameter, namely day of the week, has higher information gain than any available weather parameter and has also significant impact on the number of paint errors. Therefore it would be also suitable to analyze impact of other common production parameters and external influences to evaluate their impact on the number of paint errors.

Acknowledgments. This publication is the result of implementation of the project VEGA 1/0272/18: "Holistic approach of knowledge discovery from production data in compliance with Industry 4.0 concept" supported by the VEGA.

References

1. Ustundag, A., Cevikcan, E.: Industry 4.0: Managing the Digital Transformation. Springer, Heidelberg (2017)
2. Marz, N., Warren, J.: Big Data: Principles and Best Practices of Scalable Realtime Data Systems. Manning Publications, Shelter Island (2015)
3. Geng, H.: Internet of Things and Data Analytics Handbook. Wiley, Hoboken (2017)
4. AVEVA Group plc: Wonderware. https://www.wonderware.com. Accessed 11 Jan 2019
5. RapidMiner: Data Science Platform. https://rapidminer.com. Accessed 11 Jan 2019
6. Penny, W., Frost, D.: Neural networks in clinical medicine. Med. Decis. Making **16**, 398 (1996)

Support to Early Diagnosis of Gestational Diabetes Aided by Bayesian Networks

Egidio Gomes Filho, Plácido R. Pinheiro$^{(\boxtimes)}$, Mirian C. D. Pinheiro,
Luciano C. Nunes, Luiza B. G. Gomes, and Pedro P. M. Farias

Universidade de Fortaleza - UNIFOR, Fortaleza, Brazil
egidio.filho@gmail.com, {placido,caliope,porfirio}@unifor.br,
lcominn@uol.com.br, luizab2g@gmail.com
https://www.openpublish.eu

Abstract. Gestational Diabetes Mellitus (GDM) is one of the major diseases that affect pregnant women. On average, about 7% of pregnant women are affected by this disease. The consequences of non-treatment for the mother vary from the problems usually caused by Type 1 or 2 diabetes - such as dizziness, weight gain, hyperglycemia - to complications at the time of delivery. For the fetus, it can cause exaggerated weight gain, hypoglycemia, jaundice, type 2 diabetes, and even fetal death. Therefore, early diagnosis is important to indicate adequate follow-up and treatment in a timely manner. In this context, we carried out the structuring of the diseases that are manifested in concomitance or that are opportunized by the favorable environment caused by the evolution of undiagnosed Diabetes, through Bayesian Networks, with emphasis on Naive Bayes, based on data from a Health Plan Operator which covers eleven Brazilian states. Thus, the identification of these diseases and their respective symptoms can be used to support the early diagnosis of GDM.

Keywords: Gestational diabetes · Bayesian networks ·
Comorbidities of GDM

1 Introduction

The early diagnosis of diseases has become an ally in improving the quality of life of people, as well as in the generation of economic growth in the countries, thus being of great relevance to society. The earlier a disease is diagnosed, the sooner its treatment can be started by avoiding millions of early deaths [1,2].

Diabetes Mellitus refers to a chronic metabolic disorder due to an absolute or relative lack of insulin [3]. There are three major types of diabetes, Type 1 diabetes (destruction of β Cells, and usually insulin deficiency); Type 2 diabetes (progressive defect in insulin secretion due to its resistance); and gestational diabetes (intolerance to carbohydrates of varying degrees of intensity, diagnosed for the first time during gestation, that may or not persist after childbirth [4].

Gestational Diabetes Mellitus - GDM generates problems for both the mother (pregnancy-induced hypertension and pre-eclampsia, cesarean delivery,

© Springer Nature Switzerland AG 2019
R. Silhavy (Ed.): CSOC 2019, AISC 985, pp. 360–369, 2019.
https://doi.org/10.1007/978-3-030-19810-7_36

polyhydramnios, type 2 diabetes) and for the baby, namely macrosomia and organomegaly; fetal hypoxia and NRDS (Newborn Respiratory Distress Syndrome); hypoglycemia; polycythemia; hyperbilirubinemia; hypocalcemia; type 2 diabetes [5–7].

The risk factors considered by the American Diabetes Association are age ≥25 years; overweight/obesity (BMI > 25 kg/m2); family history (first degree); history of altered glucose metabolism; obstetric history: gestational loss of repetition, GDM, polyhydramnios, macrosomia, fetal/neo-natal death without a specific cause, fetal malformation, neonatal hypoglycemia, respiratory distress syndrome; ethnicities at risk: Hispanic, Asian, African and Native American [8].

According to [9], it is necessary to diagnose and control Gestational Diabetes early in order to minimize the risks, however, the Oral Glucose Tolerance Test (OGTT) is not adequate in the first trimester of pregnancy, since hyperglycemia induced by pregnancy is not always evident.

Therefore, this research was mainly motivated by the need for early diagnosis of Gestational Diabetes Mellitus, reducing the impacts for mother and baby; due to the fact that GDM was present along with other pathologies in 76.0% of the women [10]; and by the need of identifying an appropriate method to support the diagnosis of GDM in the first trimester of pregnancy, since OGTT is inadequate.

In these circumstances, this study presents the diseases, based on the medical history of pregnant patients, recovered from a health insurance company which offers coverage to eleven Brazilian states, in the period between January 2004 and December 2009, which emerge simultaneously or have their manifestation facilitated by the appearance of GDM as a way of supporting its early diagnosis, through the application of Bayesian Networks.

The article is structured as follows: Sect. 2 reviews the comorbidities of gestational diabetes within specific literature, Sect. 3 lists some works related to Gestational Diabetes Mellitus in the area of computer science, Sect. 4 presents a brief description of Baysean Networks, while Sect. 5 analyzes the medical history of pregnant women applied to Bayesian Networks, and finally, Sect. 6 presents the conclusions of the article, as well as the proposition of future work.

2   Comorbidities of Gestational Diabetes

In a study with 928 pregnant women, only 12.2% of the patients did not present complications during pregnancy, and the mothers had an average of 2.4 intercurrences throughout the period [11]. In this context, as reviewed in the literature, diseases that may occur concomitantly with GDM or have their manifestation favored by the onset of GDM are described below.

2.1   Threatened Miscarriage and Hemorrhage in Pregnancy

According to [12] the installation of GDM greatly affects the gestational environment for the fetus, increasing the congenital malformations by three times and preterm deliveries by ten times. For [13], miscarriage, macrosomia and perinatal mortality are more frequent in women who have developed GDM.

In this study, 76% of pregnant women who had GDM also had other pathologies, 16.0% had premature amniorrexis and another 16.0% had placental abruption [10]. In her research on diabetic pregnant women, [7] reported that 0.7% of pregnant women with GDM miscarried, whereas 1.4% had premature amniorrexis and 0.7% resulted in preterm labor.

2.2   Primary and Gestational Hypertension

It was observed by [14], the existance of complications in 14.1% of the pregnancies analysed, from which the hypertensive diseases of pregnancy (Pre-Eclampsia and Gestational Arterial Hypertension) had a prevalence of 5.9%.

On the other hand, [7] in their evaluation of the frequency of maternal-fetal complications in diabetic pregnant women, identified a prevalence of hypertension in 11.2% of cases and SHD, Specific Hypertension Disease in Pregnancy, in 9.8%. According to [10], systemic hypertension is the most prevalent pathology (18.0%) associated with gestational diabetes.

2.3   Abdominal and Pelvic Pain

The biomechanics of pregnant women are altered during the second and third trimesters of pregnancy, due to the growth of the uterus, weight gain and breast size, reasons that favor the displacement of the woman's center of gravity, causing accentuated lumbar lordosis, promoting pelvic anteversion and modifying the support base, factors that can cause pain and discomfort in the mentioned areas, as well as falling responsible for 17–39% of maternal traumas and 3–7% of fetal deaths [15,16]. [17] affirms that, with the advancement of gestational age, the lumbar and pelvic pains become accentuated.

2.4   Vagina and Vulva Infections and Cervix Inflammation

GDM is an endocrine condition that can alter vaginal pH, allowing the overgrowth of microorganisms, facilitating infections of the genital tract. The overall infection rates among 200 women screened for GDM were 68.5% and 40 patients had at least two diseases [18]. The research carried out by [7] pointed out that 8.4% of the pregnant women diagnosed with GDM had vulvovaginitis.

2.5   Fetal Problems or Pregnancy-Related Conditions

Pregnant women with poorly controlled GDM may develop polyhydramnios, the excess of amniotic fluid, greatly increasing the abdominal perimeter, and it may be accompanied by dyspnea or preterm delivery due to overdistension of the uterus, associated with maternal hyperglycemia and fetal macrosomia [13].

In addition to the fetal problems cited, maternal hyperglycemia can compromise the fetus with congenital anomalies, stillbirth, RDS - respiratory distress syndrome and neonatal hypoglycemia [13]. In the study of [7] fetal problems among pregnant women diagnosed with GD included hypoglycemia with 48.6%, jaundice with 25.4%, macrosomia with 24.6%, prematurity with 19% of fetuses.

2.6 Cystitis

As mentioned by [19], people with diabetes frequently suffer from bacterial infection in the urinary tract, since diabetic neuropathy can lead to damage to the genitourinary system; the presence of hyperglycemia favors the growth of bacteria; and the deficiency of immune function prevents the patient from self protecting against the growth of bacteria.

According to [10], 8% of the women studied with GDM presented urinary infection. In addition, to [7] the most frequent complications in diabetics are the infections (22%), from which the urinary tract infection corresponds to 11.9%.

3 Gestational Diabetes Mellitus in the Area of Computer Science

A two-stage study was designed by [20]: pre-processing data and application of supervised learning algorithm viz ID3, Naive Bayes, C4.5 and Random Tree to data sets, using the software WEKA with the focus of improving the diagnosis of gestational diabetes with the application of datamining techniques (classification), highlighting that Random Trees achieved the best accuracy and the lowest error rate.

A hybrid model for the creation of a specialist system applying Bayesian Networks, Multicriteria Decision Support Methodologies and structured knowledge representations in rules of production and probability (Artificial Intelligence) for diagnosis of type 2 diabetes, from parameters of greater influence was proposed by [21].

In [22] was proposed a methodology to identify Gestational Diabetes in pregnant women, modeling a specialist system to diagnose GDM using FeedFoward Neural Network architecture.

A predictive model for predicting diabetes was developed by [23] using SMOTE (Synthetic Minority Over-sampling Technique) and Decision Tree Classifier. The proposed system has 2 stages: in the first one an unbalanced die is removed using SMOTE and in the second the diabetes is diagnosed using Decision Tree classifier. The classification accuracy obtained is 94.7013%, with attributes such as age, fasting blood sugar and prandial blood pressure, waist measurement, BMI.

A Bayesian network and decision tree was implemented in the MATLAB software by [24]. The comparison of the two methods indicates that the Bayesian model is superior in relation to accuracy and precision, for the attributes number of pregnancies, glucose tolerance test, blood diastolic pressure, triceps skin fold thickness and others.

Naive Bayes, with Genetic Algorithms, also in the PINA Indian database, was applied by [25]. Firstly using *Naive Bayes* for the classification of all attributes. After that, the Genetic Algorithm was applied for the selection of attributes and then *Naive Bayes* was used again, on the attribute selection outcome. In the second part of the process, that is, after the selection of the Genetic Algorithms, the following attibutes remained are glucose tolerance test, serum insulin, body mass index and patient age (years).

Machine Learning was approached by [26] for the early detection of diabetes. Three classification algorithms were used: Decision Tree, Support Vector Machine (SVM) and Naive Bayes. The experiments were also done at the Pima Indians Database. The results showed that Naive Bayes was executed with an accuracy of 76.30%, while Decision Tree obtained 73.82% and SVM 65.10%.

A classifier using Bayesian networks was proposed to predict whether a person is diabetic or not [27]. The classification consisted of 3 classes: 0 - without diabetes, 1 - pre-diabetes and 2 diabetes. The variables used were the same ones considered in the Pima Indians Database, but the glucose test was separated in fasting and not fasting. The result was 99.51% accurate, demonstrating that Bayesian networks is a promising technique for this type of dataset.

4   Bayesean Network

When dealing with chance or with the lack of accuracy, sufficient or unexplored information, the use of methods such as reasoning with probability becomes appropriate. In any given experiment there is a degree of uncertainty as to whether an analyzed event (from sample space) will occur. It is possible to define as probability a number between 0 and 1, or 0% and 100%, associated with the chance that a certain event occurs [28].

It is named conditional probability, or *a posteriori*, the probability of a B event occurring, since another A event occurred or not. As opposed to conditional probability, it is named Unconditional Probability, or *a priori*, the probability of any A event occurring, when there is no association between its occurrence, and the occurrence or not of another B event, prior to the A event. With the Probability Theory one can define the Bayes Rule. In this context, Bayesian Networks works with Bayesian probability, applying the Bayes' theorem.

5   Application of Bayesian Networks in the Early Diagnosis of Gestational Diabetes

Information was collected from the medical care of pregnant patients, during their pregnancies, from a database of a health insurance operator, with coverage in eleven Brazilian states. The patient's disease code, according to the International Registry of Diseases, version 10, was used as reference and input of the study, from all appointments comprising this information registered in the medical care guide (for payment/administrative purposes), for the period between January, 2004 and December, 2009.

The data were segregated into two groups, considering the attribute associated with Gestational Diabetes, namely:

– GDM Positive: pregnant patients who have a record of one of the events of ICD 10 - O24 Diabetes mellitus during pregnancy; ICD 10 - O24.4 Diabetes mellitus

arising during pregnancy and ICD 10 - O24.9 Diabetes mellitus in pregnancy, unspecified excluding ICDs associated with existing previous diabetes.
- GDM Negative: pregnant patients who have a record of one of the ICD events related to childbirth and who do not have any ICD related to diabetes.

The analysis of patients diagnosed with Gestational Diabetes and with at least one of the 10 most frequent ICD concomitant to GDM results in 255 pregnant women. The list of ICD, the number of concomitances and the probability of incidence are shown in the Table 1.

Table 1. Pregnant women with the most frequent diseases and with GDM

ICD	Description	Amount	Probability
O20.0	Threatened abortion	69	0.270588
O20.9	Hemorrhage of early pregnancy, unspecified	67	0.262745
I10	Essential hypertension (primary)	51	0.200000
O36.9	Maternal care for fetal problem, unspecified	49	0.192157
N72	Inflammatory disease of cervix uteri	45	0.176471
R10.4	Unspecified abdominal pain	39	0.152941
N76.8	Other specified inflammation of vagina and vulva	39	0.152941
O13	Gestational [pregnancy-induced] hypertension	39	0.152941
O26.9	Pregnancy related conditions, unspecified	34	0.133333
N30.9	Cystitis, unspecified	26	0.101961

As a way of identifying the posterior probability of diagnosing GDM based on the existence of comorbidities, bayes' rule was applied to the generated data, which was defined by the Eq. 1.

$$P(B|A) = \frac{P(A|B) \times P(B)}{P(A)} \tag{1}$$

given the Eq. 2,

$$P(A|B) = \frac{P(A, B)}{P(B)} \tag{2}$$

In order to simplify the construction of the Bayesian network, it is assumed that there is no relation between the predictor variables, ie, the input events, or the comorbidities associated with GDM - Naive Bayes. The next step, is to define what are the values of the terms of this equation for each disease cited:

$P(A|B)$: conditional probability or *a posteriori* probability, in Table 1;
$P(A)$: unconditional probability or *a priori* probability, pregnant women with one of the 10 selected ICDs (with or without GDM);
$P(B)$: which is the probability of a person at the base having Gestational Diabetes, or 255 pregnant women.

Once the values have been identified, the Bayes' theorem must be applied, in order to discover the relevance of each of these diseases in the diagnosis of GDM. For instance, in order to calculate the subsequent probability of a pregnant woman presenting the disease of code N72, Inflammatory disease of the cervix, one must follow the Eq. 3:

$$P(B|A) = \frac{P(A|B) \times P(B)}{P(A)} = \frac{0.176471 \times 0.161392}{0.188608} = 0.151006; \qquad (3)$$

Thus, the posterior probability of a patient developing GDM was calculated, since it was diagnosed with some of the most significant comorbidities selected by the study, resulting in the Table 2.

Table 2. Bayes' theorem applied to the most relevant diseases

ICD	Description	Bayes' theorem
O20.0	Threatened abortion	0.188011
O20.9	Hemorrhage of early pregnancy, unspecified	0.265873
I10	Essential hypertension (primary)	0.196911
O36.9	Maternal care for fetal problem, unspecified	0.350000
N72	Inflammatory disease of cervix uteri	0.151006
R10.4	Unspecified abdominal pain	0.098734
N76.8	Other specified inflammation of vagina and vulva	0.161157
O13	Gestational [pregnancy-induced] hypertension	0.278571
O26.9	Pregnancy related conditions, unspecified	0.165049
N30.9	Cystitis, unspecified	0.168831

As a form of normalization of the data generated by the Bayes' rule, the Eq. 4 was used, altering the scale of the values.

$$\frac{P}{100} = \frac{TB - Min(Value(x))}{Max(Value(x)) - Min(Value(x))} \qquad (4)$$

where the disease with the lowest value is R10.4 (0.0987) and the disease with the highest value is O36.9 (0.3500), according to Eq. 5:

$$\frac{P}{100} = \frac{TB - Value(R10.4)}{Value(O36.9) - Value(R10.4)} = \frac{TB - 0.0987}{0.3500 - 0.0987} = \frac{TB - 0.0987}{0.2513} \qquad (5)$$

Where TB is the Bayes Theorem of the Table 2 and P is the normalized value to be found. The result of applying the Eq. 5 in the Table 2 produced the Table 3.

Table 3. Normalized Bayes' theorem applied to the most relevant diseases

ICD 10	Bayes' theorem	Normalized Bayes theorem
O20.0	0.188011	35.530800
O20.9	0.265873	66.518700
I10	0.196911	39.073000
O36.9	0.350000	100.000000
N72	0.151006	20.803679
R10.4	0.098734	0.000000
N76.8	0.161157	24.843350
O13	0.278571	71.572508
O26.9	0.165049	26.392116
N30.9	0.168831	27.897543

Diseases related to GDM have been highlighted in the literature, as well as analyzed in a database, resulting in the identification of the ones with the highest incidence. Afterwards, the posterior probability of a patient to present GDM was calculated through bayes' rule, from which some of the most significant comorbidities selected were diagnosed. At the end, a function was created in order to transform the values obtained, on a scale from 0 to 100, as a form of normalization of the data.

6   Conclusions and Future Work

GDM is a public health problem that requires rapid diagnosis, preferably within the first weeks of gestation in order to avoid negative risks for the mother and fetus. Current medical protocols using Oral Glucose Tolerance Test are not feasible in the first months of gestation, indicating the need for new diagnostic options.

Techniques of bayesian networks were applied to calculate the posterior probability of a patient to have GDM, from the diagnostic of the most significant comorbidities highlighted in the literature and of higher incidence in the database.

The suggestions for future work proposals are to use the structured model in the public health network, adapting it to mobile platforms, thus, it could cover remote areas of the country, improving the prenatal consultation protocol, and therefore the health of the mother and fetus. The use of other databases to obtain the relevance of the diseases, in which recalibration of the weights would allow the improvement of the final result of the model. Finally, we strongly recommend the use of other methodologies in the definition of the diagnosis of GDM [29–33].

References

1. World Health Organization: Chronic disease prevention a vital investment. World Health Organization (2005)
2. Australian Institute of Health and Welfare (AIHW): Prevention of cardiovascular disease, diabetes and chronic kidney disease: targeting risk factors, vol. 118 (2009)
3. Ranciaro, R.M.D.C., Mauad-Filho, F.: Effects of glucose ingestion on maternal-fetal circulation. Braz. J. Gynecol. Obstet. **28**(12), 693–699 (2006). http://www.scielo.br/scielo.php?script=sci_arttext&pid=S0100-72032006001200002. Accessed 15 Sept 2018
4. Schmidt, M.I., Reichelt, A.J.: Consensus on gestational diabetes and pre-gestational diabetes. Braz. Arch. Endocrinol. Metab. (ABE&M) **43**, 14–20 (1999). ISSN 1677-9487
5. Rehder, P.M., Pereira, B.G., Silva, J.L.P.E.: Gestational and neonatal outcomes in women with positive screening for diabetes mellitus and oral glucose tolerance test - 100 g normal. Braz. J. Gynecol. Obstet. **33**(2), 81–86 (2011). ISSN 0100-7203
6. Funes, L.S., Santos, R.C., Parada, C.M.G.D.L.: Understanding the meaning of pregnancy for diabetic pregnant women. Lat. Am. J. Nurs. **12**(6), 899–904 (2004)
7. Montenegro Jr., R.M., et al.: Maternal-fetal evolution of diabetic pregnant women followed at HC-FMRP-USP in the period 1992–1999. Braz. Arch. Endocrinol. Metabol. **45**(5), 467–474 (2001). ISSN 1677-9487
8. Bolognani, C. V.: Waist circumference in the prediction of gestational diabetes mellitus. Master's dissertation, Paulista State University, Faculty of Medicine of Botucatu, p. 55 (2011). http://hdl.handle.net/11449/99257. Accessed 15 Aug 2018
9. Georgiou, H.M., et al.: Screening for biomarkers predictive of gestational diabetes mellitus. Acta Diabetol. **45**(3), 157–165 (2008). https://doi.org/10.1007/s00592-008-0037-8. Accessed 10 Aug 2018. ISSN 1432-5233
10. Neta, F.A.V., et al.: Assessment of prenatal profile and care of women with gestational diabetes mellitus. In: Nursing Network Magazine of the Northeast, vol. 15, no. 5, pp. 823–831 (2014). http://www.redalyc.org/articulo.oa?id=324032944012. Accessed 15 Oct 2018. ISSN 1517-3852
11. Varela, P.L.R., Oliveira, R.R.D., Melo, E.C.: Intercurrences in pregnancy in Brazilian puerperas treated in the public and private health systems. Latin Am. J. Nurs. **25**, 1–9 (2017). http://www.redalyc.org/articulo.oa?id=281449566112. Accessed 23 Oct 2018
12. Zaupa, C., Zanoni, J.N.: Diabetes mellitus: general aspects and diabetic neuropathy. Arch. Health Sci. UNIPAR **4**(1), 19–25 (2000)
13. Menicatti, M., Fregonesi, C.E.P.T.: Gestational diabetes: physiopathological aspects and treatment. Arch. Health Sci. UNIPAR **10**(2), 105–111 (2006)
14. Fernandes, M.D.P.: Risk factors for the development of gestational diabetes and associated maternal and neonatal complications (2016). https://repositorio-aberto.up.pt/bitstream/10216/89007/2/170941.pdf. Accessed 23 Sept 2018
15. Conti, M.H.S., Calderon, I.M.P., Rudge, M.V.C.: Musculoskeletal discomforts of gestation - an obstetric and physiotherapeutic view. Female **31**(6), 531–535 (2003)
16. Dunning, K., et al.: Falls in workers during pregnancy: risk factors, job hazards, and high risk occupations. Am. J. Ind. Med. **44**(6), 664–672 (2003). http://dx.doi.org/10.1002/ajim.10318. Accessed 15 Sept 2018
17. Mann, L., et al.: Biomechanical changes during the gestational period: a review. Motriz **16**(3), 730–41 (2010)

18. Tristão, A.D.R.: Active search and treatment of lower genital tract infections in pregnant women with positive screening for gestational diabetes: maternal and perinatal repercussions (2008). https://repositorio.unesp.br/bitstream/handle/11449/104165/tristao_ar_dr_botfm_prot.pdf?sequence=1&isAllowed=y. Accessed 4 Sept 2018

19. Ferreira, R.C., Barros, C.E.D., Braga, A.L.: Profile of urinary tract infection associated with altered glycemia rate. Braz. J. Clin. Anal. **48**(4) (2016)

20. Nagarajan, S., Chandrasekaran, R.M., Ramasubramanian, P.: Data mining techniques for performance evaluation of diagnosis in gestational diabetes. Int. J. Curr. Res. Acad. Rev.: IJCRAR **2**(10), 91–98 (2014)

21. Menezes, A.C., Pinheiro, P.R., Pinheiro, M.C.D., Cavalcante, T.P.: Towards the applied hybrid model in decision making: support the early diagnosis of type 2 diabetes. In: Lecture Notes in Computer, vol. 7473, pp. 648–655 (2012). https://www.researchgate.netpublication/262166315_Towards_the_Applied_Hybrid_Model_in_Decision_Making_Support_the_Early_Diagnosis_of_Type_2_Diabetes. Accessed 03 Jan 2019

22. Lakshmi, K.V., Padmavathamma, M.: Modeling an expert system for diagnosis of gestational diabetes mellitus based on risk factors. OSR J. Comput. Eng. (IOSR-JCE) **8**(3), 29–32 (2013)

23. Mirza S., Mittal, S., Zaman, M.: Decision support predictive model for prognosis of diabetes using smote and decision tree. Int. J. Appl. Eng. Res.: IJAER **13** (2018)

24. Mohtaram, M., Mitra, H., Hamid, T.: Using Bayesian network for the prediction and diagnosis of diabetes. Bull. Environ. Pharmacol. Life Sci.: BEPLS **4**(9), 109–114 (2015)

25. Choubey D., Paul, S.K.S., Kumar, S.: Classification of pima Indian diabetes dataset using naive bayes with genetic algorithm as an attribute selection. In: The International Conference on Communication and Computing Systems (ICCCS-2016), pp. 451–455 (2016). https://doi.org/10.1201/9781315364094-82

26. Sisodia, D., Sisodia, D.S.: Prediction of diabetes using classification algorithms. Procedia Comput. Sci. 1578–1585 (2018). https://doi.org/10.1016/j.procs.2018.05.122

27. Kumari, M., Vohra, D.R., Arora, A.: Prediction of diabetes using Bayesian network. Int. J. Comput. Sci. Inf. Technol.: IJCSIT **5**(4), 5174–5178 (2014)

28. Walpole, R.E., et al.: Probability & Statistics for Engineers and Scientists, 8th edn. Pearson Education, Upper Saddle River (2007)

29. Castro, A.K.A.D., Pinheiro, P.R., Pinheiro, M.C.D., Tamanini, I.: Applied hybrid model in the neuropsychological diagnosis of the Alzheimer's disease: a decision making study case. Int. J. Soc. Humanistic Comput. **1**, 331–345 (2010)

30. Carvalho, D., Pinheiro, P.R., Pinheiro, M.C.D.: A hybrid model to support the early diagnosis of breast cancer. Procedia Comput. Sci. **91**, 927–934 (2016)

31. Pinheiro, P.R., Tamanini, I., Pinheiro, M.C.D., Albuquerque, V.H.C.: Evaluation of the Alzheimers disease clinical stages under the optics of hybrid approaches in verbal decision analysis. Telematics Inform. **35**, 776–789 (2017)

32. Nunes, L.C., Pinheiro, P.R., Pinheiro, M.C.D., Filho, M.S., Nunes, R.E.C.: Toward a novel method to support decision-making process in health and behavioral factors analysis for the composition of IT projects teams. Neural Comput. Appl. **2018**, 1 (2018)

33. Gomes-Filho, E.: Heterogeneous methodology for supporting the early diagnosis of gestational diabetes, Master's dissertation - University of Fortaleza (2018)

Time Series of Workload on Railway Routes

Zdena Dobesova[1(✉)] and Michal Kucera[1,2]

[1] Department of Geoinformatics, Faculty of Science,
Palacky University, 17. listopadu 50, 779 00 Olomouc, Czech Republic
zdena.dobesova@upol.cz, misakucera@gmail.com
[2] Railway Infrastructure Administration,
Nerudova 773/1, 779 00 Olomouc, Czech Republic

Abstract. The article presents the processing of time series of the workload on railway routes in the Czech Republic. The data for railway stations and signal blocks on routes were processed. The aim is to describe some typical railway stations form the point of structure and workload changes. Both passenger and freight trains are recorded. The descriptive data contains the monthly aggregation of count and weight for passenger and freight trains. Monthly-length correction of data was processed before the evaluation of the time series. Examples of time series for selected stations show that passenger trains are mainly stationary time series otherwise the freight trains are non-stationary time series with a trend. Some stations have a sessional component of series in data about freight trains. In that case, it is possible to predict the time series from old previous data.

Keywords: Time series · Railway workload · Aggregation · Prediction

1 Introduction

The Railway Infrastructure Administration (SŽDC in Czech) is a state organisation that administrates the railway infrastructure in the Czech Republic. The administration collects the amount and weight of passing trains to monitor the workload of railway routes. The monitoring points are situated at mostly all railway stations and important points like signal blocks. The total number is nearly 3 350 monitoring points.

The Railway Infrastructure Administration needs this monitoring for evaluation of workload on rail routes, for economic assessment and technical maintenance. The data are collected primarily for the need of rental payments by individual carriers. Moreover, that data could be used as an important source for the calculation of the noise pollution of the population [1] and other tasks. The number and weights of the passing trains affect the operating load and technical condition of routes.

The presented investigation tries to describe the measured data from the point of time. The analysis of time series was the base idea. Firstly, selected stations were compared: the station located on the main railway route (transit corridor) and station on a regional route. Subsequently, some selected station were analysed to find if some seasonal effect exists during the year or if some unexpected deviations exist.

© Springer Nature Switzerland AG 2019
R. Silhavy (Ed.): CSOC 2019, AISC 985, pp. 370–380, 2019.
https://doi.org/10.1007/978-3-030-19810-7_37

2 Data and Methods

The Railway Infrastructure Administration provided the data in the followed structure for presented research. The attribute data contains this structure: time, the name of the station, the indicator of the type of train (passenger, freight, maintenance etc.), the number of trains, number of railway cars in train set, the average length of the train set, and the average number of axels. Only two descriptive attributes were taken for investigation – the number of trains and the weight of the train for selected measuring points. The monitoring data are accessible in monthly aggregation for each measuring point. The monitored time was three years: 2016, 2017 and 2018.

2.1 Calendar Effect and Monthly-Length Adjustments of Time Series

The data are affected by calendar influences. The are several calendar influences like different lengths of the month, different numbers of the weekend in a month, different working days in a month and movable feasts and state holidays in a year. This irregularity could have surprising consequences. It is necessary to clean the data before further processing. In the case of monthly aggregated data, the influence of different length of the month was discovered in the railroad workload. It is called monthly-length effect [2]. Especially, month February, as a shortest in length, is visible in the data (Fig. 1). There are declinations (minimum values) of the weight of train set in each February 2016, 2017 and 2018 in comparison with other months in the year.

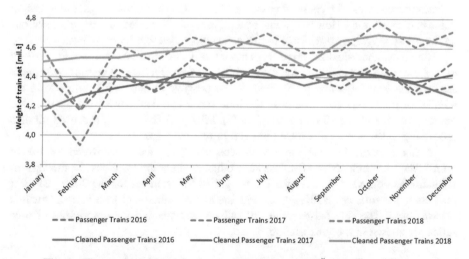

Fig. 1. Time series with origin data and cleaned data for Česká Třebová station

There are two possible solutions for recalculation and clearing data [3]. The first is recalculation to year with 360 days with constant length of month 30 days. The second solution is recalculation for the punctual length of the year, and it is 365 or 366 days. The equation for recalculation values according to the real length of each month are [4]:

$$Ynew_t = \frac{Y_t}{N_t} * \overline{N},$$ (1)

where $Ynew_t$ is a new value in a month t, Y_t is origin value in a month t, N_t is a count of days in a month t and N_t is the average length of the month in the year. In the year 2016 the average length of the month was 30.5 days (leap-year) and 30.41667 days in the year 2017 and 2018. All origin values in time series were recalculated according to Eq. (1). The monthly-length effect was removed from data. The differences of values are visible at Fig. 1 where both origin values (dot lines) and cleaned values (solid lines) are depicted. The Česká Třebová station was taken for a demonstration of time series in 2016, 2017 and 2018. The values are the total weight of the passenger train set in each month. The time series of cleaned values are more smoothed than origin values. Origin data contain more oscillation caused by the different length of the months.

2.2   Steps of Data Processing

After cleaning data, the data of various stations were displayed in graphs to investigate time series. The first step was a visual analysis. Visual analysis is the recommended initial step in analyzing the time series [5]. The first was comparing the number of passenger trains and freight trains. Subsequently, the weight of the train sets was compared. Some big differences exist. The count of passenger train set is many times higher than freight trains in most stations. However, some exception exists. We select some typical stations on the transit corridor, on regional routes and stations near border crossings.

The visual analysis discovered the variance of values in years and months. We try to find the trend and cycles in the selected time series. Because some railway stations are near the limit of capacity [6], there is not possible to expect a strong rise in the number of trains set. To decompose the time series and detect the trend we used the moving average with a 12-month window [3]. Microsoft Excel 2016 was used for data processing. Also, the graphs were prepared in this software.

In minor cases, the time series with seasonal cycles were discovered for freight trains. They are influences of industrial production in some localities. In the case of Rýmařov station, we try to predict the weight of the train set for the 2019 year. For prediction, the software WEKA v. 3.8 with the implementation of Holt-Winters method [7] was used. The selected examples of railway stations and the description of time series are shown in the next section.

3   Results

3.1   Comparison of Passenger and Freight Trains

The count and weight of train sets depend on the type and locality of the railway station. To imagine differences, the short overview is in Table 1. There are monthly average values of weight and count of train sets separately for passengers and freight trains in 2016. The Česká Třebová is a station located on the transit corridor from Prague to Ostrava. The frequency and amount of traffic are very high. The count of passenger trains is nearly seven times higher than freight trains. Despite that, the average weight of passenger train sets is only two times higher. The next examples show that in all case the average count of passenger trains is higher than the count of freight trains. The average weight of passenger trains is also higher than the weight of the freight train set, but the difference is not so high than in the case of a count of trains. It is evident that freight transport is much more massive.

Table 1. Comparison of average count and weight of the passenger and freight trains in 2016.

Station	Type and location	Weight of passenger trains [t]	Weight of freight trains [t]	Count of passenger trains	Count of freight trains
Česká Třebová	Transit corridor	4 377 056	2 287 381	13 489	2 084
Břeclav	Border crossing	2 199 604	5 953 243	6 614	5 194
Cheb	Border crossing	836 381	577 939	3 892	1 155
Rýmařov	Regional route	12 487	11 724	518	34
Senice na Hané	Regional route	15 747	445	560	2

One exception was presented in Table 1, and it is the border crossing Břeclav. There is the weight of freight trains higher than the weight of passenger trains (two times higher). The reason is transport to neighbour countries Slovakia and Austria. Also, the count of freight trains in Břeclav is higher in comparison with Česká Třebová station. The second presented a border crossing is Cheb station in Table 1. The dominance of average weight and count of freight trains is the same as other domestic stations. It is evident that the transport of freight is high to neighbouring Germany.

Table 1 presents two stations on a regional route; there are Rýmařov and Senice na Hané. The numbers are lower in comparison to previous stations located on main routes. Both stations have a low amount of workload. In the case of terminal station Rýmařov, the average monthly weight of passenger trains and freight trains is nearly the same (around 10% difference). It is evident that there is relatively huge transportation of freight. It was an impulse to investigate the changes in transport in that station.

For all presented stations the graphs of time series were constructed. The example of Česká Třebová station is presented in Fig. 2. There are visible the same high difference between passenger trains and freight trains in 2016, 2017 and 2018 years.

The fluctuation of the weight and count is not high in the category of passenger trains. More fluctuations are in weight in the category of freight trains. The question was if the workload is stationary or non-stationary time series in Česká Třebová station. The time series is stationary if it does not have a trend or seasonal effects, in addition the average is the same in selected periods [4]. The stationarity was checked by calculation the mean and variance for each year. Comparison of the means and variance showed that the times series are stationary in case of freight trains. Moreover, according to recommendation [8], the histograms were constructed for all six times series (weight and count for passenger and freight trains). The histograms have the bell curve-like shape of the Gaussian distribution for freight trains.

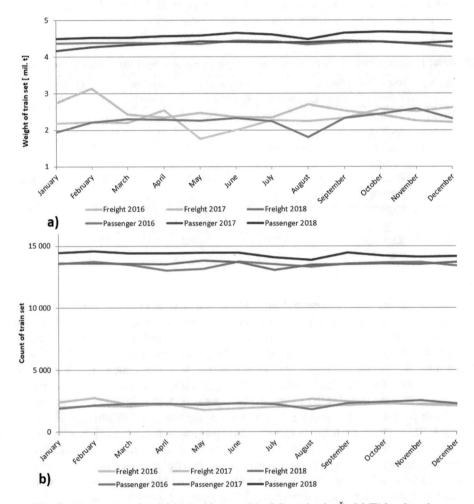

Fig. 2. Time series of weight (a) and count (b) of the trains in Česká Třebová station

The time series are stationary in case of the freight trains. It means that the workload in this station on the corridor has no trend and seasonal effect for a freight train. The time series of passenger train weight has a small increasing trend, especially influenced by data in 2018. So weight and count passenger trains are non-stationary time series in Česká Třebová station.

The Břeclav station is mentioned in Table 1 and the previous text. It is an example of the station where is an inverse portion of the passenger to freight trains from that majority of stations in the Czech Republic. Figure 3 shows the time series of train weight for Břeclav station. The weight of freight trains (light colours) is higher than the weight of passenger trains (dark colours). The oscillation of the freight train weight is higher than oscillation of passenger train weight (like Česká Třebová station). The passenger train weight is nearly stationary time series. The time series of freight trains is non-stationary time series. The decomposition of the time series for freight trains is presented in the next Sect. 3.2.

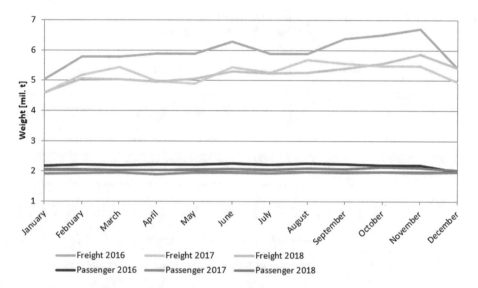

Fig. 3. Time series of the weight of train sets in Břeclav station in three years

3.2 Decomposition and Seasonal Part in Time Series

Time series data can exhibit a variety of patterns, and it is often helpful to split (decompose) a time series into several components, each representing an underlying pattern category. Three types of time series patterns exist - trend, seasonality and cycles [5, 7].

If an additive decomposition is assumed then it is described by Eq. (2):

$$y_t = S_t + T_t + R_t, \tag{2}$$

where y_t is the data, S_t is the seasonal component, T_t is the trend-cycle component, and R_t is the remainder component, all at period t.

Alternatively, a mõultiplicative decomposition is described by Eq. (3):

$$y_t = S_t \times T_t \times R_t \tag{3}$$

The additive decomposition is the most appropriate if the magnitude of the seasonal fluctuations, or the variation around the trend-cycle, does not vary with the level of the time series. When the variation in the seasonal pattern, or the variation around the trend-cycle, appears to be proportional to the level of the trend of time series, then a multiplicative decomposition is more appropriate.

After the visual analysing and the counting of variations, the additive model is more appropriate for time series of investigated workload on railway routes. Namely, the time series of weight are non-stationary on the contrary of stationary time series of train count (both passenger and freight train). The trend has various fluctuation in three-year time series. Only some time series have an evident seasonal part. The remainder component is mostly present in the time series. Two examples of interesting decompositions of time series are presented: Břeclav and Rýmařov stations.

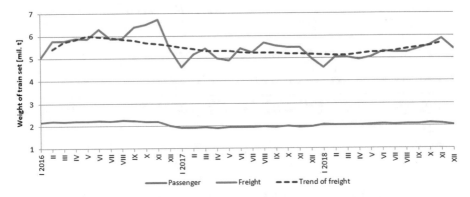

Fig. 4. The trend in time series of weight for freight trains in Břeclav station

To decompose the time series the classical method of moving averages was used. For smoothing data with expected seasonal component, the period for smoothing is recommended to the length equal the length of the season. In the case of monthly aggregated data, the length is 12 [3]. Figure 4 is a graph where the trend calculated by a central moving average by this equation:

$$T_t = \frac{y_{t-6} + 2 * \left(y_{t-5} + y_{t-4} + y_{t-3} + y_{t-2} + y_{t-1} + y_t + y_{t+1} + y_{t+2} + y_{t+3} + y_{t+4} + y_{t+5}\right) + y_{t+6}}{24}$$

(4)

The equation was applied for time t_7 to t_{29} for Břeclav station. The start and end of time series are calculated by shorter moving averages (3, 5 and 7 element windows). The weight of freight train has a stable declining trend in 2017, and increasing trend in 2018 with remainder component R. There is no repetition of the seasonal part except a regular decrease in December and January for each year.

The interesting decomposition is for Rýmařov station. In Fig. 5 is a graph of weight both for the freight (light colours) and passenger trains (dark colours). The weight of passenger trains is nearly stationary series. Otherwise, freight trains have a strong seasonal part. The lower amount of weight is in the winter. The trend increases in the spring with the highest value in summer (from May to September) followed by decreasing in the autumn. The series in 2016 and 2017 are more similar during the year than time series in 2018. The weight of freight trains in winter is also high in 2018; there is not so big difference in winter-spring time and summertime in 2018. All presented time series have a decrease in December that is probably caused by Christmas holidays and non-working days.

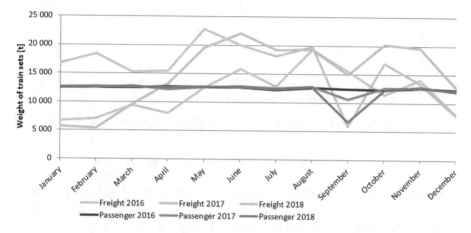

Fig. 5. Time series for Rýmařov station with a seasonal component

The interpretation of the strong seasonal part is followed. The railway station Rýmařov is a service station for the large wooded area. It is located in the Jeseníky Mountains in the Bruntál District, Moravian-Silesian Region. The surrounding area is typical for high activities in logging, cutting and preparing the timber. Rýmařov station is a terminal of the regional railway route. The timber industry is a reason for this seasonal component of a time series of weight. Because the count of freight trains has not this noticeable seasonal cycle, the length of train sets must be much longer in summer than in winter season. The weight of train sets in the year 2018 has noticeably hight values during all year. It is probably caused by high logging and cutting timber after bark beetle calamity and windfallen of trees.

Moreover, there is visible declination in September 2018. It could be caused by a temporary reconstruction of the route. It is supported by declination both freight and passenger trains. This declination was also found in the count of freight and passenger trains. The data about the count of trains verified the hypothesis about the reconstruction of the railway route.

3.3   Identification of Unusual Observations

Time series could also reveal some extraordinary situation like some reconstruction on the rail route. Figure 6 shows the time series for Senice na Hané station. It is a regional route with a small number of trains. Especially, freight trains have some zero values of a count in some months.

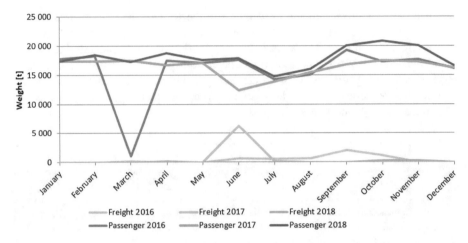

Fig. 6. Time series for Senice na Hané station with the identification of an outlier

The high decline of passenger trains is in spring of the 2016 year. In comparison with the situation in spring 2017, it is an unexpected situation. There are two possible reasons. The first reason could be wrong source data. The second could be a temporary reconstruction of the route. The second is valid for Senice na Hané station in March. Visual analysis of time series is a very illustrative way to reveal some unusual situation or outliers in data. The automatic identification of unusual observation could be a hint for some extraordinary situation. The exclusion of this data is necessary before finding a seasonal component. Otherwise, the results of decomposition could be misleading. Besides the visual analysis of time series, a boxplot and statistics are other options how to find any outliers or unusual observation in data.

3.4   Predictions of Train Weight

On the base of old data, it is possible to predict data for the future. The software WEKA v. 3.8. was used for prediction. WEKA is a collection of machine learning algorithms for data mining tasks that is freely accessible for use. The additional package *time series forecasting* by Mark Hall was installed into WEKA [9]. The package implements the Holt-Winters triple exponential smoothing method for prediction. The Holt-Winters method, implemented in that package, needs minimally three-year long time series to predict time series for 12 months in future.

The experimented time series for Rýmařov station had a strong seasonal part in spring and summer when the weight of freight train increased caused by the timber industry. The interpretation of data is depicted in the chapter about the decomposition of time series. The input parameters for Holt-Winter method were set like these - value smoothing factor 0.1, trend smoothing factor 0.2 and seasonal smoothing factor 0.1 for data from Rýmařov station. Figure 7 presents the fluctuation of time series from 2016 to 2018. Last part of the graph is a prediction for 12 months - the year 2019 (dot line). The prediction also predicts the seasonal increase in weight in summer time (Fig. 7). The prediction is wrongly influenced by decreasing in September 2018 due to the

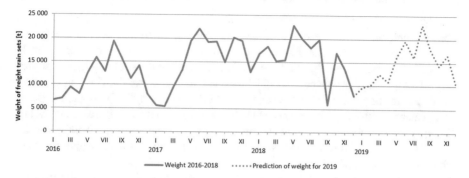

Fig. 7. Prediction of weight for freight trains at Rýmařov station for 2019

reconstruction of the route. Also, the data is influenced by the hight amount of cutting timber after calamity in 2018. The longer and older data in time series would be more precise in the prediction for the future.

4 Conclusion

The paper presents time series analyzes as the first analysis of the huge data set of the railway infrastructure measuring in the Czech Republic. Before evaluating the monthly time series, it is necessary to clear the data from the monthly-length effect as shown in the time series for Česká Třebová station. The presented analyses show that greater seasonal fluctuations during the year can be expected in freight transport than in passenger transport, which is given by valid timetable. Conversely, freight rail transport depends on customer orders for shipment. An analysis in a more detailed time series (for a day, a week, days of month) cannot be performed because the data is aggregated and provided only in monthly summaries.

It is already apparent from the examples that the stations differ in the number and weight of the trains during the year. Therefore, we plan to concentrate on analyzing the data at the border crossings to all neighbouring countries of the Czech Republic. The article presents the time series of Břeclav station near the border crossing, which is passed by freight trains to the neighbouring countries Slovakia and Austria.

It is planned to continue by further analysis of the data. One of the planned tasks is a classification of stations according to the annual load by number and weight both passenger and freight trains. Classification means to find all stations with similar workloads thus belong to one class (group) of similar stations. It is a task of clustering of time series [10].

It will also be interesting to compare the predictions made for 2019 with actual measured values in future. In order to evaluate and interpret variations and trends in time series, it will be necessary to continue consultation with rail experts to interpret data better. Data sometimes are influenced by temporary reconstruction of routes performed by a decrease of number and weight of train sets.

The presented time series analysis will also be used as examples of real-world data from the practice for the teaching at the study branch Geoinformatics at the Palacký University in Olomouc. The author of the article has a positive experience with the application of the knowledge that the teacher has acquired in solving practical problems. This practical knowledge could be added in the syllabus. This experience is mentioned in the article about the relational database design for the botanical garden plant database (BotanGIS project) [11]. Time series analysing is lectured in Data Mining for Geoinformatics. Students are familiar with the software WEKA for data mining and tasks. As a result, next year will be added some practical examples into the syllabus of subject Data Mining.

Another planned research is to compare rail workload with the state of surface and rail wear. SŽDC measures the technical condition (wear and tear) with a special moving measuring vehicle. These data are further used to evaluate rail damage and maintenance planning. The data obtained by technical vehicle could strongly relate with the weight of trains monitored in point stations. This type of research could be valuable for discovering the dependencies in data and prediction of technical maintenance.

Acknowledgement. This article has been created with the support of the student project IGA_PrF_2019_014 of the Palacky University Olomouc.

References

1. SŽDC: Annual Report 2017. SŽDC (2018)
2. Cleveland, W.S., Devlin, S.J.: Calendar effects in monthly time series: modeling and adjustment. J. Am. Stat. Assoc. **77**, 520–528 (1982)
3. Litschmannová, M.: Introduction to the time series analysis. VŠB-TU, Faculty of Electrical Engineering, Department of Applied Mathematics, Ostrava (2010)
4. Hančlová, J., Tvrdý, L.: Introduction to the time series analysis. VŠB-TU, Ostrava (2003)
5. Křivý, I.: Analysis of Time Series. University of Ostrava, Ostrava (2012)
6. Cestr, A.: Railway infrastructure for serviceability of Ustecky Region. In: Conference Železniční dopravní cesta. SŽDC, Ústí nad Labem (2018)
7. Hyndman, R.J.: Forecasting: Principles and Practice. Monash University, Australia (2018)
8. How to Check if Time Series Data is Stationary with Python. https://machinelearningmastery. com/time-series-data-stationary-python/. Accessed 15 Jan 2019
9. Hall, M.: Time Series Analysis and Forecasting with Weka. http://wiki.pentaho.com/display/ DATAMINING/Time+Series+Analysis+and+Forecasting+with+Weka. Accessed 10 Jan 2019
10. Samé, A., Chamroukhi, F., et al.: Model-based clustering and segmentation of time series with changes in regime. Adv. Data Anal. Classif. **5**, 301–321 (2011)
11. Dobesova, Z.: Teaching database systems using a practical example. Earth Sci. Inform. **9**, 215–224 (2016)

Hybrid Models of Solving Optimization Tasks on the Basis of Integrating Evolutionary Design and Multiagent Technologies

L. A. Gladkov$^{(\boxtimes)}$, N. V. Gladkova, and S. A. Gromov

Southern Federal University, Taganrog, Russia
leo_gladkov@mail.ru, nadyusha.gladkova77@mail.ru

Abstract. The paper is devoted to the problem of building hybrid intelligent systems for solving multi-objective optimization problems. The authors present the definition of a hybrid system, and the main problems and tasks of its development. The main idea is that integration of methods of computational intelligence and multiagent systems (MAS) can be promising and useful for developing intelligent systems. The paper describes the concepts of designing agents, multi-agent systems, and the design process with elements of self-organization (interaction, crossing, adaptation to the environment, etc.). The authors propose a method of forming child agents as a result of the interaction of parent agents, develop various types of crossover operators, and present the idea of creating agencies (families) as units of the MAS evolving. To implement the proposed ideas, hybrid fuzzy-evolutionary models of forming agents and agencies based on the use of fuzzy coding principles are created and described in the paper. The authors developed a software system to support evolutionary design of agents and multi-agent systems for estimating the effectiveness of the hybrid approach. The results demonstrate the effectiveness of the proposed approach.

Keywords: Multi-agent systems · Evolutionary computing · Hybrid models · Fuzzy genetic algorithms · Computational intelligence

1 Introduction

To date the relevant problem in terms of designing information systems and technologies in different branches of social and economical human activity is intellectualization. Intellectualization of information systems is a process of implementing functions which are usually performed by human. Such functions can include analysis and decision making in terms of incomplete, fuzzy or inconsistent input information; finding, evaluating and interpreting new nontrivial useful dependencies in large scopes of input information [1].

In fact, creation of mathematically justified and accurate models and methods is either inacceptable economically or impossible to implement. In that case, information systems on the basis of integrated fuzzy hybrid mechanisms and models can be a well-balanced compromise solution.

© Springer Nature Switzerland AG 2019
R. Silhavy (Ed.): CSOC 2019, AISC 985, pp. 381–391, 2019.
https://doi.org/10.1007/978-3-030-19810-7_38

Using methods of computational intelligence for solving tasks of knowledge processing, storing and retrieval allows us to work with poorly formalized information and obtain needed scientific basis.

Synergetics is a new multidisciplinary research area which studies common dependencies of transition from chaos to order and vice versa in open non-linear systems of a different nature (technical, economic, social, etc.). Synergetics is based on similarity of mathematical models in disregard of different nature of the described systems. This feature differs synergetics from the other research fields appearing at the intersection of disciplines, taking the subject from one discipline and the methods from another one [2].

Today one of the most effective instruments for intellectualization of information systems is using hybrid mathematical models combining the benefits of different computational intelligence methods such as evolutionary computation, fuzzy logic, neural network models, bioinspired algorithms of swarm intelligence, multiagent technologies.

Integration of different research fields, absence of clear borders of subject area and the use of the suitable tools allow us to consider the computational intelligence as the relevant multidisciplinary research field.

2 Problem Statement

Hybrid systems are composed of different elements (components) united to achieve the stated goals. Integration and hybridization of different methods and information technologies allow us to solve complex problems which cannot be solved by separate methods and technologies. In terms of integrating heterogeneous information technologies the synergetic effects are of the higher order than while integrating different models in the same technology.

In hybrid architecture uniting several paradigms, the effectiveness of one approach can compensate the weakness of another one. Combining different approaches, we can avoid the shortcomings of separate ones. This effect is referred as synergetic effect [3, 4].

Unlike the natural ones, the artificial systems created by human are not usually able to develop, self-organize, change its structure dynamically or adapt to changing external environments. Thus, the main task of the developers of new information systems and technologies of knowledge processing is to include the ability to use the natural experience obtained by the previous generations, adapt and self-organize into the developed systems. In that regard, the use of methods and approached inspired by the nature, such as evolutionary computations, swarm intelligence, multiagent systems in the created decision support systems is well founded [5–9].

Developed information systems as well as search and decision support systems can be classified conceptually as combined artificial systems, i.e. human-created systems combining artificial and natural subsystems [10]. They also can be considered as goal-directed systems which function are based on the desirability factors [11].

The model of evolutionary systems level can be represented as follows [12]:

$$SYS = (GN, KD, MB, EV, FC, RP), \quad (1)$$

where *GN* denotes the genetic origin (creating the initial set of solutions); *KD* denotes the living conditions; *MB* denotes the exchange phenomena (evolutionary and genetic operators); *EV* denotes the evolvement (evolution strategy); *FC* denotes functioning; *RP* denotes the reproduction.

Another promising branch of science includes methods combining fuzzy management rules and bioinspired search opportunities. Two main approaches to using such hybrid methods are [13]:

1. Using evolutionary algorithms for solving optimization tasks and searching in terms of fuzzy, ambiguous or insufficient information about the object, parameters and criteria of the solved problem; using GFRBS (genetic fuzzy rule-based systems). Hybrid systems are used for training and setting different components of the fuzzy rules system: automated generation of GFRBS knowledge bases, its testing and output functions adjustment [14, 15].
2. Using methods based on fuzzy logic for modeling structure and operators of evolutionary algorithms, along with management and adaptation of evolutionary algorithms parameters [14, 15].

To control and changing parameters of evolutionary algorithms dynamically, the system is supplied with fuzzy logic controller (FLC), which changes search parameters dynamically by using the experience and knowledge of the experts of the subject area to avoid the preliminary convergence [16].

The rule-generation system on the basis of knowledge and reasoning performs logical derivation, which is transformed into the controlling action after defuzzification. Changing the algorithm parameters leads to changing the search process and getting new results which transforms from the state variable into fuzzy sets in the fuzzification block [17].

3   Development of Hybrid Systems on the Basis of Multi-agent Approaches

Another promising approach to organization of structure of the modern intelligent knowledge management systems is using the multi-agent architectures [15].

According to different information systems, the agent can be interpreted differently. The multi-agent system can be considered as a population of simple independent agents. Each agent implements itself in local environment and interacts with other agents. Relations between the agents are horizontal, and global behavior of the agents is based on the fuzzy rules.

One of the main problems of creating effective intelligent systems is creating the agent's program implementing the agent's function transforming the input actions into the output reactions.

To date there are different approaches and methods of developing artificial agents and multi-agent systems (MAS). Among them we can note methods of bottom-up design based on the agents' roles and interaction including Gaia, MASE, PASSI, TROPOS, etc.; methods of top-bottom design in multi-dimensional space of criteria.

Classical methods require building a set of models to define the multi-agent system specification. Each model includes its components and the relations between them. The designed model are classified as internal and external. External models are referred to the system level of description: the main components are represented by the agents, and the interaction is described with the use of relations of inheritance, aggregation, etc. Thus, we develop the abstract structures of the agents. Internal models are developed for each class of the agents and describe the internal structures of the agents: their opinions, purposes, plans, etc. [17].

There are two main types of the external models: an agent model and an interaction model defining the ways of communication between the agents. The agent model is divided into models of the agents' classes and models of the agents' examples. These models define the agents' classes, their possible implementations interrelated by the relations of inheritance, aggregation, etc. Agents' classes define agents' attributes setting opinions, purposes and plans of the agent.

The purpose of the agent model is to describe different types of the agents existing in the systems. Types of the agents are defined by a set of roles. Thus, the developer can propose to unite different similar roles into a single type of agents. The main criteria here is effectiveness of implementation: the developer intends to implement the optimization of decisions. Uniting several roles into a single type is one of the ways of reaching this kind of effectiveness [17].

The example of such decision is represented by the situation, when the computational resources, required for each agent, are very large. Then, the number of agents needs to be reduced by uniting several roles into a single agent. In that context we need to reach a reasonable compromise between the simplicity of understanding functional characteristics of the agent and the effectiveness of its implementation.

The model of agents' interaction includes description of services, interrelations and obligations between the agents. It consists of sets of protocols defined for each inter-role interaction. Protocol is considered as a set of following attributes:

- purpose: short description of meaning of interaction, i.e. 'requesting information', 'giving task';
- initiator: role, which is responsible for interaction initiation;
- respondent: roles to be interacted with;
- inputs: information used by initiator to start interaction;
- outputs: information given by respondent during interaction.

Herewith, it is suggested that the protocol implementation causes a series of interactions.

This scheme is defined formally, separating from the concrete schemes of implementation (direct series of steps). Such consideration of interactions means that the main focus is on the purpose and nature of interaction rather than the clear scheme of communication.

The internal model representing the opinions and plans of a concrete class of agents is considered as the extension of the object-oriented models (opinions and purposes) and dynamical models (plans) [17].

Except the described models, we can also develop the following types of models:

- models to describe the tasks, which can be solved by the agents (initial purposes, ways of decomposition, methods for solving, etc.);
- organization models (e.g. description of the agents' communication or characteristics of the organization, where the MAS is to be implemented);
- communication models, which specify the characteristics of the partners human-computer interface.

However, almost all well-known technologies require pre-defining the functions and agents' types. They are based on quite strict pre-set protocols of communication. In fact, they do not consider different mechanisms of self-organization, evolution, and cooperation of the agents in the MAS. Thus, it is more relevant to create a new class of methods for designing agents in the MAS, which would be based on using bionic methods and models, and the ideas and technologies of evolutionary design particularly. Evolutionary design (ED) of artificial technical system is considered as targeted using of computer models of evolution at each stage of system development. Such approach is at the intersection of design theory and self-organization theory. Each kind of self-organization assumes cooperation of the agents in the MAS and adaptation of the agents to the environment in a certain evolution scheme. There are different approaches to evolutionary design of the agents in the MAS, which are based on different models of evolution [18]. Natural base for classification of concepts and strategies of ED can be represented by analyzing the external and internal reasons of evolvement of the agents or the MAS.

In terms of external reasons of evolvement, evolutionary design of the MAS is considered as a process of its evolutionary adaptation to the environment. The environment here is the source of evolution of the designed system and its main driving force. Thus, the main direction of evolvement of the designed system is to provide correspondence to current conditions of the environment. This can be achieved through direct adaptation of the system to the environment.

In particular, the starting point of evolution can be related to critical conditions of the environment. Such conditions can disturb the functioning of the MAS and its agents. In that situation mutation can help the agent to survive and adapt to the changed conditions. This class of mutations is the most promising and focused on reducing functional incompetence.

The internal reasons of the MAS change are considered in the MAS itself. They can be related to their motivation, adaptation for achieving the common goal, etc.

Let us consider the agent's evolutionary design as the processes of formation of its genetic variation and evolutionary adaptation to the environment. That means that ED is defined as the process of formation and deployment of the agent's genotype and phenotype. The agent's genotype corresponds to the whole hereditary (genetic) information inherited from the parents. The agent's phenotype consists of a set of structures of the agent defined by contextual rules, which appear as a result of genotype development in the environment. In that regard, it is often required to process qualitative fuzzy information and consider different strategies and computer models of evolution.

Let us introduce the conception of evolutionary design of the agents and the main ways of its implementation. Table 1 demonstrates the correspondences between the concepts in evolutionary modeling and multiagent system theory.

Table 1. Comparison of the concepts

Evolutionary modeling	Agent and multiagent systems theory
Gene	Parameter (property) of the agent
Chromosome	Set of parameters (properties)
Specie	Agent
Family (2 parents and 1 child)	Agency
Population	Evolutionary multiagent system

The problem of evolutionary design (ED) of the artificial systems can be formulated as follows [17]:

$$ED = \langle E, K, O, Q \rangle, \tag{2}$$

where E denotes a set of evolution models; K denotes a set of criteria of ED; O denotes a set of objects of ED; Q denotes a set of the procedures of ED.

Evolutionary theory, evolutionary modeling and fuzzy logic allow as to develop the algorithm defining the principles of the agents' interaction. The agents has parameters which are defined in the interval of [0, 1]; thus, using fuzzy logic we can modify genetic operators and the mutation operator in the algorithm. Resulting from the crossover algorithm of the parents agents we obtain the child agents which form the family (agency). To demonstrate the agency and the common structure of the multiagent system we propose to use graphs.

It seems clear that the process of development of agent and multiagent systems is composed of abrupt jumps as well as gradual changes which can last for the several generations' life.

Creation of the common theory of the agents' and the MAS's theory assumes investigation of essential problems including the following:

- to analyze the common reasons and factors of evolution of the agents and the MAS's;
- to research the mechanisms of adaptation of the agents to the environment and its changes;
- to define the reasons and mechanisms of diversity of the agents' and agency's types;
- to investigate the methods and tools for simulation modeling of the evolutionary processes.

There are several well-known evolution models which are popular to use in the computer science [10]. Such models describe individual aspects of the evolution. For the final choice of the common evolution scheme and model used for the agent theory problem it is required to study these problems and to investigate the other modern evolutionary research.

There are also distinguished the macroevolution and microevolution levels. In terms of biology, microevolution is referred to the intraspecific level, and macroevolution is referred to interspecific level. The differences of microevolution and macroevolution in relation to the agent and MAS's theory are demonstrated in Table 2.

Table 2. Difference between macroevolution and microevolution

Process	Microevolution	Macroevolution
Selection unity	Agent	Virtual community (a system of the MAS's)
Variability source	Mutation/crossover	Formation of a virtual community
Selection type	Natural selection in the MAS	Selection of different types of the MAS's
Self-reproduction type	Reproductive rate	Virtual community formation rate
Evolution mechanisms	Genetic drift	Phylogenetic drift

Recent experience has shown that using homogeneous methods, which correspond to a single scientific paradigm cannot always be effective for solving complex problems. The hybrid architecture unites several paradigms and allows us to compensate the shortcomings of one method by the benefits of another one. Combining several approaches, we can avoid the weaknesses of each of them individually. Thus, one of the leading trends defining the development of modern computer science and computer-aided design is expansion of integrated and hybrid systems. Such systems are composed of different elements (components) united to achieve the stated goals. Integration and hybridization of different methods and information technologies allow us to solve complex problems which can not be solved by some separate methods and technologies. In terms of integrating heterogeneous information technologies the synergetic effects are of the higher order than while integrating different models in the same technology.

The choice of a technology to process the initial information depends on the peculiarities of the solved problems, the number of quantitative and qualitative parameters describing the problem and the level of its investigation. Therefore, it is required to define the conditions of using each of the considered technologies and to develop methods and algorithms to be adapted for solving the concrete problems in the subject area.

4 Model Description

To form the child agent, we propose a model based on analysis of possible interaction types between parent-agents in the process of evolutionary design. The population of agents is considered to be an evolving multi-agent system (EMAS) with a set of parameters. In such connection, a modified genetic algorithm is used for building the model. Genetic algorithm finds structures of effective interaction between the agents and the forms of the child agent. The genetic algorithm plays the role of a superior coordinator, imposing restrictions on the activity of the entire population of the agents. Analysis of the results of imposing these constraints allows us to accumulate positive properties in the population and to form the structures of agencies and child agents that are most suitable for specific conditions. The choice of genetic operators depends on the types of interactions between the agents. In this model, we consider four possible types of interaction, which are represented as schemes for the generation of children, i.e. crossing operators. The model is organized in such way that only two agents can participate at a time, which means that each new agent (child agent) has exactly two parent agents. There are necessary triggering conditions for each of the four interaction schemes, i.e. not every pair of agents can interact. The normalized genotypes must satisfy operation conditions of a given scheme, only then a pair of agents has an opportunity to begin interaction.

Let there be some population of agents (evolving multi-agent system) EMAS = $\{A_1, \ldots, A_n\}$, where A_i is the i-th agent. Each i-th agent of the system is characterized by a set of parameters $(p_{i1}, \ldots, p_{ir})$, one part of which is inherited, and the other is formed during the agent's own activity. Let us denote a set of parameters transmitted by inheritance as $a_{i1}, \ldots, a_{im}$. The aggregate of these parameters forms the genotype of the agent. The phenotype of an agent is related to the definition of various rules for its interaction with the environment.

In general, the population has G generations. There are distinguished two types of agents for each generation: parent agents denoted as A_{pi} and child agents denotes as A_{nk}. Together, parent agents and child agents form a special case of the agency, called the family. A pair of successive agencies (three generations of agents) forms the minimal unit (elementary state) of EMAS.

Each agent in the agency has its own genotype and phenotype. Let us assume that the agent's genotype consists of 2 genes that carry information about the state of the agent, and the possibilities of its interaction with other agents (to form a child). This information is characterized by the corresponding resource parameters - the RESg agent's shared resource and the resource used to create the child (included in the child) RESb, while RESb < RESg.

Each pair of parents generates at least one descendant, i.e. in fact, the evolutionary operator (the crossing operator) is realized:

$$evo : A_{pi} \times A_{pj} \rightarrow A_{nk}, \tag{3}$$

where $A_{pi} \times A_{pj}$ is the interaction of a pair of parent agents belonging to a multitude of parents; A_{nk} is a descendant agent. In this case, evo $\in$ EVO, where EVO is the set of evolutionary operators (crossing operators).

The crossing operator reflects the pattern of interaction between agents. The paper suggests four schemes of possible interactions, each scheme has its own unique set of response conditions:

$$EVO = \{As, Comb, SelU, Mer\}, \tag{4}$$

Let us present the following notations given in the formula (4).

- *As* denotes the operator of crossing of the "association" type. Parents agents have approximately the same common resource and participate on a parity basis in creation of a new agent. The stimulation to cross between both agents is internal here and the resource of the newly formed child agent is less than the resource of each parent.
- *Comb* is a combination operator. One of the parent agents takes priority because it owns a significantly larger resource, and the resource of the child agent is in the range between the resources of the parent agents, while the "weak" parent is stimulated to cross from external side.
- *SelU* is a selective association operator. Elite parent agents are chosen to create a new strong agent, with the original agents giving a large amount of resources to the parent agent, but the remaining resources are above the survival level.
- *Mer* is an operator of the "merge" type. The resource of the child agent is greater than each of the parent agents, and the remaining resource of the parent agent is below a subsistence minimum. Therefore, they die, and the stimulation to cross is external.

Thus, for the generation of a new agent, it is necessary to have a pair of parent agents that satisfy certain conditions for implementing the interaction.

In general, the sequence of actions is the following:

1. To define the main criteria (conditions) for the evolutionary design of agents.
2. To implement fuzzy coding of resource characteristics of agents.
3. To implement agents' mutation. In the process of vital activity of the MAS agents, a gene mutation of parent agents occurs.
4. To select a pair of parent agents from the source MAS and establish relations between them on the generation G_i.
5. To cross parent agents and form a child agent.
6. To form a family (agency). To create units of evolutionary MAS.
7. To select a pair of parent agents and establish relations between them on the generation G_{i+1}.
8. To check the stop condition fulfillment. The value of the current generation counter G_i is compared to the number of simulated Gmax generations introduced in the initial simulation conditions. If the value of the counter is greater than the value of Gmax, the simulation process is terminated.

To implement the presented ideas of the evolutionary design of agents and MAS and to carry out the experiments, the authors developed a software system. The results of the experiments are represented as the quantitative characteristics that show which agent interaction schemes are involved in the process of forming the child agent and what agencies are formed when generating the child agent.

To determine the average number of crossing agents types, we used such fixed input parameters:

- the number of individuals in the initial population is 100;
- the threshold for the probability to obtain an additional resource by an individual is 0.2;
- the threshold for the probability to exclude the consumption of a common resource is a - 0.2.

After determining the average values of the generated structures (agencies), we change the number of simulated generations: Gmax = 10, Gmax = 30, Gmax = 50. The test results are shown in the Fig. 1.

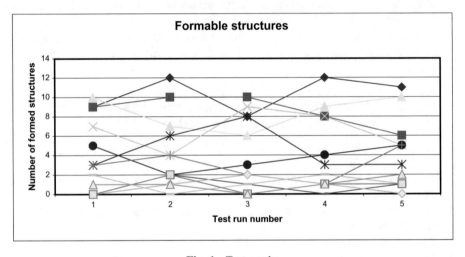

Fig. 1. Test results

5 Conclusion

The paper presents a model of forming child agent, based on the analysis of possible types of interaction between parent agents in the process of evolutionary design. The authors considered the population of agents to be an evolving multi-agent system (EMAS) and proposed a modified genetic algorithm. The task of the genetic algorithm is to find effective structures of interaction between agents and the formation of a descendant agent. The genetic algorithm plays the role of a higher coordinator, imposing restrictions on the activities of the entire population of agents. Analysis of the result of the imposition of these restrictions allows us to accumule positive properties in the population and form the structure of agencies and child agents that are most suitable for specific conditions.

Acknowledgment. This research is supported by the grant from the Russian Foundation for Basic Research (project # 18-07-01054, 19-01-00715).

References

1. Russel, S.J., Norvig, P.: Artificial Intelligence: A Modern Approach. Prentice Hall, Upper Saddle River (2003)
2. Haken, H.: The Science of Structure: Synergetics. Van Nostrand Reinhold, New York (1981)
3. Haken, H.: Synergetics, An Introduction: Nonequilibrium Phase Transitions and Self-Organization in Physics, Chemistry, and Biology, 3rd edn. Springer, New York (1983)
4. Luger, G.F.: Artificial Intelligence: Structures and Strategies for Complex Problem Solving, 6th edn. Addison Wesley, Boston (2009)
5. Michael, A., Takagi, H.: Dynamic control of genetic algorithms using fuzzy logic techniques. In: Proceedings of the Fifth International Conference on Genetic Algorithms, pp. 76–83. Morgan Kaufmann (1993)
6. Lee, M.A., Takagi, H.: Integrating design stages of fuzzy systems using genetic algorithms. In: Proceedings of the 2nd IEEE International Conference on Fuzzy System, pp. 612–617 (1993)
7. Herrera, F., Lozano, M.: Fuzzy Adaptive Genetic Algorithms: design, taxonomy, and future directions. J. Soft Comput. 7(8), 545–562 (2003)
8. Gladkov, L.A., Kureichik, V.V., Kureichik, V.M.: Genetic algorithms. Phizmatlit, Moscow (2010)
9. Redko, V.G.: Evolutionary cybernetics. Nauka, Moscow (2001)
10. Gladkov, L.A., Kureichik, V.V., Kureichik, V.M., Sorokoletov, P.V.: Bioinspirated methods in optimization. Phizmatlit, Moscow (2009)
11. Prangishvili, I.V.: Sistemnyy podkhod i obshchesistemnye zakonomernosti. SINTEG, Moscow (2000)
12. Borisov, V.V., Kruglov, V.V., Fedulov, A.S.: Nechetkie modeli i seti. Goryachaya liniya – Telekom, Moscow (2007)
13. Gladkov, L.A., Gladkova, N.V., Leiba, S.N.: Hybrid intelligent approach to solving the problem of service data queues. In: Proceeding of 1st International Scientific Conference "Intelligent information technologies for industry", IITI 2016, vol. 1, pp. 421–433 (2016)
14. Gladkov, L.A., Gladkova, N.V., Legebokov, A.A.: Organization of knowledge management based on hybrid intelligent methods. In: Software Engineering in Intelligent Systems. Proceedings of the 4th Computer Science On-Line Conference 2015 (CSOC 2015), Vol 3: Software Engineering in Intelligent Systems, pp. 107–113. Springer International Publishing (2015)
15. Gladkov, L.A., Gladkova, N.V., Gromov, S.A.: Hybrid fuzzy algorithm for solving operational production planning problems. In: Advances in Intelligent Systems and Computing. Proceedings of the 6th Computer Science On-Line Conference 2017 (CSOC 2017), Vol 1: Artificial Intelligence Trends in Intelligent Systems, vol. 573, pp. 444–456. Springer International Publishing (2017)
16. King, R.T.F.A., Radha, B., Rughooputh, H.C.S.: A fuzzy logic controlled genetic algorithm for optimal electrical distribution network reconfiguration. In: Proceedings of 2004 IEEE International Conference on Networking, Sensing and Control, Taipei, Taiwan, pp. 577–582 (2004)
17. Tarasov, V.B.: Ot mnogoagentnykh sistem k intellektual'nym organizatsiyam. Editorial URSS, Moscow (2002)
18. Tarasov, V.B., Golubin, A.V.: Evolyutsionnoe proektirovanie: na granitse mezhdu proektirovaniem i samoorganizatsiey. Izvestiya TRTU. Tematicheskiy vypusk « Intellektual'nye SAPR » , № 8(63), pp. 77–82 (2006)

Artificial Intelligence Tools for Smart Tourism Development

Tomáš Gajdošík(✉) and Matúš Marciš

Faculty of Economics, Matej Bel University, Tajovského 10,
975 90 Banská Bystrica, Slovakia
{tomas.gajdosik,matus.marcis}@umb.sk

Abstract. The recent development and use of information technologies in tourism lead to innovations that have revolutionised and automated almost every phase of tourist journey. In order to respond to these changes, the smart tourism concept emerged, providing real-time solutions, advanced analytics and enhancing tourist experience. The volume of created data and the need of real-time interactions and results challenge the tourism sector and open the door for the use of artificial intelligence (AI). However, there has been a lack of academic research on AI and its connection to smart tourism development so far. Therefore, the aim of the paper is to review artificial intelligence tools used in tourism and to identify its role in smart tourism development. The article focuses on best practice examples and case studies that examine the use of AI phenomena in tourism.

Keywords: Artificial intelligence · Smart tourism · Machine learning · Tourist experience

1 Introduction

Artificial intelligence is revolutionising almost every sector of economy as it enables computers to make autonomous decisions leading to more effective processes. Due to the fast-paced development of tourism, the adoption of artificial intelligence is inevitable as it helps in service delivery and value creation processes [1]. Tourists are much more demanding personalisation and have more digital experience. The challenge is to integrate their real-time interaction with a tailor-made experience. Moreover, tourism businesses need to analyse a large volume of data and react within the shortest time to ensure their competitive position.

The artificial intelligence can help to understand the tourist's needs, as well as respond to challenges of tourism businesses. It can provide special, tailor-made experiences, and even promote opportunities to better explore a tourism destination. The new AI tools could also ensure that tourism providers are more efficient, innovative and sustainable [2].

© Springer Nature Switzerland AG 2019
R. Silhavy (Ed.): CSOC 2019, AISC 985, pp. 392–402, 2019.
https://doi.org/10.1007/978-3-030-19810-7_39

2   Artificial Intelligence in the Context of Smart Tourism

The smart concept has emerged as a result of the rise of information technology and the need for sustainability. It is mainly based on information technologies that integrate hardware, software and network technologies to provide real-time awareness of the real world and advanced analytics to help people make more intelligent decisions about alternatives, as well as actions that will optimise business processes and business performances [3]. These technologies trigger innovation and lead to higher competitiveness, while ensuring a sustainable development [4, 5].

As tourism is highly dependent on information technologies [6, 7] and in the last years these technologies have been so tightly knitted into the fabric of the travel experience and management of tourism product [8], the smart phenomenon has also penetrated into the tourism sector. Information technologies in smart tourism development should enhance tourist experience giving all the related real-time information about the destination and its services in the planning phase, enhance access to real-time information to assist tourists in exploring the destinations during the trip and prolong the engagement to relive the experience by providing the descent feedback after the trip [9]. Within these smart technologies, AI is becoming a promising way how to make intelligent human-like decisions. Although AI has existed since the past three decades, it is the power achieved by processing and storage technologies due to rapid advancement of Moore's law that makes a difference in the way processes are being automated [10]. Artificial intelligence has been structured from the beginning of the field into four different areas. (1) problem solving, (2) knowledge representation and knowledge-based system, (3) machine learning and (4) distributed artificial intelligence [11].

In the tourism sector, machine learning (ML) is the most applicable area of AI, as it is focused on predictive and perspective analytics. It involves educating the algorithm in the sense of combining learning from experience, learning form data and following the instructions. Machine learning needs data, so the algorithm can be trained and thus to continue the self-improvement process. Subsequently, big data has become a key component of the information technologies infrastructure in smart tourism. Unlike the traditional data, big data refers to large growing datasets that include heterogeneous formats and has a complex nature that requires powerful technologies and advanced algorithms [12]. Traditional analytic systems are not suitable for handling big data [13], while machine learning functions in a complex environment and takes into consideration many variables. Based on McKinsey report [14] the potential impact of AI in tourism can double what is achievable using traditional analytic methods, amounting to between 7 and almost 12% of total revenue for the sector (Fig. 1).

Its capabilities lie in understanding, reasoning and learning. A key differentiator between AI and traditional analytical techniques is the possibility of applying unstructured big data in audio, video, image or text formats. AI and machine learning in smart tourism context should collect, process and utilise big data along all phases of tourist journey. However, there is a lack of academic research on AI and its connection to smart tourism development so far. Moreover Navío-Marco et al. [15] call for the investigation of the potential of AI in travel, tourism and hospitality industries and the immediate satisfaction through digital technologies, where human-computer integration plays a major role.

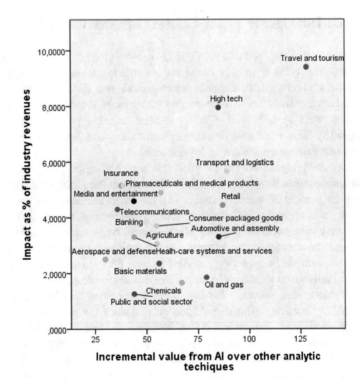

Fig. 1. The impact of artificial intelligence on various sectors of economy.

3   Materials and Methods

The aim of the paper is to review artificial intelligence tools used in tourism sector and to identify its role in smart tourism development. In order to examine the use of artificial intelligence in tourism, best practice examples as multiple case studies are chosen. The best practice has become a popular qualitative research method in business and tourism domain, as it describes leading cases as role models. The case study methodology is suitable both in tourism studies and in the field of information science, when technology is dynamic, changing and newly implemented [16]. The presented study uses multiple cases studies to fully and complexly examine the phenomenon of artificial intelligence.

4   Results

The smart tourism phenomenon implies the use of information technologies during all phases of the tourist journey. Nowadays the possibilities of information search about destinations and their services in the planning phase is overwhelming, leading to the need of personalised recommendations. The massive use of technologies in planning and especially staying phase leads to the demand of real-time interaction with service providers. Therefore the chat bots and virtual assistants are started to be used by tourism businesses. The evaluation phase is focused on tourist feedback, which is many times expressed on social media (Fig. 2).

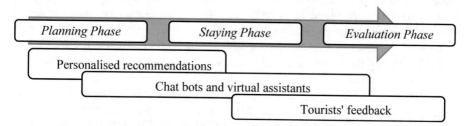

Fig. 2. The use of artificial intelligence during tourist journey

The amount of data created and the need for real-time solutions challenge the tourism sector. The AI tools are welcomed and are starting to be used by tourism businesses, as well as destinations. The applications of AI in tourism consist of the in-built mechanism and algorithms that enable to predict potential interest of tourists, personalise the tourism product and analyse the feedback. Therefore, they are valuable during all phases of tourist journey.

4.1 Providing Personalised Recommendations for Tourists

A tourist journey begins with the information search, decision-making and the booking process itself. From the tourist point of view, the most significant are search engines (e.g. Google), word of mouth marketing, hotel website, internet distribution systems (IDS, e.g. booking.com) and online travel agencies (OTA, e.g. Expedia). Moreover, tourists are also searching for information on social networks (e.g. Facebook), in travel agencies, on destinations' web and review sites (e.g. TripAdvisor). In the last years the significance of sharing economy platforms (e.g. Airbnb) and meta search engines (e.g. Trivago) has been rising steadily, leading to the massive creation of big data.

With the help of AI, these digital footprints of each tourist allow to understand needs, budget and travel preferences, and suggest the right offer at the right time. AI provides automated, personalised and intelligent travel services. It enables to learn the behaviour, choices, and preferences to the travellers and provides a personalised product. Recommender systems use, within the unsupervised learning, the cluster behaviour prediction to identify the important data necessary for making a recommendation. These personalisation techniques are particularly relevant in travel recommender systems. Travel recommender systems suggest products and provide tourists with relevant information to facilitate their decision making in a complex environment of the Internet. It provides personalised travel recommendations based on the traveller socio-demographic features (e.g. age, gender and interests), trip characteristics (leisure, business) or location.

There are many types of recommender systems, which can be content-based (content filtering, based on item content analysis), knowledge-based (knowledge record about users, items and needs), memory-based (database of users known preferences) or based on collaborative filtering (uses similar users' information to give recommendation) [17]. Among the collaborative filtering, association rule mining is becoming a promising way how to overcome problems of scalability or increasing recommendations utility. It provides stronger recommendations to tourists based on their past history and their patterns of co-occurrence across the digital footprints. The recommendations are generated by finding the strongest association rules of a tourist's frequent items and previously unknown items.

Companies like Expedia, Booking.com and TripAdvisor have already experimented in this area by using AI for customer recommendation service. In order to provide personalised recommendations and improve the anti-fraud algorithms, TripAdvisor has developed a custom Big Data platform. It uses Hadoop to store and process web log data, SQL Servers to report against aggregated data, Hive to query the data and put it into tables and machine learning to continuously improve the site experience. From ML techniques it uses Stanford's maximum entropy classifier to classify tourists' reviews. This classifier "read" a review, score its helpfulness and decide whether the review should be automatically rejected, accepted and published or queued for human interaction. It also uses agglomerative clustering within NLP to find clusters of phrases that has similar theme and thus provide recommendation for tourists searching specific theme. An example of classifying hotels in New York based on previous comments are in Fig. 3 [18].

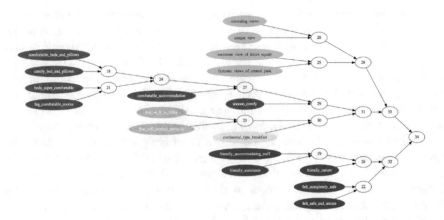

Fig. 3. Example of hotel classification in TripAdvisor based on tourists' comments

4.2   Interacting with Tourists Using Chat Bots and Virtual Assistants

Tourists are nowadays spending more and more time in the digital environment, which push the tourism businesses to move into digital services. The progress in information technologies allows virtual service agents (bots or virtual assistants) to enhance customer experiences through real-time interactions. These experiences start in a planning phase, but are more used in the dynamic context of the trip – in the staying phase. Virtual service agents serve as 24/7 customer care service and are based on machine learning algorithms, most of which fall in the supervised machine learning category, which uses trained data to develop an algorithm for classifying new examples. Training a supervised machine learning system involves providing it with representative inputs and corresponding outputs. The example of training a bot is shown in Fig. 4.

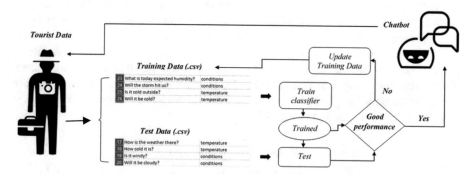

Fig. 4. Example of training a bot for tourism purposes

With the proper pre-programming, bots can empower the tourist experience, starting from automated reminders before the arrival to suggesting nearby activities in the destination. However, they can also be used post-trip, to send feedback forms. The use of bots has expanded rapidly among tourism enterprises, such as airlines, travel agencies and hotel booking services. Dutch airline KLM was an early adopter of chatbot technology. Their chatbot BB (Blue Bot) is based on DigitalGenius platform, which core are deep learning algorithms, trained on customer service data and integrated directly into the existing software. Once enabled, the platform automates and increases the quality and efficiency of customer service and supports conversations across text-based communication channels like email, chat, social media, SMS and mobile messaging. When an agent receives a customer question, the AI suggests an answer. The agent decides whether the proposed answer is accurate, adjusts it if necessary and sends the response. The AI system learns based on what the agent does. With Google Assistant's bot users receive tips on how to pack bags for a flight based on a destination.

Another example is Bold360, an AI-integrated chatbot and live-agent software that uses natural language processing (NLP) to answer user questions in a conversational and helpful manner. Bold360 uses proprietary algorithms to break down unstructured language into recognizable inputs and consider contextual information to understand the customer's natural language. The AI technology can understand hidden information, handle complex sentences that contain a lot of information and retain information from previous questions to make conversations flow naturally. Bold360 was deployed by Thomas Cook travel agency to deliver personalised travel solutions and provide consistent responses across multiple markets.

Moreover, except of bots, virtual assistants are also used in a tourism sector, namely hospitality. IBM´s Watson Assistant uses NLP that lets customers talk to the hotel the way they talk to a friend. The assistants can also deliver tailored responses based on previous behaviours and preferences. With data, skills and AI, the virtual assistant can synchronise with guests' calendars, interact with their homes, make recommendations or use any number of third-party applications. The property can also develop unique skills that extend the solution of their loyalty program, payment system and entertainment programs. Also Amazon Alexa is being used by several hotels (e.g. Marriott) as smart room equipment. Guests can verbally control many aspects of lighting, temperature and audio-visual components of a room using voice commands.

The above mentioned examples show the rise of conversational artificial intelligence. These conversations are mostly formed by unstructured data consisting from either voice, text, images or videos (Table 1). Therefore, it becomes critical to capture and understand the intent before generating the right response using machine learning models deployed on conversational AI systems.

Table 1. The use of conversation types in tourism

Conversation type	Technology components	Types of model used	Examples in tourism
Voice	Voice identification, Voice-to-text conversation	Convolutional neural networks (CNNs), Deep neural networks (DNNs), Generative adversarial networks (GANs), Boosted trees	Personal Guide SmartEcoMap, Voice assistant Google Assistants, Amazon Alexa
Text	Language detection, Language understanding	Recurrent neural networks (RNNs), Support vector machines, Naïve Bayes classification, Boosted trees	Chatbots, Sam, Kayak, Bold360, Booking.com's Assistant, Mezi, Hipmunk
Face to face	Object (logo, face, product) detection and classification, Action recognition, Cross-camera person reidentification	Transfer learning, CNNs, Artificial models with GANs	Watson-enabled robot concierge, Dash the room service robot, Robotic butler A.L.O

4.3   Evaluating Tourists' Feedback

Traditionally, the evaluation phase has been left to the tourist with his photos, videos and souvenirs physically shared among friends. With the massive use of social media, sharing the experiences in forms of status updates, comments, photos and videos is gaining the importance and significantly affects the reputation. AI can be applied to customer feedback as well, as both ML and NLP can be used to make feedback analytics more effective.

Tourists' comments on review platforms, OTAs and blogs can be analysed using text analytics, especially by opinion mining and sentiment detection. AI tools can highlight frequently used words, distinguish sentiment, as well as look at the correlation of certain words (Fig. 5). Machine learning can be used to make predictions based on historical feedback data. NLP can address sentiment behind feedback and collect text together to quickly uncover patterns and trends within the feedback [19].

If a tourist expresses a frustration on a social media, the AI tool analyses the tourist's intent and the context to automatically reach out with real-time interventions that are most likely to deliver a positive impact. These interventions could range from providing additional information, helping the tourist to understand the situation to more options that can meet the tourist's requirements.

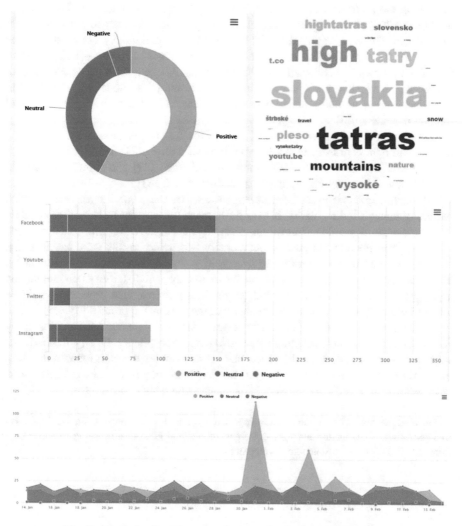

Fig. 5. Example of sentiment analysis on the destination High Tatras

For example, AI platform Metis can help tourism providers digging into customer feedback such as surveys and reviews, measure performance and instantly discover what really matters to guests. The city of Cork uses Citibeats, that can instantly and visually analyse data, compare opinions and feelings shared e.g. on social media to get genuine insights on visitors. With the Citibeats segmentation of amounts of data, positive and negative visitors' emotion, this destination is able to focus resources more efficiently.

5   Conclusion

The applications of AI in tourism reduce time taken to complete repetitive tasks, while improving the accuracy of processes and leading to real-time outcomes. The artificial intelligence tools can be used along all phases of tourist journey to enhance tourist experiences. Thanks to digital footprints of each tourist, enormous amounts of data are present about everything that is relevant to different stages of travel — before, during and after a journey and collected in different formats. AI can present near-real-time data processing, integrating and sharing using complex analytics, modelling, optimisation and visualisation to make better operational decisions relevant both for users and tourism providers [20].

However, the main concerns of AI include the fear of job loss, social acceptance and the willingness of tourists to adopt and interact with AI [21] as well as the privacy and individuality of users to share more of their data [22] and tourism providers' cultural data silos which may result in isolation of data [23].

Moreover, research in artificial intelligence shows that complex reasoning, which requires precision and regularity, is hard for humans but easy for machines, while tasks that require generalisation, perception, creativity and interacting with real world are relatively easy for humans but computationally expensive. In this vein, any task that can be described by an algorithm or is repetitive, can be outsourced by technology. Skills like pattern recognition, recombinant innovation, multi-sensory communication and development of creative solutions to previously unimagined problems would be left for humans [24]. This leads to the fact that the personal approach, which is crucial in tourism, will not be replaced. Conversely, artificial intelligence supports it by facilitating the work and providing more place for real hospitality. Finally, based on the case studies presented, it can be stated that AI contributes to smart tourism development, by enhancing overall tourist experience and providing businesses and destination with better results and lead to overall sustainable competitive advantage.

Acknowledgments. The research was supported by the research project VEGA 1/0809/17 Reengineering of destination management organizations and good destination governance conformed to principles of sustainable development.

References

1. van Doorn, J., Mende, M., Noble, S.M., Hulland, J., Ostrom, A.L., Grewal, D., Petersen, J.A.: Domo arigato Mr. Roboto: emergence of automated social presence in organizational frontlines and customers' service experiences. J. Serv. Res. **20**(1), 43–58 (2017). https://doi.org/10.1177/1094670516679272

2. Tussyadiah, I., Miller, G.: Perceived impacts of artificial intelligence and responses to positive behaviour change intervention. In: Pesonen, J., Neidhardt, J. (eds.) Information and Communication Technologies in Tourism 2019, pp. 359–370. Springer, Cham (2019). http://doi.org/10.1007/978-3-030-05940-8_28

3. Washburn, D., Sindhu, U., Balaouras, S., Dines, R.A., Hayes, N.M., Nelson, L.E.: Helping CIOs understand "smart city" initiatives: defining the smart city, its drivers, and the role of the CIO, Cambridge (2010)

4. Aina, Y.A.: Achieving smart sustainable cities with GeoICT support: The Saudi evolving smart cities. Cities **71**(2016), 49–58 (2017). https://doi.org/10.1016/j.cities.2017.07.007

5. Janković, S., Krivačić, D.: Environmental accounting as perspective for hotel sustainability: literature review. Tour. Hosp. Manag. **20**(1), 103–120 (2014)

6. Benckendorff, P., Sheldon, P., Fesenmaier, D.: Tourism Information Technology. CABI International, Oxfordshire (2014)

7. Buhalis, D.: eTourism: Information Technology for Strategic Tourism Management. Pearson Education Limited, Edinburgh (2003)

8. Xiang, Z., Tussyadiah, I., Buhalis, D.: Smart destinations: foundations, analytics, and applications. J. Destin. Mark. Manag. **4**(3), 143–144 (2015). https://doi.org/10.1016/j.jdmm.2015.07.001

9. Buhalis, A., Amaranggana, A.: Smart tourism destinations enhancing tourism experience through personalisation of services. In: Tussyadiah, I., Inversini, A. (eds.) Information and Communication Technologies in Tourism 2015, pp. 377–389. Springer, Cham (2015). http://doi.org/10.1007/978-3-319-14343-9_28

10. Adiki, S.: Impact of big data and artificial intelligence on economy. In: Mulay, A. (ed.) Economic Renaissance in the Age of Artificial Intelligence. Business Expert Press, New York (2019)

11. Torra, V., Karlsson, A., Steinhauer, H.J., Berglund, S.: Artificial intelligence, pp. 9–26. Springer, Cham (2019). http://doi.org/10.1007/978-3-319-97556-6_2

12. Oussous, A., Benjelloun, F.Z., Ait Lahcen, A., Belfkih, S.: Big data technologies: a survey. J. King Saud Univ. Comput. Inf. Sc. (2017) http://doi.org/10.1016/j.jksuci.2017.06.001

13. Bibri, S.E., Krogstie, J.: The core enabling technologies of big data analytics and context-aware computing for smart sustainable cities: a review and synthesis. J. Big Data **4**(1), 38 (2017). https://doi.org/10.1186/s40537-017-0091-6

14. McKinsey Global Institute: Notes from the AI frontier: insights from hundreds of use cases (2018)

15. Navío-Marco, J., Ruiz-Gómez, L.M., Sevilla-Sevilla, C.: Progress in information technology and tourism management: 30 years on and 20 years after the internet - Revisiting Buhalis & Law's landmark study about eTourism. Tour. Manag. **69**(May), 460–470 (2018). https://doi.org/10.1016/j.tourman.2018.06.002

16. Neuhofer, B., Buhalis, D., Ladkin, A.: A typology of technology-enhanced tourism experiences. Int. J. Tour. Res. **16**(4), 340–350 (2014). https://doi.org/10.1002/jtr.1958

17. Hatami, H., Soleimani, B., Ziafat, H.: Recommendation systems based on association rule mining for a target object by evolutionary algorithms. Emerg. Sci. J. **2**(2), 100–107 (2018)

18. https://www.tripadvisor.com/engineering/using-nlp-to-find-interesting-collections-of-hotels/

19. Alaei, A.R., Becken, S., Stantic, B.: Sentiment analysis in tourism: capitalizing on big data. J. Travel Res. (2017). http://doi.org/10.1177/0047287517747753

20. Gretzel, U., Sigala, M., Xiang, Z., Koo, C.: Smart tourism: foundations and developments. Electron. Markets (2015). https://doi.org/10.1007/s12525-015-0196-8

21. Alexis, P.: R-Tourism: introducing the potential impact of robotics and service automation in tourism. Econ. Sci. Ser. **17**(1), 211–216 (2017)

22. Berger, H., Dittenbach, M., Merkl, D.: Activation on the move: querying tourism information via spreading activation. LNCS, vol. 2736 (2003)

23. Ninaus, G.: Using group recommendation heuristics for the prioritization of requirements. In: Proceedings of the Sixth ACM Conference on Recommender Systems - RecSys 2012 (2012). https://doi.org/10.1145/2365952.2366034

24. Sigala, M.: New technologies in tourism: from multi-disciplinary to anti-disciplinary advances and trajectories. Tour. Manag. Perspect. **25**, 151–155 (2018). https://doi.org/10.1016/j.tmp.2017.12.003

Author Index

© Springer Nature Switzerland AG 2019
R. Silhavy (Ed.): CSOC 2019, AISC 985, pp. 403–404, 2019.
https://doi.org/10.1007/978-3-030-19810-7

Printed in the United States
By Bookmasters